T0399758

Biogas Technology

About the Editors

Dr. Snehasish Mishra, Associate Professor, PhD (Life Sciences), School of Biotechnology, KIIT Deemed University has research exposure in the areas of microbiology, biotechnology, limnology, environmental biotechnology, renewable energy, bioprocess engineering and food sciences. He has more than two decades of research experience, including international exposure. He heads the Govt. of India sponsored MNRE's Biogas Development and Training Centre since 2009-2010. With two years industry experience and 14 years academic experience, he has significantly contributed in research, development and extension of waste management, aquaculture, bioenergy, environmental sciences research. He has authored research and popular articles, and book chapters, with more than 46 publications in peer reviewed journals and books of national and international repute. He is regularly involved from time-to-time in the short-term trainings funded by DST, DBT and MSME for skill-development and livelihoods of the farming community through biotechnological interventions. He is a reviewer and editorial member with a number of research journals. He has also been an active participant in biotech entrepreneurship and industry-academia interface. For his continued contribution to the cause of science he has received several accolades and recognitions from academies, societies and associations. He has successfully guided, and continues to guide, numerous PhDs and MPhils, MTechs and MScs.

Dr. Tapan K. Adhya, Professor, School of Biotechnology, Kalinga Institute of Industrial Technology (KIIT) deemed University, Bhubaneswar - 751024, also is the Director, South Asia Nitrogen Centre, New Delhi. He is a global name in the field of microbial interactions in the biogeochemical cycles. His previous assignment has been as the Director, National Rice Research Institute, Cuttack, Odisha. Using tropical paddy as the model ecosystem, the focus of research by Prof. Adhya is on sustainable management of resource poor tropical soils to maintain high productivity with minimum environmental impact and investigations into sediment microflora for C- & N-cycling. Prof. Adhya initiated field-level research on greenhouse gas emission from tropical Indian paddy, and also identified strong mitigation options. In recognition of his research achievements, he has been awarded with the fellowship of all the three major science academies of the country. Dr. Adhya is the editorial board member of 6 journals published by Springer-Nature publications and has published more than 170 peer-reviewed journal papers with total citations of more than 4,850 with an H-index of 36 and i10 index of 101. He has also edited three books published by Elsevier and Springer-Nature. http://biotech.kiit.ac.in/faculties/Tapan-Kumar-Adhya.html

Dr. Sanjay K. Ojha completed his PhD in Biotechnology from KIIT University, Bhubaneswar before doing his postdoctoral research at the Department of Biophysics, AIIMS, New Delhi. Currently, he is a Research Scientist in Pandorum Technologies Pvt Ltd., collaborating with LV Prasad Eye Institute, Hyderabad. Dr Sanjay did his PhD on 'Feasibility and Enhancement of Biomethanation of Kitchen Refuse with Special Mention of the Amylolytic Potential of *Bacillus aryabhattai* KIIT BE-1' focussing on identifying and applying amylase enzyme from a novel bacteria for enhanced biomethanation from solid wastes. He worked as the Project Officer at the Biogas Development and Training Centre (Ministry of New & Renewable Energy, New Delhi) where he skilled on initiating and managing labs, with generous support of KIIT-TBI. He acquired knowledge on innovation ecosystem and entrepreneurship particularly related to biogas sector. Additionally, he has worked on several projects where he gained experience in molecular biology, biophysics, fermentation and other related technologies. In the course, he trained and guided several MScs and MTechs. He has published many papers in the national and international peer reviewed journals, and has authored numerous book chapters. Along with research, he is actively involved in teaching, and worked as a visiting professor in Vinoba Bhave University, Hazaribag. He has been felicitated by numerous scientific bodies/societies in recognition of his scientific contributions. He is a Lifetime member of Societies such as the Biotech Research Society of India (BRSI), and the Association of Microbiologists of India (AMI). Dr Sanjay is in editorial board member of research journal like International Journal of Contemporary Microbiology.

Biogas Technology

Editor in Chief
Snehasish Mishra
School of Biotechnology
Kalinga Institute of Industrial Technology (KIIT)
Bhubaneswar - 751 024 (Odisha)

Associate Editors
Tapan K. Adhya
School of Biotechnology
Kalinga Institute of Industrial Technology (KIIT)
Bhubaneswar - 751 024 (Odisha)

Sanjay K. Ojha
Pandorum Technologies Pvt Ltd.
Collaborating with LV Prasad Eye Institute
Hyderabad

CRC Press is an imprint of the
Taylor & Francis Group, an informa business

NEW INDIA PUBLISHING AGENCY
New Delhi – 110 034

First published 2021
by CRC Press
2 Park Square, Milton Park, Abingdon, Oxon, OX14 4RN

and by CRC Press
6000 Broken Sound Parkway NW, Suite 300, Boca Raton, FL 33487-2742

CRC Press is an imprint of Informa UK Limited

British Library Cataloguing-in-Publication Data
A catalogue record for this book is available from the British Library

Library of Congress Cataloging-in-Publication Data
A catalog record has been requested

ISBN: 978-0-367-70201-4 (hbk)

Central Agricultural University

Lamphel Pat, Imphal 795004, Manipur

Prof. S. Ayyappan
Chancellor

172, Shreepadam, 5th Main
Avalahalli, BDA Extension
Bengaluru 560085, Karnataka
Mob. 91-9582898989
sayyappan1955@gmail.com

Foreword

Easy access to energy and energy security are the backbone of any societal development. This has been listed among the 17 Sustainable Development Goals (SDGs) adopted by the UN General Assembly in September 2015 as the 2030 Agenda for Sustainable Development. SDG 7 which speaks of 'Affordable and Clean Energy', aims to ensure access to affordable, reliable, sustainable and clean energy for all. Present global overdependence on fossil fuels in the energy mix, with its attendant environmental impacts coupled with resource constraint issues, has raised serious concern over its continued use.

In this backdrop, Biogas has positioned itself as an important component in the energy mix. Backed by technological advances, the scope of biogas generation from conventional sources has been widened to include newer resources like food and agro-industrial wastes. Making biogas technology more user-cum-environment friendly is expected to substantially contribute towards national energy security while converting 'waste to wealth'. This has become imperative with the recent flagship 'GOBAR (Galvanizing Organic Bio-Agro Resources) – DHAN' (2018) initiative of the Govt. of India. The scheme focuses on managing and converting cattle dung and solid waste from farms and fields to useful compost, biogas and bio-CNG. It will also help in keeping villages clean and generate energy while increasing income of farmers and cattle herders.

The evolving biogas sector in terms of application of latest technology has made it simple to adopt, broader in application, and accurate in lab testing. Ministry of New and Renewable Energy (MNRE), Govt. of India through its regional Biogas Development Technology Centres (BDTCs) has intensively supported technology development in this sector. The KIIT-BDTC started its operation in the year 2009-10. Since its inception, it has been aggressively involved in research, development, extension and dissemination activities related to Biogas Technology. KIIT-BDTC has five states namely, Odisha, Andhra Pradesh, Telangana, Jharkhand and Bihar under its jurisdiction.

The KIIT-BDTC team has taken this unique initiative to collate latest information in this critical area in this book that is well-diversified to include basic science to application methodology in a single volume. With renewed emphasis on alternate energy source to diversify the energy mix of the country, biogas technology deserves a closer look. I am sure that this book will serve as a ready reckoner in this subject and supporting country initiative on biogas research.

January 7, 2019

(S. Ayyappan)

Preface

Over dependence on the fossil fuels to meet the ever-increasing energy demand of the modern world, with its attendant negative impact on the environment and global climate change, has awakened the conscience of researchers and policy planners alike for the search of climate neutral and environmentally benign non-fossil fuel based resources. Biogas is a renewable energy resource that can be an alternative solution for the world's insatiable energy demands while helping in managing waste and reducing the greenhouse gas (GHG) emissions. It is also regarded as carbon neutral as the carbon in biogas comes from organic matter (feedstock) that captured this carbon from atmospheric CO_2 over a relatively short timescale.

Apart from recycling the organic matter, especially wastes - both agricultural and municipal, biogas generation also helps capture N, a valuable agricultural input, present in the refuse as it gets biogasified through conversion into the nutrient-rich biodigestate with a narrow C:N ratio that makes it a useful manure. Thus the technology remains an important cog in the wheels of a circular bio-economy with a zero-waste based social infrastructure. Although the technology of biogasification involving biomethanation of organic residues is more than half a century old, the basic technology has failed to capture the fancy of the research community for its commercial exploitation including technology upgradation and fine-tuning for the use of alternate feedstock as well as downstream process technologies for effective use of this important bioprocess for a thoroughly 'circular' approach.

This book has been written and compiled to collate latest information on biogas technology to help readers, researchers and extension workers alike to understand the fruitful exploitation of the process. It has fourteen chapters, primarily in three major categories, the first category dealing with the basic biomethanation process including its ecology, microbiology, biochemistry and molecular biology (*viz.*, Methanogenesis-Microbiology and ecology by Nayak *et al.*; Biochemistry and molecular biology of biomethanation by Mishra *et al.*, Methane driven biogeochemical processes in terrestrial ecosystem by Bharati *et al.*, Bioenergetics and evolutionary relationship in biomethanation with respect to *Mcr* by Srichandan *et al.*), the second category dealing with the evolution of the technology in Indian/global scenario from the lab to the land (*viz*., Application of methanation as an alternate energy technology by Rana and Nanda; Biogas generation from food wastes and its purification technology by Das; Biomethanation of rice straw - Feasibility assessment by Maheshwari *et al*.; Biogas from dairy effluents by Satpathy and Das; Valorization of pineapple wastes for biomethane generation by Sarangi and Nanda; Optimised process considerations for enhanced kitchen refuse biomethanation by Malesu and Singh; Electrofermentation in aid of bioenergy and its industrial application by Kumar *et al*.), and the last category dealing with the economics of the technology (*viz*., Industrial applications of anaerobic digestion by Mohanty and Das; Biomethanation under biphasic conditions: Success story of Nisargruna biogas plant by Mehetre and Kale; Improved biogas plants and related technologies: A status report by Sooch; Socioeconomic analyses of biogas technology towards the upliftment of rural India by Paikaray and Mishra). All the various known and active names in this field of research and development have put their hearts and minds into their contributed chapters. The additional details provided in the Annexures (*viz*., Model bankable scheme for biogas commercialisation venture; Frequently asked questions in adopting biogas technology; Common terminologies in biogas research; Glossary of abbreviations and symbols frequently used in biogas research; and Prominent global entities in biogas R&D and commercialisation) double the usefulness of the compilation. The Volume should contribute to this field of research and development in this sector by igniting the young research minds and educating the experienced minds alike involved, we are confident.

Finally, the Editors take pride to congratulate and express their gratitude to all the valuable contributions received from various corners, including the administrative and technical support received from the Publisher. The Editors are open to the various constructive comments and criticism from the valued readers so that the same would help considering and incorporating in newer editions.

Editors

Contents

Contributors

Archana Kumari
Department of Biotechnology
Bodoland University, Kokrajhar-783370
Assam, India

Amar Kumar Das
Department of Mechanical Engineering
Gandhi Institute For Technology (GIFT)
Gramadiha Gangapada
Bhubaneswar – 752 054, India

Beom Soo Kim
Department of Chemical Engineering
Chungbuk National University
Cheongju-28644, South Korea

B. Baliyar Singh
Department of Biotechnology, CET, BPUT
Bhubaneswar, Odisha, India

B.B. Mishra
Department of Microbiology
OUAT, Bhubaneswar, Odisha, India

Biswajit Paikaray
BDTC, Bioenergy Lab
School of Biotechnology, KIIT
Bhubaneswar, Odisha, India

D.M. Das
Scientist, Agricultural Engineering
KVK, OUAT, India

D.P. Samantaray
Department of Microbiology
OUAT Bhubaneswar
Odisha, India

Haragobinda Srichandan
Mineral Resources Research Division
Korean Institute of Geoscience and Mineral
Resources Yuseong-gu, Deajeon
South Korea

Department of Material Science and Engineering
Chungnam National University
Daejeon, 300-764, South Korea

K. Bharati
Indian Institute of Soil Science
Berasia Road, Nabibagh, Bhopal – 462 038
Madhya Pradesh, India

K. Chandrasekhar
Department of Civil Engineering
Yeungnam University
Gyeongsan-38451, South Korea

M.K. Mohanty
College of Agricultural Engineering
& Technology
OUAT, Bhubaneswar

Neha Ahirwar
Indian Institute of Soil Science, Berasia Road
Nabibagh, Bhopal – 462 038
Madhya Pradesh India

Puneet K. Singh
BDTC, Bioenergy Division
School of Biotechnology, KIIT University
Bhubaneswar, Odisha-751024, India

Prakash Kumar Sarangi
Directorate of Research
Central Agricultural University
Imphal, Manipur, India

Pranav Kshirsagar
Bioenergy Group MACS-Agharkar
Research Institute, GG Agarkar Road
Pune – 411004, Maharashtra, India

Prashant K. Dhakephalkar
Bioenergy Group
MACS-Agharkar Research Institute
GG Agarkar Road
Pune – 411004, Maharashtra, India

Preseela Satpathy
Department of Environment and Sustainability
Institute of Minerals and Materials
Technology, Bhubaneswar, Odisha, India

Prasun Kumar
Department of Chemical Engineering
Chungbuk National University
Cheongju – 28644, South Korea

Puneet K. Singh
Bioenergy Lab
Biogas Development and Training Centre
School of Biotechnology
KIIT University, Bhubaneswar
Odisha - 751024, India

Ritesh Pattnaik
School of Biotechnology
KIIT Bhubaneswar, Odisha-751024, India

Rachita Rana
Department of Chemical and Biological
Engineering, University of Saskatchewan
Saskatoon, Canada

Snehasish Mishra
BDTC, Bioenergy Lab
School of Biotechnology
KIIT Bhubaneswar, Odisha – 751 024, India

Sanjay K Ojha
Department of Biophysics, All India
Institute of Medical Sciences
New Delhi – 110 024, India

S.R. Mohanty
Indian Institute of Soil Science
Berasia Road, Nabibagh, Bhopal – 462 038
Madhya Pradesh, India

S.K. Nayak
Department of Biotechnology, CET, BPUT
Bhubaneswar, Odisha, India

Sonil Nanda
Lassonde School of Engineering
York University, Toronto, Canada

Sneha Tapadia-Maheshwari
Bioenergy Group
MACS-Agharkar Research Institute
GG Agarkar Road
Pune – 411 004, Maharashtra, India

S.T. Mehetre
Nuclear Agriculture and Biotechnology
Division, Bhabha Atomic Research Centre
Mumbai – 400 085, India

S.P. Kale
Nuclear Agriculture and Biotechnology
Division, Bhabha Atomic Research Centre
Mumbai – 400 085, India

Sarbjit Singh Sooch
School of Renewable Energy Engineering
Punjab Agricultural University
Ludhiana – 141004, Punjab, India

Tapan Kumar Adhya
School of Biotechnology, KIIT
Bhubaneswar, Odisha – 751 024, India

Trupti Das
Department of Environment and Sustainability
Institute of Minerals and Materials
Technology, Bhubaneswar, Odisha, India

Vijay K. Malesu
Bioenergy Lab, Biogas Development and
Training Centre, School of Biotechnology
KIIT University, Bhubaneswar
Odisha - 751024, India

Adopting Biogas Technology – Frequently Asked Questions

Below is provided a compilation of 111 questions that may frequently occur in the mind of the users/potential users, researchers, extension workers, relevant line-department officials, bankers, and anyone who has an interest in biogas technology. These FAQs are categorised into various headings, *viz.*, 'A. Fundamentals of biogas technology', 'B. Functional characteristics of biogas digester and its operation', 'C. Biogasifiable substrates and their management', 'D. Slurry management and its manurial benefits', 'E. Off-grid power (electricity) generation', 'F. Environmental, economic and other benefits of biogas technology', and 'G. Technical issues in biogas technology'.

A. Fundamentals of biogas technology

1. What is Biogas?

Biogas is an energy product derived from the decomposition of organic materials such as animal dung in the absence of oxygen making a mixture of gases (methane, carbon dioxide, and small amounts of other gases), the gas of utility being methane. This can be used directly for cooking and lighting, or to generate electricity. Biogas is an eco-friendly renewable energy source found abundantly hidden in cattle dung, kitchen wastes, agro residues and municipal biodegradable wastes, virtually all sorts of organic refuse.

2. What is the composition of biogas?

Methane, carbon dioxide, and hydrogen sulfide are in the main constituents of biogas. It contains methane (CH_4), carbon dioxide (CO_2), water vapours, and traces of hydrogen sulphide (H_2S) and nitrous oxide (NO_2). Generally, biogas contains 55%-75% CH_4, 44%-24% CO_2, and 1% or less of the mixture of other gases. Reports suggest the percent composition of biogas as, methane (CH_4): 50–75, carbon dioxide (CO_2): 25–50, hydrogen sulphide (H_2S): 0–3, H_2: 0–1, N_2: 0–10, and water vapours and nitrous oxide (NO_2).

3. Specify the role of each gas ingredient in biogas?

The methane (CH_4) gas is largely responsible for combustion, whereas carbon dioxide (CO_2) acts as a fire extinguisher. Similarly, H_2S is a pungent smelling gas and corrosive in nature to the devices that runs on biogas. Further, the traces of NO_2 and NH_3 likely to be present are relatively obnoxious in nature if present in large quantity.

4. Is biogas a kind of renewable energy?

Yes, anaerobic digester technology is employed globally to source renewable energy. Biogas from biodigester is comprised primarily of methane that can be used as a fossil fuel alternative. This renewable natural gas can substitute for fossil fuel natural including for heating, cooking, driving. Biogas can also be used as fuel to make clean electricity. All of these options provide an opportunity to turn organic 'waste' into a valuable renewable energy resource in a sustainable manner.

5. What is the 'waste to wealth' concept?

Generating money from the various forms of the so-called waste, or saving money through proper waste management is the basis of the concept of 'waste-to-wealth'. In an agro-food industry, for instance, the digester can be used as a primary waste treatment unit where the biogas is used to compensate some energy cost in the plant, thereby reducing the size of the secondary waste treatment and the incurred cost.

6. What source biogas could be generated from?

Biogas is commonly made from animal manure, sludge settled from wastewater, and at landfills containing organic wastes. However, biogas can also be made from almost any feedstock containing organic compounds, both wastes and biomass (energy crops). Carbohydrates, proteins and lipids are all readily converted to biogas. Many wastewaters contain organic compounds that may be converted to biogas including municipal wastewater, food processing wastewater and many industrial wastewaters. Solid and semi-solid materials including plant or animal matter can be converted to biogas.

7. What is the working principle of a biogas plant (anaerobic digester)?

Biogas is produced as a result of anaerobic fermentation of the dead organic biomass in the presence of water, and the detritivorous microbes. Anaerobic fermentation by a specialised group of microbes (methanogens) produces the methane gas.

8. How does a biodigestor work?

The life and soul of an anaerobic biodigester is the 'microscopic' bacteria that live, constantly multiplying inside the digester, and breakdown the material fed into it. The bacteria require a sufficient amount of organic feedstock, vitamins, supplements, and some water to live. The bacteria also would love to thrive in a warmer environment, and a relatively stable pH levels.

9. What is the calorific value of biogas?

The calorific value of biogas is about 6kWh/m^3; it corresponds to about half a litre of diesel. The net calorific value depends on the efficiency of the burners or appliances. Methane is the valuable component in the biogas fuel.

10. What are the factors influencing the choice and adoption of biogas technology?

The plant installation cost, multiuse as a household fuel, level of income and education, promotion of the technology and beneficiary awareness are some of the factors that influence the choice and adoption of biogas technology.

11. Is biogas explosive?

While in unfortunate situations, explosions can occur, biogas itself is relatively safe. Under normal operating conditions, low pressure and low temperature, biogas is an uncompressed, wet gas, and tests have shown that it can simply be burnt off (or oxidised with a flare) in a controlled setting. When natural gas, whose main component is methane, builds up without proper ventilation, it can become flammable and cause explosions. A deadly example is, methane caused the 2010 coal-mine explosion in West Virginia killing 29 miners.

12. Can biogas be used as an alternative to fossil fuels? How?

Methane is the principal gas in biogas. Methane is also the main component in natural gas, a fossil fuel. Biogas can be used to replace natural gas in many applications including, cooking, heating, steam production, electricity generation, as a vehicular fuels, and as a pipeline gas.

13. Can I make/use biogas at home or at my place of business?

Biogas can be made at home or at a business from food waste, yard and grass-trimmings, and other organic solid wastes. However, efficient use of biogas is more readily accomplished at larger scales. A typical home-based (household) biodigester might be useful in an hour cooking per day on biogas from domestic wastes.

14. How much does biogas generation cost?

Current price for natural gas is around $7/1000ft^3$ or around Rs. 450. Depending on the application, this is very similar to current estimates for the cost of biogas production.

15. Is there a wholesale price for renewable natural gas (RNG) sold for vehicular fuel?

Typically, one could receive the avoided cost of cogeneration of commodity plus negotiated percentage of value realised from sale (of the renewable identification number; RIN, attached to the gallon equivalent of fuel) and the low carbon fuel standard (LCFS) credit associated with the generated gas quantity. A fixed, wholesale price for RNG as vehicular fuel doesn't exist; it may vary month to month, region to region and country to country. Vehicular fuel seller would risk (or somehow hedge) his exposure to erratic RIN, LCFS and NG commodity pricing (based on current market force) to offer a fixed price to producer to create RNG.

16. What is Biomethane/BioCNG?

Biomethane/BioCNG means pipeline-quality gas derived from organic material. Though not derived from fossil fuels, it is identical in properties to natural gas. Biomethane/BioCNG is produced from biogas after cleaning/upgrading to meet natural gas pipeline specifications, by removing the interfering gases like CO_2 and H_2S to obtain almost pure (90-98%) CH_4. Biomethane/BioCNG is injected into the gas network or compressed for use in vehicles, then.

17. Give an idea on the industrial biogas plants.

Anaerobic process is largely used to treat industrial wastes and wastewaters for more than a century now; anaerobic process treats organically-rich industrial wastewaters before disposal. Energy and environmental concerns (strict environmental legislations) in recent years have increased interests in managing organic industrial wastes by treating anaerobically. With advanced treatment technologies worldwide (the Europe leading the way), diluted industrial wastewater is digested. Technologies to treat food-processing, pharma and agro-industrial wastewaters are standardised. Industries treating wastewaters likewise are: •Food processing, *e.g.*, vegetable canning, milk and cheese manufacturing, potato processing, slaughterhouse; •Beverage industry, *e.g.*, breweries, soft drinks, distilleries, coffee, fruit juices; and •Organic products industries, *e.g.*, paper and board, rubber, biochemicals, starch, pharmaceuticals.

18. What is the biogas programme of the concerned Indian Ministry?

The New and Renewable Energy (MNRE) Ministry implements National Biogas and Manure Management Programme since 1981-82 in India, setting up family-type biogas plants. From 2018, the programme is revamped as New National Biogas and Organic Manure Programme.

19. What is the amount of subsidy in terms of CFA released to a beneficiary?

Due to its renewable nature and the potential to replace fossil fuels, the MNRE encourages by providing 50% subsidy to install household/community biodigester. The subsidy receivable by a beneficiary of a 2m^3 plant is Rs. 9000. For bigger (> 25m^3) plants, the beneficiary gets 40% of CFA to construct and 50% subsidy on machineries through channel partners.

20. What is a family-type biogas plant?

A family-type biogas plant is 2–6 m^3 size wherein household organic wastes like cattleshed refuse, sewage wastes, and kitchen refuse is fed to generate biogas for in-house consumption.

21. How many household level (family-type) biogas plants is setup so far in the country?

As per 2011 census, a total of 4.31 million such biogas plants have been setup in the country.

B. Functional characteristic of biogas digester and its operation

22. What is 'anaerobic digestion' of biomass, to generate biogas?

Anaerobic digestion is where microbes living in places without oxygen breakdown organic, biodegradable material over time and convert to biogas and organic manure; anaerobic means 'no oxygen'. It is a (four-step) series of interdependent biological process wherein microbes breakdown biodegradable material in an oxygen-free or oxygen-limited environment. Thus, microbial fermentation/degradation taking place in the absence of oxygen inside a bioreactor is the anaerobic digestion. One of the end-products is the combustible biogas that generates heat, electricity and/or can be processed into renewable natural gas and transportation fuels.

23. What are the stages in anaerobic digestion/fermentation?

The four interdependent and complementary stages/steps involved in anaerobic biodigestion are, hydrolysis, acidogenesis, acetogenesis, and methanogenesis.

24. What is methanogenesis?

Methane-producing bacteria decompose low molecular weight organic compounds to make CH_4 as end-product, utilising H_2, CO_2 and acetic acid. Methane-producing microbes occur in anaerobic environments (such as, marine sediments, ruminant stomach, marshes, etc.). They are obligatory anaerobes and sensitive to environmental change. In contrast to the acidogenic and acetogenic bacteria, methanogens belong to the archaea group with heterogeneous morphology and numerous common biochemical and molecular properties that distinguish them from all other bacteria, the main difference being the bacterial cell wall makeup.

25. What are anaerobic digesters?

Anaerobic digesters are specially-designed insulated tanks used to facilitate the anaerobic digestion process under controlled environment to maximise biogas production. An anaerobic digester is a large, air-tight container or tank providing an oxygen-free environment. The tank is filled with organic material and is maintained at an optimum temperature for anaerobic bacteria to digest the material. Depending on what you put into it, the contents (or substrates or raw materials) can be wet or dry. Thus, a biodigester is an enclosed anaerobic chamber wherein the specialised bacteria breakdown or 'digest' organic material fed into it.

26. How big is an anaerobic digester?

The size of the digester depends on how much organic material will be fed into the digester, and how quickly the digester can breakdown the organic material. Thus, the size could be from a large refrigerator to a small building.

27. What are the functional characteristics of a biodigester?

A fixed-dome plant is a digester with a fixed, non-movable gas holder, sitting on top. When gas is produced, the slurry displaces into the compensation tank. Gas pressure complements to the volume of gas stored and the height difference between the slurry level in digester and the slurry level in compensation tank. A fixed-dome biogas plant costs relatively low; it's simple as no moving part exists. As there is no rusting part, the plant is expected to have longer life (>20 years). As plant is constructed underground, it saves space and protects from physical damage. Being underground, the digester is thermo-regulated; day/night thermo-stability positively influences the bacteriological processes. However, construction of a fixed-dome plant is labour-intensive. As fixed-dome plant construction needs skilled masonry (creates local employment), they should be constructed under the supervision of experienced 'biogas' technicians.

28. What are the different components of a bio-digester?

The various components are, the **inlet** where the substrates enter the plant; the **outlet** where the substrate byproduct eject as slurry; the **gate valve** for transmission of gas; the **headspace** to maintain optimised/calibrated gas level inside digester; the **dome**, the outer circular part as insulating chamber; and the **digester** where anaerobic process of methanation occurs.

29. What is the production capacity of a biodigester?

True measure of biogas production is difficult as it varies with material type, temperature and acidity/alkalinity of the mixture inside the tank. Also, workmanship defects like construction engineering, pipeline leakage, biogas dome leakage, biogas storage balloon leakage, leakage in biogas handling equipment (blower or pressurising unit) etc. could be responsible.

30. What is the estimated installation/commissioning period for a biodigester?

The installation period typically depends on the size and the type (floating- or fixed-dome) of plant install. Generally it takes 45-60 days to install a plant above 1000 kg/day, including the additional conditioning period to 'charge' the digester. Smaller (like household) prefabricated ready-to-install plants are available that can be installed in a day. Considering consumption pattern of the users and the volume of available inputs, 1-85m^3 are popular in India.

31. Which are the MNRE-recommended biodigester models popular in India?

Various MNRE-approved digester models that are popular in India are Deenabandhu, Pragati, Konark, Janata, KVIC, TERI model, low-cost polyethylene model and ferro-cement digester.

32. What is the minimum retention period of a family-type biogas plant?

Estimated applicable values for mesophilic fermentation of feed are, 20–30 days for cattle dung, 15–25 for pig manure, 20–40 for chicken manure, and 50–80 for manure-plant material mixture. The digester microflora washes out faster than it reproduces with shorter retention time, practically stalling fermentation; this issue occurs in agricultural biodigester rarely.

33. Explain about the operation of a family-type biogas plant.

A major operation of the plant is daily substrate feeding with an approximate 1:1.5 solid to liquid proportion. Feeding rate and its composition needs to be constant as far as practicable; mind that all in = all out. Avoid overfeeding, abrupt feed change, foaming; don't add a substrate with very low pH and considerably high protein. Maintain continuous mixing; mix as necessary and as little as possible. Choose an appropriate temperature and keep it constant.

34. How can the biogas thus generated be measured?

It can be measured with help of a biogas flow-meter. A common flow-meter type installed in biogas production plants is the insertion type thermal mass meter fitted at the smallest and largest gas pipes under pressure. They are usually of stainless steel though other materials can be used, and variations of a traditional design supplied when the application requires it.

35. How much biogas can be stored at a time?

Theoretically there is no limit for the storage but in practice it is advisable to have a storage capacity that can take care of around 12 hrs of biogas generation. Remember that, too high accumulation of the generated biogas for too long a time may build-up the inside pressure considerably thereby affecting the durability of the digester dome.

36. Does this technology reduce odour? If so, by how much?

Anaerobic digesters can reduce odour of 'smelly' organic wastes by 80% with adequate manure management. The anaerobic digestion process appears to transform most of the volatile odorous compounds into biogas. Therefore, adopting this technology and burning the captured biogas in co-generators reduces odour.

37. Are the biodigesters cold or hot?

To keep the right bacteria alive and multiplying, many larger digesters are kept at or above 30–38°C (86–100°F). Digesters can also work at temperatures that are both lower and higher than this. As the digester bacteria are temperature-sensitive, cooler digesters take more time to breakdown the feedstock, while hotter ones may break it down more quickly.

38. What is the unit cost of a 2m^3 Deenabandhu model biogas plant?

The unit cost of a 2m^3 Deenabandhu model biogas plant is Rs 18,000.00.

39. How quickly will biogas be ready for use after generation?

Biogas can be used directly either for thermal applications or electricity generation or as a vehicular fuel (after up-gradation) immediately after its generation.

40. Why does biogas need purification for industrial use?

Since H_2S and NO_2 are hazardous to the generators and other instruments fuelling the biogas even in little quantity, their removal is necessary for industrial as well as household CNG use.

41. What is a scrubber?

It is a device that extracts H_2S from the raw biogas during purification.

42. What if the biogas doesn't fill inside the biogas dome/storage 'balloon'?

If the biogas doesn't fill inside the biogas 'balloon', please check the biogas outlet from the biodigester and the pipeline which connects the biogas balloon and the bio-digester. If there is enough pressure to break the water column inside the gas fluid separation chamber then the gas can easily escape from the chamber and will not be filled in the biogas balloon.

43. If biogas production has dropped quite a bit, how to figure out the reason behind it?

There are a number of factors that can affect biogas production in a biodigester. These are:

- Check for a leak somewhere. If there is no problem with water levels, check the gas-fluid separation chamber and its water levels. Check leakages in the balloon, and biogas pipeline.
- If temperature reaches below 20°C, biogas production drastically decreases. Provide a heating system to biodigester if this is the case.
- The digester pH should be as close to neutral (7.0) as possible. As anaerobic process in a digester produce acids, the most common pH problem is that of acidity. If a litmus test shows a low pH below 7.0, add small amount of lime (grounded lime stone). Never exceed a lime concentration of 500mg; excess lime may not solubilise, and may harm the bacteria.
- Take care not to introduce problematic chemical (like soap and detergent solutions) into the tank. Avoid using dung to feed digester of livestock that's on antibiotic or other medication.
- Cement and plastic cause no harm to the mixture in the tank, but metals should be avoided for use in the tank, or any of the tubing through which the biogas travels.

C. Biogasifiable substrates and their management

44. What are the raw materials required to run a biogas plant?

The raw material (substrate) to run a biogas plant could be any organic substrate, including animal dung, poultry wastes, plant wastes, human excreta, industrial and domestic wastes including kitchen waste, agro-residue and municipal biodegradable waste. Inorganic material like rock, floating dirt, plastic, metal cans and glass can't be anaerobically digested, and thus remain accumulated in digester after inadvertently reaching there during the digester feeding.

45. What should not be put into a digester?

Undesirable materials for a biodigester include those that can't easily be digested or those that inhibit gas production. For example, wood has a high lignin content and cannot easily be digested; it would create sedimentation issues and impact digester functioning. Materials containing heavy metals would inhibit the gas-producing bacteria. Inorganic materials (plastics, metals, glass, etc.) would create mechanical issues. Also, biodegradable forms that have a tendency to float should also be avoided as that may encourage hard scums to develop at the content-headspace interface inside the biodigester thereby dropping the performance.

46. What is a biodegradable/organic waste?

Biodegradable waste is a type of waste that can be broken down, in a matter of weeks or few months, into its base compounds by microbes and other living beings, regardless of what those compounds may be.

47. What is an organic material?

Organic material is sourced from an organism that was living at a point of time, and that which can decay. Wasted or spoiled food, plant clippings, animal manure, meat trimmings, and sewage are some common examples of such organic material used in anaerobic digestion. In contrast, inorganic material includes things like rocks, dirt, plastic, metal and glass.

48. What is a domestic organic waste?

It is all the various forms of organic wastes generated in a family holding, from kitchen to the washroom. Kitchen waste, for instance, includes the vegetable choppings, rice water, food leftovers and other organic wastes generated from a kitchen.

49. What are the uses of biogas generated through night soil treatment?

Biogas generated through night soil treatment can be utilised for any application similar to biogas generated from other waste materials. Biogas generated from night soil treatment plants may have a little higher quantity of hydrogen sulphide (H_2S) than that produced from other sources; better to filter/scrub before using as biogas.

50. What are the suitable locations of night soil based plants?

Any public, private institutions, housing colonies, flats and individual houses are the suitable locations could be an iseal location for night-soil based plants. However, social issues need to be addressed first.

51. How does biodigestion help in wastewater treatment?

Biodigestion is largely used to treat primary and secondary sludge from municipal wastewater treatment. The system is applied in many countries in combination with advanced treatment systems where the biodigestion is used to stabilise and reduce the final amount of sludge. Most engineering companies providing sewage treatment systems have also the capability to provide biodigestion systems. In European countries, between 30 and 70% of sewage sludge is treated by biodigestion, depending on national legislation and priorities. The biodigested sludge effluent can be used as agricultural manure or for energy production by incineration. There are countries where the effluent is still disposed on landfill sites, an environmentally degrading practice.

52. How can biodigestion help in Municipal solid waste (MSW) treatment plants?

In many countries, municipal solid waste is collected as mixed stream and incinerated in large power plants or disposed on landfill sites. This practice is actually a waste of energy and nutrients, as most of the organic fraction could be source separated and used as biodigestion feedstock. Even bulk collected wastes can be further processed and used for biogas production. In recent years, source separation and recycling of wastes received increasing attention. As a result, separate fractions of MSW are now becoming available for more advanced recycling treatment, prior to disposal.

The origin of the organic waste is important in determining which treatment method is most appropriate. Kitchen waste is generally too wet and lacks in structure for aerobic composting, but provides an excellent feedstock for biodigestion. On the other hand, woody wastes contain high proportions of lignocellulosic material are better suited for composting, as pre-treatment is necessary in order to be used for biodigestion. Utilisation of source separated organic fraction of household waste for biogas production has a large potential and several hundred biodigestion plants, processing organic fraction of MSW, are in operation around the world. The aim is to reduce the stream of organic wastes to landfills or even to incineration and to redirect them towards recycling.

53. Specify the name of different microbes responsible for production of biogas?

a. Hydrolysis: Biochemical reaction where the feed solubilise, and large polymers convert to simple monomers (*e.g.*, *E. coli*, *Lactobacillus*, *Bifidobacterium*, *Streptococcus*)

b. Acidogenesis: Biological reaction where simple monomers convert to volatile fatty acid (VFA) (*e.g.*, *Acetobacter*, *Syntrobacter*, *Syntrophomonas*)

c. Acetogenesis: Biological reaction where VFA convert to acetic acid, CO_2, and H_2 (*e.g.*, *Clostridium aceticum*, *Acetobacter woodii*, *Clostridium termoautotrophicum*)

d. Methanogenesis: Biological reaction where acetates convert to CH_4 and CO_2, while H_2 is consumed (*e.g.*, *Methanobacter*, *Methanomicrobia*, *Methanococcus*)

There are about 50 methanogens species which do not form a monophyletic group, although all methanogens belong to Archaea.

54. What is volatile fatty acid (VFA)?

The stability of the anaerobic digestion is reflected by the concentration of an intermediate product VFA produced during acidogenesis; it is up to six carbon atoms chain (*e.g.*, acetate, propionate, butyrate, lactate). Digester instability is attributed to the accumulating VFA that drops the pH in most cases; VFA accumulation must exceed a certain level before it reduces the pH value significantly. At such point, the VFA concentration in the digester could be high enough to severely affect the process. VFA accumulation may not always drop the pH value due to the inherent buffering capacity of the digester contents, *e.g.*, animal manure has a surplus of alkalinity. Experience shows that two different digesters can differently behave with similar accumulated VFA; same VFA concentration can be optimal for a digester but inhibitory for another. Microbial composition varying between digesters could be a plausible reason for this. Thus, VFA concentration is not a stand-alone process monitoring parameter.

55. What is kitchen waste?

Vegetable choppings, rice water and other organic wastes generated from a kitchen.

56. Any comparison of the biogas potential of the various agricultural residues?

Agro substrate	Gas yield (l/kg VS)	Methane (%)
Wheat straw	200-300	50-60
Rye straw	200-300	59
Barley straw	250-300	59
Oats straw	290-310	59
Corn straw	380-460	59
Elephant grass	430-560	60
Sunflower leaves	300	63
Algae	420-500	47
Sewage sludge	310-470	-

57. Any comparison of the biogas potential of various forms of cattle dung?

Cattle dung	Gas yield (l/kg VS)	Methane (%)
Pig	340-550	65-70
Cow	90-320	65
Poultry	310-620	60
Horse	200-300	42
Sheep	90-310	48
Barnyard	175-280	51

58. Which materials yield the most biogas?

When put into a digester, fats, oils and greases and food waste create most biogas. For this, many dairy farms add local food scraps in their digesters to enhance biogas production.

59. Does something specific have to be grown to feed into an anaerobic digester?

Agricultural producers grow crops (energy crops) as feedstock to produce renewable energy. Crops like corn and grasses are used to feed anaerobic digesters for profitable biogasification, but most organic waste materials like manure, and food scraps work equally well.

60. How long does it take to break down the organic material in the digester?

It varies based on the type of organic material you feed into the digester. Simpler organic compounds, such as simple sugars, fats and proteins, will digest fairly quickly. More complex organic compounds may take 30 plus days to completely digest, especially fibrous materials like cellulose, the major constituent of paper, paperboard, and card stock and of textiles made from cotton, linen, and other plant fibers). The operating temperature of the digester, and the digester design also have significant impact on the breakdown process.

61. What is biomethanation potential of feedstock substrate?

Biomethanation potential is the amount of methane (expressed as litre/kg VS) that can be produced from an unit gram of a biomethanation substrate.

62. What is the biogas potential of a kitchen waste/refuse?

It shall vary between the compositions of the kitchen waste. However, a litre of vegetable residue (leaf choppings, potato scrapping, etc.) shall yield an average of 330-360L gas/Kg VS, with a minimum 42% CH_4 gas.

63. What role macro- and micro-nutrients (trace elements) and toxic compounds play?

Microelements (trace elements) like iron, nickel, cobalt, selenium, molybdenum or tungsten are important in microbial activity. The considered optimal ratio of macronutrients carbon, nitrogen, phosphorus and sulphur (C:N:P:S) is 600:15:5:1. Insufficient provision of nutrients and too high digestibility of the substrate may disturb and/or inhibit the process.

Toxic compounds may reach the digester together with the feedstock, or are generated during the process. Application of threshold values for toxic compounds is difficult, as these are often bound by chemical processes on one hand, and the capacity of anaerobes to adapt to ecological conditions including the presence of toxic compounds within a limit, on the other.

D. Slurry management and its manurial benefits

64. Besides biogas, what comes out of the digesters?

The liquid slurry resulting from anaerobic digestion process is a byproduct of biodigestion, referred as the digestate. The solid in it is separated, composted, and sold to local gardeners, landscapers and farmers. Some farmers use these as bedding for the cattle and sell off to neighbours. The liquid obtained while separating the solids is organically-rich which can be used by farmers as nutrient-rich manure reducing the use of the chemical fertilisers (cutting the agricultural inputs costs).

65. What is a digestate?

Digestate is a naturally-produced biofertiliser, an important output of a biogas plant, to use in growing plants. It is also recognised by organic farming organisations as a natural organic fertiliser. Digestate recycling will reduce the amount of artificial chemical fertilisers used by local farmers.

66. How useful is the digestate from a biogas plant as manure?

It is rich in nitrogen and phosphorous and, thus, is excellent as manure. Digestate is more homogenous, compared to raw slurry, with an improved N-P balance. It has a declared content of plant nutrients, allowing accurate dosage and integration in fertilisation plans of farms. It contains more inorganic nitrogen, easier accessible to the plants, than untreated slurry. N-efficiency increases considerably and nutrient losses by leaching and evaporation will be minimised if digestate is used as fertiliser in conformity with good agri-practices. For optimum utilisation of digestate as fertiliser, the same practice criteria are valid, like in the case of utilisation of untreated slurry and manure: •Sufficient storage capacity (minimum 6 months), •Restricted season of application as fertiliser (during vegetation), •Amount applied per hectare (according to fertiliser plan), •Application technique (immediate incorporation and minimum nutrient losses).

Due to higher homogeneity and flow property, digestate penetrates in soil faster than raw slurry. Nevertheless, application of digestate as fertiliser involves risks of nitrogen losses through ammonia emissions and nitrate leaking. In order to minimise these risks, some simple rules of good agricultural practice must be respected, such as: •Avoid too much stirring of digestate before application, •Application of cooled digestate, from the post storage tank, • Application with dragging pipes, dragging hoses, direct injection in soil or disk injectors, • Immediate incorporation in soil, if applied on the surface of soil, • Application at the start of the growing season or during vegetative growth, • Application to winter crops should be started with 1/3 of the total N requirement, • Optimum weather conditions for application of digestate are: rainy, high humidity and no wind.

67. How to manage slurry in family-type and industrial-scale biogas plants?

Industrial intervention may be realised in the form of outsourcing of such huge amount of biomass for recycling and generation of biogas through the Public Private Partnership (PPP) mode. Interaction with the industries to this effect will lead to the production of green energy thereby minimising environmental menace. In order to make it commercially viable and technically acceptable, the technology business incubation (TBI) units may be promoted in urban areas inviting young research scholars, experts and consultants to develop the state of art technology as well as the formulation of marketing strategies for such outputs.

68. Any strategy to economically empower rural folks by effective slurry utilisation?

Commercialisation of such set-up lies on Vermicompost/*Azolla*-based initiatives, where the slurry generated from the unit would be utilised for the said purposes. The District Rural Development Agency (DRDA), State Agricultural Marketing Board, SNA, BDTC and channel partners of the MNRE may be involved in this endeavour to develop markets at the regional and national levels.

69. What are the effects of digestate on soil?

Organic matter degradation occurring through biodigestion includes degradation of carbon compounds, organic acids as well as odoriferous and caustic substances. When applied on soil, these create less stressful and more suitable environment for soil microbiome compared to raw dung application. Direct measurements of biological oxygen demand of digested cattle and pig slurry show ten times less oxygen demand than the undigested slurry. As oxygen consumption reduces, so is the tendency to form anoxic soil areas, *i.e.*, oxygen-free, nitrogen containing zones. The capability to build up new soil and the humus reproduction through supplied organic matter is also higher, when compared to fertilisation with raw slurry.

E. Off-grid power (electricity) generation

70. How is Electricity Produced Using Biogas?

Internal combustion engines or power turbines can convert biogas into electrical energy.

71. What is the role of biogas in off-grid electricity production?

In off-grid technology, large size (more than 25m^3) biogas plants are primarily installed to produce more than 3kWh power. It focuses on generating electricity from biogas through diesel-run generators after dewatering of vapour and scrubbing harmful gases like H_2S. The calorific value of biogas is quite high (around 4700kcal/20MJ/55-60% CH_4 content), meaning that 1m^3 of biogas corresponds to 0.5–0.6 l diesel fuel or an energy content of about 6kWh. Due to conversion loss, 1m^3 of biogas can be converted only to around 1.7kWh, practically.

72. What is cogeneration and the mechanism involved in production of electricity?

Diesel engines operate on biogas only in dual fuel mode. To facilitate the ignition of the biogas, a small amount of ignition gas - often diesel fuel - is injected together with the biogas. Modern pilot injection gas engines need about 2% additional ignition oil. Almost every diesel engine can be converted into a pilot injection gas engine. The advantage of these motors running in dual fuel mode is that they can also use gas of low heating value. However, in such cases, they consume a considerable amount of diesel. Up to engine sizes of around 200kW, pilot injection engines seem to have an advantage over gas motors due to their slightly higher (3-4% high) efficiency and lower investment costs. Apart from these, otto meter, gasoline-based engines may also be used to start the generator. The water vapour can be reduced by condensation in gas storage or on the way to engine.

F. Environmental, economic and other benefits of biogas technology

73. What are the environmental impacts of producing biogas?

Biogas production can reduce the pollution potential in wastewater by converting oxygen demanding, organic matter that could cause low oxygen levels in surface waters, Nutrients, like nitrogen and phosphorous are conserved in biogas effluents and can be used to displace fertilisers in crop production. A biodigester helps ensure a hygienic environment not only at rural households but also nearby atmosphere since kitchen waste along with other agro-waste (prime source of biogas) on large scale can be injected into the plant apart from daily cattle dung feeding. The utilisation of the digestate reduces dependence on chemical fertilisers, the initiative consequently leading to consolidating soil nutrients/fertility sustainably. Reforestation at remote locations may be promoted. Consequently, the cleanness ensures the inhabitants/rural folks preventing from contagious and vector born diseases. Successful operation of the digester also confers the benefits of being not vulnerable to retina (cataract), cardiovascular, decentry, diarrhea and other respiratory attacks on beneficiaries and reduces the incidence of occurrence of such diseases drastically.

74. How does climate affect biogas production?

Tropical climates generally have no problems with temperature because the anaerobic bacteria thrive in higher temperatures. In a temperate climate, one may need to heat the digester during colder months. The ideal temperature for the mesophilic bacteria is 38°C.

75. Isn't methane a greenhouse gas?

Methane is a greenhouse gas that is more than 20 times as damaging to the environment as CO_2, but it has to be released into the atmosphere to have 'greenhouse effect'. That is why care must be taken to contain the biogas from anaerobic digesters, and not let it release into the air. A benefit of anaerobic digestion is its ability to capture CH_4 that would normally be emitted into the air if the organic material was left to decompose uncontrollably.

76. Does biogas contribute to climate change?

Biogas combustion produces CO_2, a greenhouse gas. The carbon in biogas comes from plant matter that fixed it from atmospheric CO_2. Thus biogas is C-neutral and doesn't add to green-house gas emissions. Also, reduced fossil fuel use due to biogas lowers the CO_2 emissions.

77. What are the environmental benefits of biogas?

Methane has a greenhouse gas (GHG) factor 20 times greater than CO_2. Capturing methane from a substrate, which would anyway be emitted to atmosphere, and utilising it for other purposes will obviously reduce the GHG impact. Biogas energy is considered to be carbon neutral since carbon emitted by its combustion comes from organic matter that fixed the carbon from atmospheric CO_2. Additionally, utilising biogas will replace fossil fuels, a main contributor of green house gases.

78. What are the uses of biogas?

Biogas Used in gas stove as fuel in rural areas, street lighting and also for running engine.

79. What are disadvantages of biogas as a fuel source?

The main disadvantage of biogas is the loss of organic wastes for compost or fertiliser.

80. Who can benefit from biogas?

Farm owners, food processors, waste managers and sewage treatment plant operators can all benefit from treating their organic waste streams with biogas technologies.

81. How to compare biogas-electricity-LPG market prices in typical Indian households?

Comparison of market prices of household consumer (2015-16) - Odisha as a case

	Biogas	Electricity bill (Rs.)	LPG billsubsidised/ open market (Rs.)	Firewood bill(Rs. 3/kg)	Kerosene billPDS/open market(Rs. 16–40/l)
Paid Monthly	18,000 (one time investment)	300 (avg)	450/650	2.5 quintals750	10 litrs160/400
Paid Annually	18,000	3600 (avg)	5400/7800	9000	1920/4800
Paid in 5 yrs	20,000 (incl. maint. charges)	18000	27000/39000	45000	9600/24000

Use of a $2m^3$ plant annually compensates at least 20% electricity bill, 80-85% saving on LPG expenditure, and 85–90% saving on firewood, and a complete replacement of kerosene

82. How do you calculate carbon credits for methane emission reduction?

Carbon credit is calculated based on the amount of methane reduced through the installation of an anaerobic digester. This approach considers:

1. How much methane was emitted before the digester was installed (or 'baseline' emission)?
2. How much methane is emitted after the digester is installed?

Baseline emissions majorly influence the carbon credit that can be claimed, and can vary a great deal for farm-based anaerobic digesters based on how manure was originally managed.

For example, storage lagoons emit much more CH_4 than manure stacks. A farm originally managing its manure in storage lagoons would likely get a higher carbon credit than a farm generating the same amount of biogas that originally managed its manure in manure stacks.

83. How biogas technology helps in eliminating pollution?

Organic municipal solid wastes are major concern of the solid waste that release leachate (liquid) and odor that contaminate ground water and air pollution after decomposition, which can utilised as a potential source of feedstock for biogas production.

84. How biogas affects human health?

Burning of fossil fuels and wood generate lot of fume, carbon dioxide and carbon monoxide that gives negative impact on human respiratory system where as biogas does not release any fumes.

85. Does methane affects negatively on human health on exposure?

At normal environmental concentrations, methane has no impacts on human health. At extremely high concentrations in an enclosed space the reduction in oxygen levels could lead to suffocation.

G. Technical issues in biogas technology

86. What is biogas technology?

The technology involved in biogas generation in anaerobic condition (absence of oxygen) is called as biogas technology.

87. Where does methane come from?

Methane is the dominant component in natural gas. Oil and natural gas industry is actually not the largest source of methane emissions. The top producer is the agricultural sector, specifically the manure, belches and flatulence that comes from livestock, mainly cattle.

88. What are natural sources of methane in environment?

One of the main sources of methane into the environment is from the natural decomposition of plant and animal matter in airless conditions. This occurs in marshes (wetland), rice paddies and the guts of animals.

89. What does bioresource mean in renewable energy concept?

Since many years, people have viewed wastes as rubbish to throw away, bury and forget. Reusing and recycling materials rather than just disposal after use is preferable due to the pressures on environment, fossil fuels and virgin materials. The concept of a bioresource facility simply means a biological process employed to extract resources by sourcing materials in wastes to use elsewhere, recycled and/or generate renewable fuel and electricity.

90. What is the energy-content of biogas?

Biogas with a methane content of 60% will have an energy content of around 4-6kWh/m^3. If utilised for electric generation, a net 2kWh of usable electricity is produced, others turning into heat energy. The energy content can be increased considerably by upgrading the biogas to Bio-CNG or Bio-Methane, removing the CO_2, providing a gas of suitable quality for injection into the national gas grid or for use as a transport fuel.

91. How much energy is in biogas?

Each cubic meter (m^3) of biogas contains 6kWh equivalent calorific energy. However, when we convert biogas to electricity in a biogas powered electric generator, we get about 2kWh of useable electricity, the rest turning into heat which can also be used as heating energy. 2kWh is enough energy to power a 100W light bulb for 20hrs, or a 2000W hair dryer for an hour.

92. Why is EPA concerned about coal mine methane?

Methane is the second most important greenhouse gas after carbon dioxide. In fact, methane is over 20 times more potent than carbon dioxide (CO_2) on a mass basis over a 100 year time period. Coal mine methane (CMM) represents wasted emissions to the atmosphere, while capture and use of CMM has benefits for the local and global environment.

93. What is the difference between coal bed methane and coal mine methane?

Coal-bed methane refers to the methane found in coal seams formed during coalification, the transformation of plant material into coal. Coal-bed methane is also known as CBM, or virgin coal seam methane or coal seam gas. It is considered an "unconventional" natural gas source. In the US, coal-bed methane is a valuable resource that accounts for about 5% of total annual US natural gas production. Coal mine methane (CMM) refers to methane released from the coal and surrounding rock strata due to mining activities. In underground mines, it is an explosive hazard to coal miners, so it is removed through ventilation systems. In abandoned and surface mines, methane might escape to the atmosphere through natural fissures or other diffuse sources. Like CBM, coal mine methane is a subset of the methane found in coal seams, but it specifically refers to the methane found within mining areas (*e.g.*, within a mining pan). CBM is the methane in coal seams that will never be mined. As CMM releases through mining activities, recovering and using CMM is considered emissions avoidance.

94. How is methane emitted from coal mines?

There are three primary sources of CMM:

- Degasification systems at active underground mines. Also commonly referred to as drainage systems, these systems employ vertical and/or horizontal wells to recover methane before mining takes place to help the ventilation system keep the in-mine methane concentrations sufficiently low (well below the explosive limit) to protect miners.
- Ventilation air methane (VAM). This refers to the very dilute methane that is released from underground mine ventilation shafts. Although it is typically less than 1% methane, it is the single largest source of CMM emissions globally.
- Abandoned mine methane (AMM). Closed mines produce emissions of low- to medium-quality gas from diffuse vents, ventilation pipes, boreholes, or fissures in the ground.

95. What is Kaveri delta coal-bed methane extraction project?

The Kaveri delta coal-bed methane extraction project is presently undertaken by Great Eastern Energy Corporation Ltd (GEECL), a private company based in Gurgaon, Haryana. The project aims to extract methane gas from coal-bed using hydraulic fracturing method of hydraulic fracturing in the Kaveri river basin. The company received licence to explore and extract CBM from

Nagapattinam, Thanjavur and Thiruvarur districts which are the major rice cultivating area of Tamil Nadu. Farmers, environmentalist and experts are opposing the project and hence it is currently suspended by the Government of Tamil Nadu.

96. How much methane is emitted from coal mines?

China leads the world in coal mine CH_4 emissions with nearly 300MMTCO_2e (over 20 billion cubic meters annually) in 2010. Other leading emitters are the US, Ukraine, Australia, the Russian Federation, and India. In 2015, global CH_4 emissions from coal mines were projected to be about 630MMTCO_2e, accounting for 8% of total global methane emissions. In 2012, US coal mines emitted about 60MMTCO_2e. Between 1990 and 2012, US emissions decreased by over 30%, largely due to the coal mining industry's increased recovery and utilisation of drained gas. Between 2005 and 2011, emissions and the amount of drained gas utilised both increased by around 10% due to increased production.

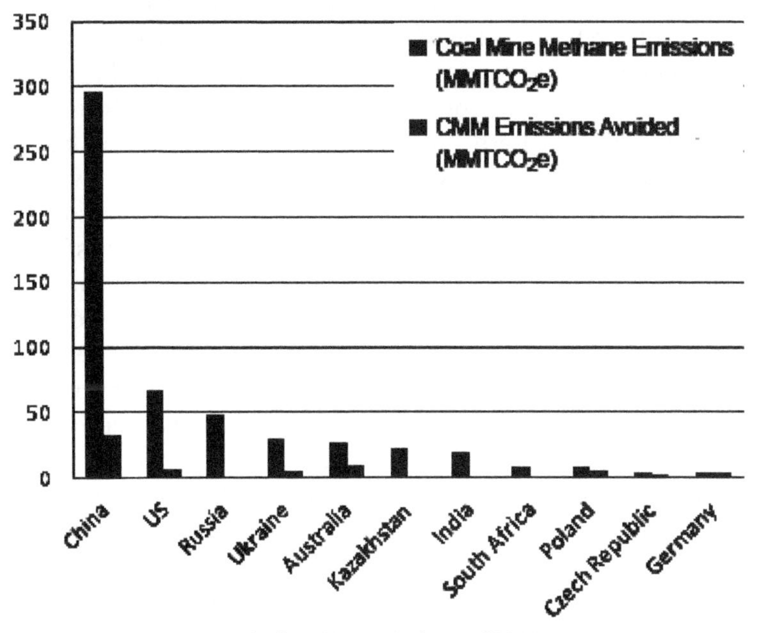

Coalmine CH_4 emissions of 2010
(*Source:* Global anthropogenic non-CO_2 greenhouse gas emissions)

97. How to find out about international developments in coal mine methane project?

CMOP is actively engaged in implementing the international Global Methane Initiative, a voluntary initiative to reduce CH_4 emissions from five key sectors: agriculture, coal mining, municipal solid waste (*e.g.*, landfills), oil & gas systems, and wastewater. The partnership focuses on near-term CH_4 abatement or recovery for use as a clean, profitable energy source.

98. Is methane available on any other planet?

After an exhaustive analysis of data obtained during 605 Martian days, NASA's Curiosity rover has confirmed the presence of methane on Mars environment which may hint that life once existed on the Red Planet. The tunable laser spectrometer in the SAM (Sample Analysis at Mars) instrument of the Curiosity robot has unequivocally detected an episodic increase in the concentration of methane in Mars' atmosphere. This puts an end to the long controversy on the presence of methane in Mars, which started over a decade ago when this gas was first detected with telescopes from the Earth, the authors from the Mars Science Laboratory (MSL) reported.Since methane can be the product of biological activity - practically all the existing methane in the Earth's atmosphere originates in this way - this has created great expectations that Martian methane could also be of a similar origin. "It is a finding that puts paid to the question of the presence of methane in the Martian atmosphere but it does pose some other more complex and far-reaching questions, such as the nature of its sources," said study co-author Francisco Javier Martin-Torres from the Andalusian Institute of Earth Sciences (CSIC-UGR) at the University of Granada, Spain. "The sources, we believe, must lie in one or two additional sources that were not originally contemplated in the models used so far. Among these sources, we must not rule out biological methanogenesis," he added. According to some current models, if there really existed methane in Mars, it would remain there for an average 300 years and during this period, it would be homogeneously distributed across the atmosphere. SAM has been detecting basal levels of methane concentration and has confirmed an event of episodic increase of up to 10 times this value during a period of 60 Martian days. The new data are based on observations during almost one Martian year (almost two Earth years), included in the initial prediction for the duration of the mission (nominal mission), during which Curiosity has surveyed about 8 kms in the basin of the Gale crater. The newly arrived MAVEN (Mars Atmosphere and Volatile Evolution) from NASA will provide continuity for the study of this subject, the US space agency said in a statement. In the near future, the Trace Gas Orbiter (TGO), jointly developed by the European Space Agency (ESA) and the Russian Space Agency (Ruscosmos) will measure the concentration of methane on Mars at a larger scale. The paper was published in the reputed journal 'Science'.

99. What is the fermentative pathway behind methane production?

In fermentative pathway, acetic acid undergoes a dismutation reaction to produce methane and carbon-di-oxide (one electron is transferred from the carbonyl function (e^- donor) of the carboxylic group to the methyl group (e^- acceptor) of acetic acid to respectively produce CO_2 and CH_4). Dismutation reaction is enzymatically catalysed. Archaea that catabolising acetate for energy are referred as *acetotrophic* or *aceticlastic*. Methylotrophic archaea utilise methylated compounds such as methylamines, methanol, and methanethiol as well.

$CH_3COO^- + H^+ \rightarrow CH_4 + CO_2$ $\Delta G° = -36$ kJ/reaction

Reduction of carbon-di-oxide to methane in the presence of hydrogen can be expressed as:

$CO_2 + 4\ H_2 \rightarrow CH_4 + 2H_2O$

Some CO_2 react with H_2 to produce CH_4 creating electrochemical gradient across membrane (used to generate ATP by chemiosmosis). Plants and algae utilise H_2O as reducing agent.

100. How is the physical appearance of methanogens?

Methanogens can be coccoid or bacilli. They are anaerobic organisms and cannot function under aerobic conditions and are very sensitive to the presence of oxygen even at trace level. Usually, they cannot sustain oxygen stress for a prolonged time. *Methanosarcina barkeri* is exceptional in possessing a superoxide dismutase (SOD) enzyme, and may survive longer than the others in the presence of O_2. Hydrogenotrophic methanogens use carbon dioxide (CO_2) as carbon source and hydrogen as reducing agent. Methanogens lack peptidoglycan but possess a cell wall of pseudopeptidoglycan. Other methanogens do not, but have at least one paracrystalline array (S-layer) made up of proteins that fit together like a jigsaw puzzle.

101. What is the role of temperature in biogas production?

Biomethanogenesis process is very sensitive to temperature. The degree of sensitivity, in turn, is dependent on the temperature range. Brief fluctuations not exceeding the following limits may be regarded as still un-inhibitory with respect to the process of fermentation.

102. What is the Hydraulic Retention Time (HRT)?

Hydraulic retention time (m^3/d) is the time duration of complete digestion of organic feed stock. This is the length of time for which a substrate is calculated to remain on an average in the digester until discharged. HRT=VR (volume of the reactor)/VD (volume of the substrate).

103. What is organic loading factor?

It indicates how many kilograms of volatile solids (VS, or organic dry matter) can be fed into the digester per m^3 of working volume per unit of time. The organic loading rate is expressed as kg VS/ (m3·d). Organic loading rate can be specified for each stage (gas-tight, insulated and heated vessel), for the system as a whole (total working volumes of all stages) and with or without the inclusion of material recirculation. Changing the reference variables can lead to sometimes widely differing results for the organic loading rate of a plant.

104. What's the role of N_2 and C/N ratio during the fermentation?

The N_2 inhibition: All substrates contain nitrogen. For higher pH values, even a relatively low nitrogen concentration may inhibit the process of fermentation. Noticeable inhibition occurs at a nitrogen concentration of roughly 1700mg ammonium-nitrogen (NH_4-N) per unit substrate. Nonetheless, given enough time, the methanogens are capable of adapting to NH_4-N concentrations in the range of 5000–7000mg/l substrate, the main prerequisite being that the ammonia level (NH_3) does not exceed 200–300mg NH_3-N per liter substrate. The rate of NH_3 dissociation in water depends on the process temperature and pH value of the substrate.

The C:N ratio: Microorganisms need both nitrogen and carbon for assimilation into their cell structures. Various experiments have shown that the metabolic activity of methanogenic bacteria can be optimised at a C/N ratio of approximately 8-20, whereby the optimum point varies from case to case, depending on the nature of the substrate.

105. What is the role of agitation in fermentation/digestion?

Many substrates and various modes of fermentation require some sort of substrate agitation or mixing in order to maintain process stability within the digester. The most important objectives of agitation are: removal of the metabolites produced by the methanogens (gas), mixing of fresh substrate and bacterial population (inoculation), preclusion of scum formation and sedimentation, avoidance of pronounced temperature gradients within the digester, provision of a uniform bacterial population density and prevention of the formation of dead spaces that would reduce the effective digester volume.

106. What are the major process-controlling parameters?

Most of the anaerobic digesters are operated either in the mesophilic (moderate) temperature range (35–37°C) or in the thermophilic (higher) temperature range (55–60°C). The pH of the slurry in the digester is maintained between 6.5 and 7.5. The typical retention time of organic matter in the anaerobic digesters varies from 2 to 30 days, depending on the type of digester and the concentration of organic matters processed. Anaerobic digesters can process liquid organic waste with a solids concentration in the range of 0.5 to 12%.

107. What is ignition temperature?

It's the minimum temperature at which a substance catches fire and starts burning.

108. What is ventilation air methane and how can it be used?

To ensure mine safety, fresh air is circulated through underground coal mines using ventilation systems to dilute in-mine concentrations of methane to levels well below explosive levels. These concentrations are regulated by mine safety authorities in each country. Typically, methane concentrations in ventilation air range from 0.1-1.0%. Ventilation air methane (VAM) refers to the very dilute methane that is released from underground mine ventilation shafts. VAM represents over half of all coal mining emissions in the United States and worldwide. With few exceptions, it is simply released to the atmosphere. It is technically possible to convert the dilute methane in ventilation air to useful energy. The high volumetric flow rate and low concentrations of ventilation air methane make it challenging to capture and utilise cost-effectively. Technologies to capture and harness the energy resource of VAM are currently being developed, demonstrated, and commercialised.

109. What is the role of ammonia in anaerobic digestion process?

Ammonia (NH_3) is an important compound, with a significant function for the AD process. NH3 is an important nutrient, serving as a precursor to foodstuffs and fertilisers and is normally encountered as a gas, with the characteristic pungent smell. Proteins are the main source of ammonia for the AD process. Too high ammonia concentration inside the digester, especially free ammonia (the unionised form of ammonia), is considered to be responsible for process inhibition. This is common to AD of animal slurries, due to their high ammonia concentration, originating from urine. For its inhibitory effect, ammonia concentration should be kept below 80 mg/l. Methanogenic bacteria are especially sensitive to ammonia inhibition. The concentration of free ammonia is direct proportional to temperature, so there is an increased risk of ammonia inhibition of AD processes operated at thermophilic temperatures, compared to mesophilic ones. The freeammonia concentration is calculated from the equation: where $[NH_3]$ and $[T\text{-}NH_3]$ are the free and respectively the total ammonia concentrations, and ka is the dissociation parameter, with values increasing with temperature. This means that increasing pH and increasing temperature will lead to increased inhibition, as these factors will increase the fraction of free ammonia. When a process is inhibited by ammonia, an increase in the concentration of VFA will lead to a decrease in pH. This will partly counteract the effect of ammonia due to a decrease in the free ammonia concentration.

110. Give a brief idea on landfill gas recovery plants?

Landfills can be considered as large anaerobic plants with the difference that the decomposition process is discontinuous and depends on the age of the landfill site. Landfill gas has a composition which is similar to biogas, but it can contain toxic gases, originating from decomposition of waste materials on the site. Recovery of landfill gas is not only essential for environmental protection and reduction of emissions of methane and other landfill gases, but it is also a cheap source of energy, generating benefits through faster stabilisation of the landfill site and revenues from the gas utilisation. Due to the remoteness of landfill sites, landfill gas is normally used for electricity generation, but the full range of gas utilisation, from space heating to upgrading to vehicle fuel and pipeline quality is possible as well. Landfill gas recovery can be optimised through the management of the site such as shredding the waste, re-circulating the organic fraction and treating the landfill as a bioreactor. A landfill bioreactor is a controlled landfill, designed to accelerate the conversion of solid waste into methane and is typically divided into cells, provided with a system to collect leachate from the base of the cell. The collected leachate is pumped up to the surface and redistributed across the waste cells, transforming landfill into a large high-solids digester.

111. What is gas chromatography?

A device which is used to calculate the presence of CH_4, CO_2, H_2S etc in a biogas sample. The measurement can be done on-line or off-line. It works on the principle of separation of the various constituent gases in a sample by providing varying injection temperature in the column.

Common Terminologies in Biogas Research

Acetoclastic methanogenesis: Production of methane from the cleavage of acetate moiety is called as acetoclastic methanogenesis.

Acetogenesis: It is the anaerobic process in which acetic acid is produced from CO_2 and an electron source (*e.g.*, H_2) *via* reductive acetyl Co-A pathway.

Acidogenesis: A biological reaction where simple monomers are converted into volatile fatty acids.

Adenosine triphosphate (ATP): ATP is a complex organic moiety with energy-rich phosphate bond that provides energy to drive many biological processes.

Aerobic microbes: An aerobic organism or aerobe is an organism that can survive and grow in an oxygenated environment.

Aerobic respiration: The process in which glucose is metabolised to CO_2 and H_2O in the presence of oxygen, releasing large amounts of ATP.

Agricultural refuse: The agricultural refuse are the agro-wastes produced in agriculture farms due to anthropogenic activities.

Anaerobic digester: The specially designed vessel and associated systems for anaerobic digestion of organic material by associated microbes.

Anaerobic digestion: Anaerobic digestion is a collection of processes by which microorganisms break down biodegradable material in absence of oxygen.

Anaerobic bacteria: Anaerobic bacteria are organisms that are capable of surviving and growing in an atmosphere of no oxygen.

Anaerobic respiration: Anaerobic respiration is respiration using electron acceptors other than molecular oxygen (O_2).

Biogas: Biogas refers to a mixture of different gases produced by the breakdown of organic matter in the absence of oxygen.

Biological Oxygen Demand (BOD): Biochemical Oxygen Demand is the amount of dissolved oxygen needed by aerobic biological organisms to break down organic material present in a given water sample at certain temperature over a specific time period.

Biomethanation: Biomethanation or methanogenesis is the formation of methane by microbes known as methanogens.

Biochemical methane potential: Biochemical Methane Potential (BMP) is a laboratory test assessing the potential biogas yield of a feedstock.

Biomethaniser: A sealed container/tank where the biological digestion of organic matter takes place for biogas production.

Carbon to Nitrogen Ratio (C/N): A carbon to nitrogen ratio is the ratio of the mass of carbon to the mass of nitrogen present in a substance.

CFA: Central financial assistance is a financial assistance provided by the central government to beneficiaries.

Chromatography: Chromatography is a laboratory technique for the separation of a mixture. **Combined heat and power plant (CHP)**: Combined heat and power is the use of a heat engine or power station to generate electricity and useful heat at the same time.

Digestate: Digestate is the material remaining after the anaerobic digestion of a biodegradable feedstock.

Electron transport chain (ETC): An electron transport chain is a series of complexes that transfer electrons from electron donors to electron acceptors via redox reactions, and couples this electron transfer with the transfer of protons across a membrane.

Emission: The production and discharge of something, especially gas or radiation.

Facultative aerobic/anaerobic microbes: A facultative aerobe/anaerobe generates ATP by aerobic respiration in presence of oxygen, but is capable of switching to fermentation or anaerobic respiration in absence of oxygen.

Farmyard manure: Farmyard manure refers to the decomposed mixture of dung and urine of farm animals along with litter and left-over material from roughages or fodder fed to the cattle.

Feedstock: Any bulk raw material constituting the principal input for an industrial process.

Fermentation: Fermentation is a metabolic process that produces chemical changes in organic substrates through the action of enzymes.

Fertilising residue material: Residual organic matter used as biofertiliser/ manure in agriculture, horticulture and forestry.

Gas chromatograph: A gas chromatograph (GC) is an analytical instrument that measures the content of various components in a sample.

Greenhouse gas (GHG): A greenhouse gas is a gas that absorbs and emits radiant energy within the thermal infrared range.

Hydraulic retention time (HRT): The hydraulic retention time is a measure of the average length of time that a compound remains in a storage unit.

Hydrogen sulphide (H_2S): Hydrogen sulfide is a colorless, flammable, extremely hazardous gas with a 'rotten egg' smell.

Hydrogenic methanogenesis: The process of methane production using CO_2 as a substrate and H_2 as an electron donor.

Hydrolysis: Hydrolysis is the chemical break down of a compound due to reaction with water.

Influent: Biomass on the in-flow side of a treatment, storage, or transfer device.

Loading rate: Loading rate defines the rate of feeding of any substrate to a reactor.

Manure: Manure is organic matter, mostly derived from animal feces except in the case of green manure, which can be used as organic fertiliser in agriculture.

Mesophilic: A moderate temperature range of 25–40°C.

Methane (CH_4): Methane is the simplest member of the paraffin series of hydrocarbons and is among the most potent of the greenhouse gases.

Methanogenesis: Methanogenesis or biomethanation is the formation of methane by microbes known as methanogens.

Methanogens: Methanogens are microorganisms that produce methane as a metabolic byproduct in hypoxic conditions.

MNRE: The Ministry of New and Renewable Energy, Government of India.

NBMMP: The National Biogas and Manure Management Programme.

NNBOMP: The New National Biogas and Organic Manure Programme.

Nutrients: A nutrient is a substance used by an organism to survive, grow, and reproduce.

Obligate aerobes: An obligate aerobe is an organism that requires oxygen to grow.

Obligate anaerobes: Obligate anaerobes are the organism which can live in environments which lack oxygen.

Oxidation: Oxidation is the loss of electrons during a reaction by a molecule, atom or ion.

Phosphorylation: A biochemical process that involves the addition of phosphate to an organic compound.

Reduction: Reduction is chemical reaction that involves the gaining of electrons by one of the atoms involved in the reaction.

Respiration: Respiration is a metabolic process common to all living things for the generation of energy.

Respiratory phosphorylation: It is a metabolic process resulting in formation of ATP by aerobic or anaerobic respiration.

Slurry: Slurry is a thin and viscous fluid mixture composed of a pulverized solid and liquid.

Substrate-level phosphorylation: Substrate-level phosphorylation is a metabolic reaction that results in the formation of ATP or GTP by the direct transfer of a phosphoryl group to ADP or GDP from another phosphorylated compound.

Terminal electron acceptor: A compound that receives or accepts a electron during oxidation.

Thermophilic: Organisms growing at a high temperature range of 50–80°C.

Total solids (TS): Total solids is a measure of the suspended and dissolved solids in water.

Volatile fatty acids: Volatile fatty acids (VFAs) are short chain fatty acids composed mainly of C_2–C_6 carboxylic acids produced in the anaerobic digestion process.

Volatile solids: Solids in water or other liquids that are lost on igniting the dry solids.

Glossary of Abbreviations and Frequently Used Symbols in Biogas Research

Description	Symbol/ Abbreviation	Unit
Absolution temperature of the gas to be compressed	T	K
Air:fuel ratio for stoichiometrically equivalent air:fuel ratio	λ	-
Amortisation per year for the CHP	KK	US$/a
Animal unit	GVE	-
Annual amortisation for concrete works	t_B	US$/a
Annual amortisation for technical equipment	t_T	US$/a
Annual amortization for technical equipment	v_T	US$/a
Annual operation time	t_s	h/a
Area for cultivation of energy plants	A	m^2
Area load	w_s	$m^3/m^2/h$
Average bioreactor volume load	B_{BR}	kgDM/m/d
Average temperature difference between heating medium and substrate	$\Delta\vartheta_{BH}$	°C
Billion	bn	-
Biogas pressure after compressing	P^2	bar
Biogas pressure before compressing	P^1	bar
Biomass concentration excess sludge	C_S	$kgCOD/m^3$
Biomass concentration in the reactor	X	$kgDM/m^3$
Bioreactor area load	B_A	kgDM/m/d
Bioreactor height	H_{BR}	m
Bioreactor volume	V_{BR}	m^3
Bioreactor volume load	B_R	$kgDM/m^3/d$, $kgCOD/m^3/d$
Breadth	B_S	m
Breathing activity	AT_4	mg O_2/gDM
Calorific value	$H_{O,N,}$ $H_{U,N}$	kWh/m^3
Capacity of the plant to deliver electrical energy	E_{el}	kW
Capacity of the plant to deliver heat	E_{th}	kW
Chemical oxygen demand	COD	-
COD value	COD	mg O_2/l
COD value of untreated sample	COD_0	mg O_2/l
Compressor throughput	V_K	Nm^3/h
Concentration of organics in the substrate	c_0	$kgCOD/m^3$

Constants	C_1,C_2	-
Cost for ignition oil	K_{OIL}	US$/a
Costs for cultivation of renewable resources	KR	US$/a
Costs for heat losses	KW	US$
Costs for power consumption	KS	US$/a
Cultivation area for maize	A_M	ha
Decanter diameter	D_D	m
Decanter length	L_D	m
Degree of decomposition determined by oxygen demand value	A_S	-
Degree of decomposition determined by the COD value	A_{COD}	-
Density of conferment	ρ_S	kg/m³
Density of heating medium	ρ_w	kg/m³
Density of Siran	ρ_{FS}	g/cm³
Density of substrate	ρ_G	kg/m³
Dew point temperature	?S	°C
Diameter o aeration pipe	D_L	m
Diameter of bioreactor	D_{BR}	m
Diameter of discharge pipe	D_{BRI}	m
Diameter of heating pipe	D_{HR}	m
Diameter of preparation tank	D_{PT}	m
Diameter of residue storage tank	D_E	m
Diameter of windings of heating pipe	D_W	m
Difference in oxygen concentration (day 1 *vs* day 5)	BOD_5	Mg O_2/l
Differences in absolute temperatures	ΔT_E, “T_A	K
Disintegration intensity	B	kJ/kg
Dry matter	DM	% or g/l
Dry matter in outflow of sludge bed reactor	$DM_{R,e}$	g/l
Economic potential	P_{econ}	kWh/a
Efficiency of the compressor	η_K	%
Efficiency of the preparation tank pump	η_{VP}	%
Efficiency to produce electrical energy	η_{el}	%
Efficiency to produce heat	η_{th}	%
Electrical power consumption of the plant	E_{Eel}	kW
Factor to increase the bioreactor volume	f_{VBR}	-
Factor to increase the residue storage tank	f_{VE}	-
Factors to increase the preparation tank	f_{VPT}	-
Feedback from the residue storage tank to the bioreactor	V_E	m³/d
Filling height for pellet sludge	H_{BP}	m
Flow of gas to the compressor	m*	M³/h
Flow rate of coferments	@$_S$	Mg/a
Flow rate of dry matter into the bioreactor	DM_{BR}	kgDM/d
Flow rate of heating medium in the pipe	V_w	m³/h
Flow rate of ignition oil	@$_{oil}$	Mg/d
Flow rate of substrate	@$_G$, @$_{G1}$, @$_{G2}$	mg/d
Flow rate of the preparation tank pump	V_{VP}	m³/h
Fraction of the investment costs without CHP for concrete works	x_B	-
Fraction of the investment costs without CHP for technical equipment	x_T	-
Gas formation within 21days	GB_{21}	nl/kgDM

Gas velocity in empty reactor	wG	$Nm^3/m^2/s$
German industrial norm	D_{IN}	-
Gibbs free energy	$\Delta G'_f$	kJ/mol
Grinding ball density	ρ_{MK}	kg/m^3
Grinding ball diameter	d_{MK}	m
Heat loss of the bioreactor	Q_{BR}	kW
Heat transfer coefficient at the wall inside the bioreactor	$(\alpha BR)_i$	$W/m^2/K$
Heat transfer coefficient at the wall inside the heating pipe	$(\alpha H)_i$	$W/m^2/K$
Heat transfer coefficient at the wall outside the bioreactor	$(\alpha BR)_a$	$W/m^2/K$
Heat transfer coefficient at the wall outside the heating pipe	$(\alpha H)_a$	$W/m^2/K$
Heat transmission coefficient of the insulation of the bioreactor	λBR	W/m/K
Height of silo	H_S	m
Height of the gas/solid separator	H_{BS}	m
Height of the preparation tank	H_{PT}	m
Height of the residue storage tank	H_E	m
Inflow rate	$V_G{}^*$	m^3/d
Inhabitant	IN	-
Insurance costs	*KV*	US$/a
Investment costs for concrete works	*KB*	US$
Investment costs for technical equipment	*KT*	US$
Investment costs for the CHP	K_{CHP}	US$
k-Factor of the bioreactor wall with insulation	k_{BR}	$W/m^2/K$
k-factor of the heating pipes	k_H	$W/m^2/K$
Length of the heating pipe	L_{HR}	m
Length of the silo	L_S	m
Local overhead costs	*KP*	US$/a
Lowest ambient temperature	?A	°C
Maintenance costs for the concrete work	*KX*	US$/a
Maintenance costs of technical equipment	*KY*	US$/a
Maintenance costs of the CHP	*KZ*	US$/a
Maximum COD value	COD_{max}	mg O_2/l
Maximum temperature difference between substrate and outside of reactor	$\Delta\vartheta_{BR}$	°C
Maximum temperature difference in substrate inside and outside of reactor	$\Delta\vartheta_{SU}$	°C
Maximum theoretical yield	E_{Rmax}	mg_{DM}/ha/a
Million	Mio	-
Minimum tolerable theoretical residence time	t_{min}	h
Molecular weight	M_E	kg/kmol
Net income from current	*G*	US$/a
Net income from fertiliser	*D*	US$/a
Net income from heat	*W*	US$/a
Newton number of an agitator	Ne_{BRR}	-
Nominal capacity of electrical power of the CHP	*E*	kW
Normal	N	-
Normal biogas density	$\rho_{BG}{}^*$	kg/Nm^3
Not specified	n.s.	-
oDM in the outflow of a sludge bed reactor	$oDM_{R,e}$	g/l
Organic dry matter	oDM	KgCOD, kgDM

Organic sludge load	B_{RoDMSB}	kg/kg/d
Outer diameter of agitator	D_{BRR}	m
Overlapping	S	mm
Oxygen uptake of untreated substrate	OUR_0	mg/l/min
Oxygen uptake rate	OUR	mg/l/min
Plant investment costs without CHP	KA	US$
Plate inclination	Á	-
Population equivalent	PE	-
Porosity	ε	%
Porosity of Siran (porous glass)	ε_{FS}	%
Pour diameter of Siran	d_{FS}	m
Power consumption of a coferment conveyor	P_{SC}	kW
Power consumption of agitator	P_{BRR}	kW
Power consumption of compressor	P_A	kW
Power consumption of the air compressor	P_K	kW
Power consumption of the pumps	P_{VP}	kW
Pressure after compressor	pK2	bar
Pressure before compressor	pK1	bar
Pressure head of the preparation tank pump	ΔP_{VP}	bar
Produced flow of biogas	@$_{BR}$	mg/d
Produced flow of biogas	V_{BR}	m^3/d
Relative density	ρ	kg/Nm^3
Required energy to heat the substrate	Q_{SU}	kW
Residence time	T	d
Residence time in the activated sludge tank	t_{TS}	d
Residence time in the activated sludge tank	v_{TS}	d
Residence time in the bioreactor	t_{BR}	d
Residence time in the preparation tank	T_{PT}	d
Residence time in the residue storage tank	t_E	d
Revolutions of an agitator	^{n}BRR	rpm
Rotational velocity of the agitator system	v_u	m/s
Running time of a coferment conveyor	t_{SC}	h/d
Silo volume	V_S	m^3
Sludge age	Θ	D
Sludge volume index	$^I{}_{SV}$	mg/l
Solar energy	E_S	kW
Special gas constant for CH_4	R_{CH4}	kJ/kg/K
Specific biogas energy	E_{spec}	kW/m^3
Specific cost for ignition oil	$K_{OILspec}$	US$/l
Specific costs for cultivation of renewable resources	KR_{spec}	US$/ha/a
Specific costs for power consumption	KS_{spec}	US$/kWh
Specific economic potential	P_{econ}	kWh/ha/a
Specific energy per volume of ignition oil	$E_{OILspec}$	kWh/l
Specific enthalpies at different stages of the process	h_1, h_2, h_3, h_4, h_5	kJ/kg
Specific heat capacity of the heating medium	c_w	kJ/kg/K
Specific heat capacity of the substrate	cSU	kJ/kg/K
Specific investment costs for biogas plant per unit volume of bioreactor	KA_{spec}	US$/$m^3$
Specific investment costs for CHP per unit capacity of electrical energy	KK_{spec}	US$/kW

Specific local overhead costs	KP_{spec}	US$/h
Specific maintenance costs for CHP	y_{CHP}	US$/a
Specific maintenance costs of technical equipment	y_T	US$/a
Specific maintenance costs of the concrete work	y_B	US$/a
Specific price for sold current	KB_{spec}	US$/kWh
Specific price for sold heat	KW_{spec}	US$/kWh
Specific surface area	O_{spec}	m^2/m^3
Specific surface area of Siran	O_{FSspec}	m^2/m^3
Specific technical potential	P_{techn}	kWh/ha/a
Specific theoretical potential	P_{theor}	kWh/ha/a
Specific work of the compressor	w_t	kJ/kg
Surface of the bioreactor where heat is lost	A_{BR}	m^2
Technical potential	P_{techn}	kWh/a
Technically usable area	A_{Dtechn}	ha
Temperature	?	°C
Temperature difference in inlet and outlet of heating medium to bioreactor	$\Delta\vartheta_H$	°C
Temperature of the heating medium at the inlet	αHE	°C
Temperature of the heating medium at the outlet	?HA	°C
Temperature of the substrate in the bioreactor	αBR	°C
Theoretical potential	P_{theor}	kWh/a
Theoretical yield	E_R	mg_{DM}/ha/a
Thickness of the insulation of the bioreactor	s_{BR}	m
Threshold time value – PEL, permissible exposure limit	TLV	-
Time for discharging the reactor content	t_{BRI}	h
Time of amortisation for the CHP	t_K	a
Time of local work	t_P	h
Time of operation of an agitator	t_{BRR}	min/h
Total available area	A_D	ha
Total energy	E_{tot}	kW
Total heat loss	Q_V	kW
Total investment costs	K, K_1, K_2	US$
Total organic carbon	TOC	ppm, mg/ l
Total organic carbon in the residue	TOC^*	% DM
Total organic carbon in the substrate	TOC	ppm, mg/ l
Total power consumption of the agitators	$(P_{BRR})_{tot}$	kW
Total power consumption of the coferment conveyors	$(P_{SC})_{tot}$	kW
Velocity in the discharge pipe	v_{BRI}	m/s
Velocity of air in aeration pipe	v_L	m/s
Velocity of gas in gas pipes	v_F	m/s
Velocity of inflow	v_G	m/h
Velocity of the heating medium in the pipe	v_H	m/s
Velocity of the substrate in heat exchanger pipes	v_W	m/s
Velocity of the upstream	v_A	m/h
Velocity of the upstream	v_A	m/h
Volume flow of co-ferment in the conveyor	V_{SC}	m^3/h
Volume of compressor pressure vessel	V_K	m^3
Volume of residue storage tank	V_E	m^3
Volume of the gas holder	V_{GS}	m^3
Volume of the preparation tank	V_{PT}	m^3

Volume rate of air in the aeration pipe	V_L	Nm^3/h
Volume rate of ignition oil	V_{Oil}	m^3/d
Volumetric flow of excess sludge	V_S	m^3/d
Wobble index, upper Wobble index, lower Wobble index	W_O, $W_{O,N}$, $W_{U,N}$	kWh/m^3
Yield of CH_4 per biomass	E_M	$kmolCH_4/kg$

1

Methanogenesis Microbiology and Ecology

S.K. Nayak[a], B. Baliyar Singh[a], *D.P. Samantaray[b,1] and B.B. Mishra[b]

[a]*Department of Biotechnology, CET, BPUT, Bhubaneswar, Odisha*
[b]*Department of Microbiology, OUAT, Bhubaneswar, Odisha*
[1]*Corresponding Author: dpsamantaray@yahoo.com*

ABSTRACT

Methanogenesis is a ubiquitous process primarily encountered in anaerobic environments. Methanogens are a group of diverse obligate anaerobic bacteria that produce methane from complex organic matter. Despite their differences in morphological characters and habitat, they show coherence in terms of methane-producing metabolic activity. Being distinctive from all other classical prokaryotes, these are grouped under another kingdom of Archea. Other distinctions of the members are based on their unique cell wall structures and membrane components, as well as differences in their highly conserved gene sequences. Methanogens likely owe their cosmopolitan status to their unique mode of energy metabolism of methane generation. They are active in various ecological niches, from rumens of animals to submarine volcanic vents, sewage digestors and landfills. Methanogenic bacteria are abundant in habitats where high affinity electron acceptors are negligible. Methane fermentation involves bacteria that obtain energy by catalysing anaerobically degradable organic matter to CH_4 *and* CO_2 *by consuming* H_2, CO_2, *and low-carbon organic acids as their substrates. The gaseous products of methanogenesis are greenhouse gases, thus are major contributors to global warming. Methanogens could be used to produce renewable carbon-neutral energy substitute, biogas. Methane cycling archaea play a major role in global methane production as energy for the future.*

Keywords: Anaerobiosis, Archaea, Methanogenesis, Methanogens

Abbreviations used

ppb	: Parts per billion
ppm	: Parts per million
Tg	: Teragram
GI	: Gastrointestinal
Pa	: Pascal
mM	: Millimolar
rRNA	: Ribosomal RNA
GDGT	: Glycerol dialkyl glycerol tetraether
nos.	: Numbers
TCA	: Tricarboxylic acid
E_h	: Redox potential

Background

Methane (CH_4) is an odourless and transparent gas that spans everywhere on this blue planet, spreading from miles below the Earth's surface to miles above it (about 6km). Historical measurements show that CH_4 concentration in the atmosphere is more than 2.5 times higher than preindustrial era, topping about 1800 ppb in recent years (Stocker *et al.*, 2014). Although atmospheric CH_4 concentration was stable during 1999–2005 (~1.8 ppm), its concentrations began to rise again at a rate of 6 ppb per year (Stocker, 2014) since 2007. Apart from water vapour and CO_2, CH_4 is the dominant causative agent of greenhouse effect. The climate warming potential of CH_4 is about 20 times stronger than that of CO_2. Natural sources like wetlands, oceans and termites account for 36% of global CH_4 emissions. Other sources include waste-landfills (16%), biomass burning (11%), rice cultivation (9%), livestock farming and burning of fossil fuels (Bousquet *et al.*, 2006). The total amount of CH_4 produced may be far greater in some important habitats where the microbial CH_4 production is known to be very significant. Methanogens estimatedly produce about one billion tonnes CH_4 annually which suggest that methane-producing bacteria play a major role in the Earth's carbon cycle. Estimates based upon the isotopic composition of the atmospheric CH_4 suggest that about 74% CH_4 is derived from recent microbiological activity (Liu and Whitman, 2008). Being a major greenhouse gas, its generation and atmospheric chemical interaction are of considerable interest, from researchers to policy makers.

On the other hand, CH_4 being highly flammable has renewed the interest as an alternative energy source to petroleum-based fuels. Methane from microbes is more eco-friendly than other bio-fuels produced from feedstock like corn ethanol, not competing with agricultural resources like land, irrigation and fertilisers. Controlled methanogenic indirectly help in reducing greenhouse effect through carbon sequestration as they use CO_2 to make CH_4. Because of their impact on the environment, roles in waste management and energy conservation, there has been considerable study on methanogens pertaining to basic microbiology, ecological diversity, metabolic reaction and genomics. Unique properties of methanogens and significant ecological adaptations have facilitated their enumeration in natural habitats. This chapter discusses the microbial diversity and physiological characteristics with respect to their ecological niches.

Ecological distribution of methanogens

Woese and associates proposed a reclassification of all living organisms in late 70s primarily into three kingdoms: Eubacteria, Archaebacteria and Eukarya. Although morphologically similar, like membrane bound organelles, multiplication by binary fission, many ribosome shapes and metabolic reaction, Eubacteria and Archaebacteria differ in cell wall composition. HA Barker, a pioneer in methanogen research, reorganised the dispersed methanogens into coherent taxonomic group among other well-characterised bacteria in 1956 based on their striking physiological characteristics. The subsequent demonstration of an array of novel coenzymes in methanogens by Balch and Wolfe (1976) provided further support to Barker's conclusions.

Methanogens likely owe their cosmopolitan status to their unique mode of energy metabolism through CH_4 generation. Never-the-less, all methanogens share certain unique and unifying physiological properties. Methanogens are non-spore forming anaerobes. Methanogens generally feed on the end-products of various eubacterial and eukaryotic fermentations and anaerobic oxidations of both complex and simple organic compounds (Jones *et al*., 1987). Even if they are metabolically restricted group, they inhabit virtually every habitat that encourages anaerobic degradation of organic compounds, including extreme niches like geothermal springs and deep-sea hydrothermal vents. Broadly, methanogen ecosystems can be grouped into three different types, such as, a) lakes and marine sediments, marshes, swamps, rice soils, sludge and digesters in which the organic matter is completely degraded, b) the intestinal tracts of ruminants and termites where the degradation process is incomplete and most intermediate products formed (*e.g.*, volatile fatty acids) is reabsorbed as nutrition, and c) extreme thermic conditions (*e.g.*, hot springs) where methanogenesis occurs as a part of the geochemical process. In all of these, the methanogens occupy the terminal niche in the transfer of electrons generated by the anaerobic degradation of organic matter. Microbial studies on methanogenesis are limited, primarily to the ecology and physiology bacteria associated with aquatic sediment, GI or anaerobic digester habitats. Table 1 lists a few of the prominently studied methanogens.

Table 1: Physiological profile of methanogens

Class	Order	Family	Genus	Substrate	Temp	pH	Typical habitat	Remark	Reference
Methano -bacteria	Methano- bacteriales	Methano- bacteriaceae	*Methanobacterium* (Eurythermic)	H_2, formate, 2-propanol, CO_2	37–45°C	7.0-8.5	Anaerobic digesters, freshwater sediments, marshy soils, rumen	Hydrogenotrop, thermo- to alkaliphile	Liu and Whitman, 2008; Jones *et al.*, 1987
			Methanobrevibacter	H_2, methanol, formate	33–45°C	6.0-8.5	Animal/human GI tract, termite gut, anaerobic digester, rice paddy, decaying woody tissues	-	Leadbetter *et al.*, 1998; Miller, 1989
			Methanosphaera	H_2, methanol	37–40°C	6.0-7.0	Animal GI tracts	-	Biavati *et al.*, 1988
			Methanothermo bacter	H_2 , formate	55–65°C	7.0-8.0	Anaerobic digester	Thermophile	Boone *et al.*, 1993
		Methano- thermaceae	*Methanothermus*	H_2, CO_2	80–88°C	5.5-7.5	Hot spring	Thermophile	Lauerer *et al.*, 1986
Methano-cocci	Methano- coccales	Methano- coccaceae	*Methanococcus*	H_2, formate	35–40°C	6.0-8.2	Marine sediment	Mesophile	Burggraf *et al.*, 1990
			Methanothermo coccus	H_2 , formate	60–65°C	6.5-8.0	Marine geothermal	Lithotroph sediment	Boone *et al.*, 1993
		Methano- caldoco ccaceae	*Methanocald ococcus*	H_2	65–94°C	5.2-7.6	Marine geothermal sediment	-	Boone *et al.*, 1993
			Methanotorris (Erst. *Methanoignis*)	H_2	45–90°C	5.0-7.5	Marine geothermal sediment	Extreme thermophile	Burggraf *et al.*, 1990

Class	Order	Family	Genus	Substrate	Temp	pH	Typical habitat	Remark	Reference
Methano -microbia	Methano- microbiales	Methano- microbiaceae	*Meth anomicrobium*	H_2, formate, alcohols	38–43°C (40)	6.0-7.8	Animal GI tracts, anaerobic digester, groundwater	Mesophile	Rivard *et al.*, 1983
			Methanoculleus	H_2 , formate	20–55°C	5.5- 8.0	Anaerobic digester, marine & fresh-water sediment, rice paddy, oil field & hot springs	Mesophile	Zabel *et al.*, 1985
			Methanofollis	H_2 , formate	35–45°C	6.3-8.8	Anaerobic digester	-	Kuhner *et al.*, 1991
			Methanogenium	H_2, formate, methanol, CO_2	15–57°C	6.4-8.1	Marine & freshwater sediment, rice paddy, animal GI tract	Thermophile, psychrophile	Maestrojuan *et al.*, 1990
			Methanolacinia	H_2	30–40°C	7.0	Marine sediment	-	Zellner *et al.*, 1989
			Methanoplanus	H_2, formate	17–41°C	6.5-7.5	Oil field	-	van Bruggen *et al.*, 1986; Liu and Whitman, 2008
		Methano- spirillaceae	*Methanospirillum*	H_2/CO_2, formate, 2-propanol, 2-butanol, CO_2	30–45°C	5.9-7.7	Anaerobic digester, marine sediment mesophile	Hydrogenotroph,	Garcia *et al.*, 2000
		Methano- corpusculaceae	*Methanocorpusculum*	H_2/CO_2, formate, 2-propanol/ CO_2	30–40°C	6.0-8.0	Anaerobic digester, freshwater sediment	Mesophile	Xun *et al.*, 1989
		Methanocalcul aceae	*Methanocalculus*	H_2, CO_2, formate	14–45°C	8.0–10.2	Soda lakes, hypersaline Siberian lake	Strict anaerobe, mesophile	Sorokin *et al.*, 2015
	Methano-	*Methanolinea* regulaceae	H_2, CO_2	20–40°C	6.5–7.4 (37)	(7.0)	Rice paddy, municipal sewage sludge	-	Sakai *et al.*, 2012

Class	Order	Family	Genus	Substrate	Temp	pH	Typical habitat	Remark	Reference
			Methanoregula	H_2, CO_2	20–40°C	4.5–5.5 (5.1)	Acidic bog	Acidophile, hydrogenotroph, ombrotroph	Brauer *et al.*, 2011
			Methanosphaerula	H_2/CO_2, formate, mesophile to mild acidophile	14–35°C (30)	4.8-6.4 (5.5)	Marshy area	Strict anaerobe, mesophile, mildly acidophile	Cadillo-Quiroz *et al.*, 2009
	Methano-sarcinales	Methano-sarcinaceae	*Methanosarcina*	H_2, CO_2, acetate, methanol & methylamine	35–60°C	5.5-7.0	Anaerobic digester, marine & fresh -water sediment, animal waste lagoon, rumen	Acetotroph/ Acetoclastic methanogen	Murray and Zinder, 1985; Sowers and Gunsalus,1988;
			Methanococcoides	(Tri/di/mono) methylamine & methanol	23–35°C	7.0-7.5	Marine sediment	Obligate methanogen, Psychrophile, mildly halophile	Liu and Whitman, 2008; Sower and Ferry, 1983
			Methanohalobium	Methaylamine ($MeNH_2$)	40–55°C	6.5-8.3	Hypersaline sediment	Obligate methanogen, halophile	Liu and Whitman, 2008
			Methanohalophilus	$MeNH_2$	35–40°C	6.3-8.0	Hypersaline sediment	Obligate methanogen, moderately halophile	Liu *et al.*, 1990
			Methanolobus	$MeNH_2$	(37)°C	6.0-7.5	Hypersaline sediment	Mesophile	Sowers and Ferry, 1985
			Methanomethyl ovorans	$MeNH_2$	20–50°C		Freshwater sediment, anaerobic digester	-	-

Class	Order	Family	Genus	Substrate	Temp	pH	Typical habitat	Remark	Reference
			Methanosalsus (Methanosalsum)	$MeNH_2$Dimethylsulfide or methanethiol	35–55°C (50)	8.2-10.3	Hypersaline sediment, salt lakes of Egypt and Kenya	Obligate methanogen, alkaliphile, halophile	Boone *et al.*, 1993; Sorokin *et al.*, 2015
			Methanimicrococcus	H_2, $MeNH_2$, methanol	20–40°C	6.7 - 8.2	Animal GI tract	-	Sprenger *et al.*, 2000
		Methano-saetaceae	*Methanosaeta* (Erst. *Methanothrix*)	Acetate, formate	35–60°C	6.0-8.0	Anaerobic digester, freshwater sediment	Obligate acetotrophic	Patel and Sprott, 1990
		Methermicoccaceae	*Methermicoccus*	Methanol, methylamine & trimethylamine	50-80°C (65)	5.5–8.0	Shengli oilfield, China	Thermophile	Cheng *et al.*, 2007
	Methano-cellales (Erst. RC-I)	Methano-cellaceae	*Methanocella* (3 spp. - *M. paludicola*, *M. arvoryzae*, *M. conradii*)	Acetate (main), H_2, formate	25–40°C (35–37)	6.5–7.8 (7.0)	Rice paddy, wetland, upland	Obligate hydrogenotroph, mesophile	Lyu and Lu, 2015
Methanopyri	Methanopyrales	Methano-pyraceae	*Methanopyrus*	H_2	85–110°C	5.5-7.0	Marine geothermal sediment	Hyperthermophile	Kurr *et al.*, 1991
Thermoplasmata	Methanom-assiliicoccales (Erst. Methanoplasmatales or RC-III)	Methano-massiliicoccaceae	*Methanomassiliicoccus*	H_2, methanol	25–45°C	7.2–8.4	Peat soil (wetland), marine ecology, termite & cockroach gut, human feaces	Obligate hydrogenotroph, mesophile, mildly alkaliphile, chemoheterotroph	Lang *et al.*, 2015; Dridi *et al.*, 2012

Erst.: Erstwhile, GI: Gastrointestinal; RC-I: Rice cluster-I; Obl: Obligate; RC-III: Rumen Cluster-III; Temp: Temperature; Figures in parentheses indicate optimum temperature and pH

Ecological diversification in methanogens

Marine sediments

Reports from stable-hydrogen isotopes studies have depicted that CO_2 reduction by H_2 is a predominant source of methanogenesis in deep marine sediments (Whiticar, 1999; Parkes *et al.*, 2007) producing 75–320Tg CH_4 per annum. However, instead of escaping to atmosphere, most methane is anaerobically oxidised to CO_2 by sulphate-reducing bacteria. Mostly in upper layers of sediments, or sulphate-reducing zones, methanogenesis is limited and accounts for less than 0.1% of total carbon turnover (Capone and Kiene, 1988).

In sediments with high organic matter input, sulphate is depleted with depth, and methanogenesis can become the predominant terminal process. In methanogenic zones, which are usually beneath the sulphate-reducing zones, the dissolved bicarbonate pool is replenished from oxidation of carbon compounds in upper sediments and can obtain concentrations greater than 100mM (Whiticar, 1999). In marine sediments, members of order Methanococcales, Methanomicrobiales and Methanobacteriales (Newberry *et al.*, 2004; Kendall and Boone, 2006) gain energy strictly by CO_2 reduction or formate oxidation. In some marine habitats, methylated compounds are generated from osmolytes of marine bacteria, algae, phytoplankton and some plants. Methylotrophic methanogens of the genera *Methanococcus*, *Methanosarcina* and *Methanolobus* (Kendall and Boone, 2006; Kendall *et al.*, 2007; Lyimo *et al.*, 2000) use these methylated compounds generated in sulphate-rich sediments to produce CH_4 in limited quantity. Acetate being the minor substrate for methanogenesis, few aceticlastic methanogens belonging to *Methanosarcina* use methylated compounds for methanogenesis (Newberry *et al.*, 2004; Elberson and Sowers, 1997; Von Klein *et al.*, 2002). Efficient utilisation of CH_4 by anaerobic methanotrophs is the reason of marine sediment being a minor source of total CH_4 budget.

Freshwater sediments and lowlands

Methanogens are generally absent from the water column of unstratified lakes and rivers as the deep waters are aerated rapidly by convection currents. However, anoxic condition in stratified lakes is observed due to slower oxygen diffusion rate between layers and structural arrangement of sediments restricting oxygen dispersion. Sediment anoxic condition coupled with low (100–200μM) sulphate concentration favours the uninhibited growth of methanogens here compared to marine sediments. In Lake Dagow sediments, the relative abundance of hydrogenotrophic methanogenesis increased from 22–38% (at depth of 0–18cm). It's observed that population of Methanomicrobiales species increased with depth, while that of Methanosaetaceae decreases. In contrast, hydrogenotrophic methanogenesis was observed only in the upper 2cm of Lake Rotsee sediments (Casper *et al.*, 2003; Falz *et al.*, 1999). Due to absence of competition in sulphate-reducing bacteria, the acetate becomes dominant substrate for methanogenesis.

In most freshwater environments, aceticlastic and hydrogenotrophic methanogens that are collectively responsible for 70% and 30% of CH_4 production respectively are isolated (Whiticar, 1999; Conrad, 1999). Here, Methanosaetaceae, Methanomicrobiaceae and Methanobacteriaceae families dominate. Acidic freshwater sediments are dominated by Methanosarcinaceae species. Moreover, methanogenesis from methylated compounds is non-significant because of the absence of methylated compounds in lacustrine conditions (Capone and Kiene, 1988).

Geothermal regions

Methanogenesis also occurs in environments where geochemical sources predominantly provide H_2 and CO_2, required for proliferation of chemolithotrophic methanogenic bacteria (Deuser *et al.*, 1973). Halophiles or thermo-acidophiles isolated from hot springs, solfataras or submarine hydrothermal vents sites are reportedly methanogenic. *Methanobacterium thermoautotrophicum* (optimal temp 65°C; Zeikus *et al.*, 1980) and *Methanothermus fervidus* (optimal temp 83°C; Stetter *et al.*, 1981) were isolated from terrestrial hot springs while *Methanococcus jannaschii* (optimal temp 85°C; Jones *et al.*, 1983), *Methanococcus igneus* (optimal temp 88°C; Burggraf *et al.*, 1990), and *Methanopyrus kandleri* (optimal temp 105°C; Kurr *et al.*, 1991) were isolated from the hydrothermal vents regions of Galapagos Rifts.

The upper photosynthetic layer (cyanobacterial mat) in terrestrial hot spring provides appropriate anaerobic condition to grow methanogenic bacteria *Methanothermus thermoautotrophicum* (Sandbeck and Ward 1982; Zeikus *et al.*, 1980). *Methanothermus* is the only methanogen isolated from solfatara fields, southwest Iceland (Stetter *et al.*, 1981; Lauerer *et al.*, 1986). Although acetotrophic methanogens are less frequent in hot springs, thermophilic *Methanothrix* sp. from Kamchatka, Russia grew on decaying algal materials (Nozhevnikova and Chudina, 1984). Microbial methanogenesis in hypersaline algal mats depend on complex interaction of salinity and methylated amines to osmoregulate in algae (King, 1988). Examples of hypersaline methanogens are *Methanohalophilus mahii* (Paterek and Smith, 1988), *Methanohalobium evestigatum* (Zhilina and Zavarzin, 1987) and *Methanosalsus zhilinae* (Kevbrin *et al.*, 1997).

Cultivated paddy fields

Rice field is a major anthropogenic CH_4 emission source, significantly (21%) contributing to the global greenhouse gas budget (Chen and Prinn, 2005). About 3-6% of photosynthetically fixed CO_2 is converted to CH_4 (Dannenberg and Conrad, 1999). Rice fields develop anoxic conditions when the area is flooded for long period of time during cultivation. In flooded rice paddies, the rapid exhaustion of O_2, NO_3^-, Ferric ion, and SO_4^{2-} with high input of plant carbon develops a favourable condition for methanogenesis. The rhizosphere methanogens are observed with distinct community structure than in the bulk soil. Most H_2-dependent methanogenesis in such ecosystems occurs as a consequence of direct interspecies hydrogen transfer between juxtaposed microbes in soil consortia (Conrad *et al.*, 1985). *Methanomicrobiaceae*, *Methanobacteriaceae*, *Methanosarcinaceae*, *Methanosaetaceae* and *Methanocellaceae* sp. (RC-I) dominate the methanogen population here (Chin *et al.*, 2004; Lu *et al.*, 2005; Krüger *et al.*, 2005).

An interesting estimation is that, the relative populations of methanogens remain constant upon flooding and seasonal drying (Schütz *et al.*, 1989; Krüger *et al.*, 2005; Lueders and Friedrich, 2000). During flooding of rice paddies, hydrogenotrophic methanogens represent 20–50% of the total prevalent methanogens while aceticlastic methanogens dominate in dry seasons (Roy *et al.*, 1997). Studies suggest that incubations of rice roots or rice-field soils with *Methanocellaceae* sp. (RC-I) prefer low H_2 concentrations and supply of amino acids and sugars at moderately high (45–50°C) temperatures. However, upto 80% of the produced CH_4 in planted soils does not reach the atmosphere but apparently oxidise in the rhizosphere (Holzapfel-Pschorn *et al.*, 1985).

Rumens and GI tracts

Enteric fermentation in farm animals is an important source of CH_4 emissions. 27% of the total CH_4 emission (annually, about 90MMT of CH_4) come from livestock farming (Bousquet *et al.*, 2006). Rumen is the primary fermentation-digestion site of a ruminant diet, and depends highly on the interaction and diversity of the microbial consortia (Apajalahti, 2005; Mackie, 2002; Weimer *et al.*, 2009). CH_4 produced in cattle and sheep was 0.46 and 0.75l/d/kg body weight respectively while it was 0.01–0.16l/d/kg body weight (5% of total CH_4 emitted) in monogastric animals.

In 1958, the first pure methanogen culture from the bovine rumen was obtained by Smith and Hungate. During the mid 80s, *Methanobrevibacter smithii* was isolated from human samples as well as from various non-ruminant animals (Miller and Wolin, 1986). Based on the culturable and metagenomic data, *Methanobrevibacter* sp. and *Methanomicrobium* sp. were the most abundant methanogens while *Methanobacterium* sp. and *Methanosarcina* sp. were fewer in number in sheep rumen (Lin *et al*., 1997; Skillman *et al*., 2004; Yanagita *et al*., 2000).

Methanomicrobium was predominant in Indian Murrah buffalo (*Bubalus bubalis*; Chaudhary and Sirohi, 2009). Few non-*Methanobrevibacter* sp., *Methanosphaera*, *Methanomicrococcus*. *Methanobacterium*, *Methanosarcina*, *Methanomicrobium* (ruminant) and *Methanomassiliicoccus luminyensis* (human faeces) are reported. Methanogens have also been isolated from terrestrial arthropods (Hackstein and Stumm, 1994) and insects and termites (Martius *et al*., 1993) guts.

The amount of CH_4 contributed by termites to atmosphere ($2–5 \times 10^{12}$gm annually) is ambiguous. *Methanobrevibacter curvatus*, *M. curticularis* and *M. filiformis* were isolated from the hindgut of termite *R. flavipes* (Kollar) (*Rhinotermitidae*) (Miller, 2001). Studies report that the variability in the number and type (composition) of methanogens in the rumen is not influenced by geographic locations although different herds and different diets could (Sundset *et al.* 2009).

Man-made digester tanks

Being a ubiquitous and anaerobic process, methanogenesis is most often associated with organic matter decomposition in microbial habitats like sewage sludge digesters. It provides alternative source of generating large quantities of renewable fuel in the form of biogas. Anaerobiosis of CH_4 from wastes depends on cooperative action of at least three different categories of metabolic reaction carried out by different microbial groups, *viz*., enzymatic hydrolysis of organic polymers into soluble organic polymers by *Bacterioides*, *Clostridium* and *Streptococcus*, anaerobiosis of monomers into acetate and H_2 by acetogenic bacteria, and finally converting acetate, H_2 and CO_2 into CH_4 by methanogens.

A wide variety of methanogens from Methanomicrobiales and Methanobacteriales are found in anaerobic digestors (McHugh *et al*., 2003). Common aceticlastic methanogens in thermophilic digestors include *Methanosarcina thermophila* and thermophilic *Methanosaeta* (Zinder, 1993). Increased VFA concentration with the increase in the loading rate decreases the pH for efficient growth of acid-tolerants like *Methanobrevibacter acididurans* (Savant *et al*., 2002). Depending on the type of digester and the feedstock, *Methanosaeta* or *Methanosarcina* dominate the methanogen population in a digester. *Methanosaeta* perform better in high feeding-rate digester, such as an upward-flow anaerobic sludge blanket (UASB), presumably due to efficient adhesion and granulation (Grotenhuis *et al*., 1991; Sekiguchi *et al*., 1999). *Methanosarcina* are more turbulence and shear sensitive, and they frequently dominate fixed-dome and unstirred digestors.

Factors influencing the distribution of methanogens

Various active organisms protect methanogens from oxic environments in natural habitats, as the aerobic and eukaryotic microbes consume O_2 from environments. Methanogenic bacteria are abundant in habitats where electron acceptors such as O_2, NO_3^-, Fe^{3+} and SO_4^{2-} are limiting. Abundance of these compounds helps other organisms to outcompete methanogens. For instance, sulphate-reducing bacteria utilise H_2 in sulphate-rich upper sediment layer at concentrations that is subminimal for the methanogens (Kristjansson *et al.*, 1982; Lovley, 1985) which is probably due to the more positive reduction potential of SO^{2-} compared to CO_2. The abundance of other H_2 and acetate consumers greatly influences the population diversity of methanogens. In Lake Constance sediments, the presence of H_2-consuming and absence of acetate-consuming sulphate-reducing bacteria resulted in 100% CH_4 production by aceticlastic methanogens exhibiting syntrophic association (Schulz and Conrad, 1996).

Few factors affecting the relative contribution of aceticlastic and hydrogenotrophic methanogens in freshwater sediments have been reported. The hydrogenotrophic methanogenesis decreases at low pH. In Lake Knaack sediments (pH 6.8), only 4% of the CH_4 production was derived from H_2/CO_2. Low pH provides a selective advantage to homoacetogens, which reduce CO_2 to acetate instead of CH_4 which in turn limits hydrogenotrophic methanogenesis.

The relative contribution of hydrogenotrophic methanogenesis decreases with temperature, due to better adaptation of homoacetogens to low temperatures (Chin *et al.*, 1999). Moreover, as H_2 production process is less favoured at low temperatures, H_2 production by syntrophic bacteria decreases. From marine sediments in particular, CH_4 emission is substantially larger if anaerobic methane-oxidising microbes would not consume more than 75% of the CH_4. Probably because of the efficient attenuation by anaerobic methanotrophs that marine sediments are only a minor source in the atmospheric CH_4 budget. Aerobic rather than anaerobic methanotrophs, which live at the interface of anoxic and oxic zones, are the important CH_4 consumers.

Association of methanogens with others

In methanogenic habitats, complex organic matter is degraded to CH_4 by the cooperative actions of different groups of anaerobes. The obvious interaction of competition is observed with other microbes utilising the electron acceptors other than CO_2, such as sulphate-reducing, denitrifying and iron-reducing bacteria. This phenomenon probably occurs because these compounds are better electron acceptors, and their reductions are thermodynamically more favourable than CO_2 reduction to methane. Besides methanogens, homoacetogens are another group of anaerobes that reduce CO_2 for energy. During acetate production or acetogenesis, CO_2 is reduced by H_2 or other substances like sugars, alcohols, methylated compounds, CO_2 and organic acids.

Acetogenesis with H_2 is thermodynamically less favourable than methanogenesis. Therefore, homoacetogens do not compete well with methanogens in many habitats, although they outcompete methanogens in some environments (like, the hindgut of certain termites and cockroaches). A possible explanation to this may lie in their metabolic versatility as well as low sensitivity to intermittent oxic environment. The organic polymers are initially degraded by specialised bacteria to simple sugars, lactate, volatile fatty acids and alcohols. These are further fermented by syntrophs and related bacteria to acetate, formate, H_2 and CO_2 which are substrates for methanogenesis. The conversion of volatile fatty acids and alcohols to acetate, CO_2 and H_2 is only favourable at hydrogen partial pressures below 10^2Pa (Zinder, 1993). When methanogens are present, hydrogen rapidly metabolises and maintains concentrations below 10Pa. Therefore, the syntrophic bacteria depend on the association with methanogens or other hydrogenotrophic organisms for energy production. Methanogens have also been found as endosymbionts in protozoa, in removing hydrogen produced by the protozoa *via* interspecies hydrogen transfer, allowing the protozoa to produce more oxidised and energy yielding products like acetate.

In some environments (systems) the growth of aceticlastic methanogens and obligate syntrophs is too slow to maintain a large population. In the rumen and colon, the acetate accumulates to concentrations of 50–100mM which is well above the concentration required for the growth of aceticlastic methanogens like *Methanosarcina* sp. These organisms do not catabolise significant quantities of acetate as their growth rate on this substrate is too slow to maintain the population in a rapid-turnover ecosystem. However, when methylamine or methanol is present, the cell numbers of *Methanosarcina* in the rumen may reach 10^5–10^6/ml as these substrates support faster growth. *Methanobrevibacter* sp. is the most commonly found CO_2 reducing methanogen in the intestinal tracts of non-ruminants.

Methanosphaera sp. has also been isolated from colonic environments; they only grow by using H_2 to reduce methanol to CH_4. Methanogenesis is considered rate-limiting, and high activity of the methanogens is important to maintain efficient anaerobic digestion and avoid accumulation of H_2 and short chain fatty acids. High levels of H_2 (>10Pa) lead to inhibition of the anaerobic fermentation and accumulation of electron sinks like lactate, ethanol, propionate and butyrate. Thus, efficient interspecies hydrogen transfer is quite important for good performance of anaerobic digestors. Temperature and pH are two main parameters influencing methanogenesis in anaerobic digestors. Slightly thermophilic temperature (50–60°C) is desired in certain anaerobic digesters to increase reaction rate and decrease the retention time. The most common hydrogenotrophic methanogens in thermophilic digesters include *M. thermoautotrophicus* and *Methanoculleus thermophilicum* (Hori *et al*., 2006).

Microbes involved in methanogenesis

The microbial activity is almost exclusively responsible for CH_4 production from both natural as well as anthropogenic systems. The methanogens (methanogenic archaea or methanoarchaea) produce CH_4 during the organic matter digestion or decomposition, anaerobically. Methanogens appear to be monophyletic. Hence, all modern methanoarchaea possess an ancient ancestor within the Euryarchaeota. The phylogeny suggested by many authors advocates the lineages represented by *Haloferax*, *Thermoplasma* and *Archaeoglobus* were derived from methanogenic ancestors (Hedderich and Whitman, 2013). Bryant termed this special group of microbes as methanogens eliminating taxonomy confusion with other microbial group, the methane-oxidising bacteria. Methanogens may no longer be a mysterious group of poorly studied microbes. Presently, 34 genera and more than 200 methanogen species have been documented. *Methanoarchaea* is truly cosmopolitan and is the only archaea that is currently culturable.

Major taxonomic groups

Methanogens taxonomically belong to the Euryarchaeota phylum of archaeal super kingdom. Morphological feature like cell shape is used as a primary property for taxonomic classification of methanogenic genera. Physiological and nutritional properties are the bases for species designation. Diverse origin of methanogens was evident from morphological and ultrastructural organisation dissimilarity, the diverse nutritional uptake and wide range of G+C contents, even if they share common mode of energy yielding metabolism. Thus, the diversity observed among CH_4-producing bacteria parallels the description for other bacterial groups like the phototrophic bacteria distinguished on energy-yielding metabolism. Taxonomically, they are classified into seven orders, Methanobacteriales, Methanococcales, Methanomicrobiales, Methanosarcinales, Methanocellales, Methanopyrales and Methanomasiliicoccales, each phylogenetically related to another as distantly as the Cyanobacteriales to the Proteobacteriales. Methanomasiliicoccales is the newest order recently discovered and grouped under class Thermoplasmata (Paul *et al*., 2012) which was confirmed by the isolate *Methanomassiliicoccus luminyensis* (Dridi *et al*., 2012). Among these orders, only the Methanosarcinales can ferment acetate to CO_2 and CH_4 and can grow as well on methanol, methylthiols or methylamines as sole energy source. On the other hand, Methanobacteriales, Methanococcales and Methanopyrales include hyperthermophilic species (Boone *et al.*, 1993; Mesle *et al.*, 2013). The characteristic of utilisation of methanogenic substrate by Methanocellales exhibited hydrogenotrophy, more or less similar to that of Methanomicrobiales. Methanomassiliicoccales have an energy metabolism distinct from other methanogens. All three *Methanomassiliicoccus* representatives from Methanomassiliicoccales genome revealed differential adaptations to the GI environment and possibly contrasted taxonomical evolutionary history.

Order Methanobacterials

The members from these orders are gram positive hydrogenotrophs (and possess double-layered cell wall) that use H_2 as electron donor and CO_2 as electron accepter to produce CH_4. Some members use formate as well as simple secondary alcohols (2-propanol). This order is divided into two families *Methanobacteriaceae* and *Methanothermaceae*. Variation in shape is observed among the species of family Methanobacteriaceae, *e.g.*, *Methanobacterium* sp. exhibit rods to filamentous shape. Most are mesophile, but few thermophile (in hydrothermal regions) and alkaliphile (in freshwater sediments) species are also found. *Methanobacterium subterraneum*, an isolate of a deep granitic groundwater is an alkaliphilic, eurythermic and halotolerant methanogen (Kotelnikova *et al.*, 1998). The only thermophillic microbe of this family *Methanothermobacter thermoautotrophicum* convert about 5% of acetate to CH_4 in 150h reducing methyl and carboxyl moieties. Extreme thermophiles isolated from volcanic springs comprise of three species (*M. fervidus*, *M. jannaschii* and *M. sociabilis*) of *Methanothermus* and are grouped under Methanothermaceae.

Order Methanococcales

Methanococcales is an order of irregular coccoid shaped methanogens, isolated from marine and coastal environments. They are mildly halophilic and chemolithotrophic microbes that use H_2 or formate as electron donor and CO_2 as electron acceptor. A single family (Methanococcaceae) and genus (*Methanococcus*) comprised all Methanococcales species earlier. Now, it have catagorised into two families of Methanococcaceae and Methanocaldococcaceae based on the optimum growth temperature as their growth varies from mesophilic to hyperthermophilic. As of now, the discovered 33 species are grouped under two genera *Methanococcus* and *Methanothemococcus*. Likewise, Methanocaldococcaceae family comprises of 21 species grouped under two genera *Methanocaldococcus* and *Methanotorris*. These usually do not have filaments, biochemically gram negative, with motility or without.

Order Methanomicrobiales

Order Methanomicrobiales has at least five families, Methanomicrobiaceae, Methanospirillaceae, Methanocorpusculaceae, Methanocalculaceae and Methanoregulaceae. Similar to the members of Methanococcales, these use H_2 or formate electron donors for methanogenesis (hydrogenotrophy), majority using formate. The use of secondary alcohols as substitute electron donors is observed in some species. Additionally, morphological diversification from cocci, rods, curved to sheathed rods is observed (Liu and Whitman, 2008). The cellular lipids contain 96% of caldarchaeol or non-cyclic GDGT and 4% mainly of diether archaeol (Koga *et al.*, 1998). Based on the habitat these are mesophiles, neutrophiles, alkaliphiles, halotolerants or natronophiles.

All the families of order Methanomicrobiales have single genus except Methanomicrobiaceae which has six genera, *Methanomicrobium, Methanolacinia, Methanogenium, Methanoplanus, Methanoculleus* and *Methanofollis* and 96 species that show diversity in morphology, physiology and phylogeny. Methanocorpusculaceae the other family of Methanomicrobiales contains a single genus of coccoid, hydrogenotrophic methanogens (*Methanocorpusculum*). Based on 16S rRNA partial sequencing, *Methanospirillum* (*Methanospirillum hungateii*) is grouped under family Methanospirillaceae as it shows significant phylogenetic diversification from other members of Methanomicrobiales. Methanoregulaceae family of this order comprises of genus *Methanolinea, Methanoregula* and *Methanosphaerula* with total 32 species.

Order Methanosarcinales

This order has three families, Methanosarcinaceae, Methanosaetaceae and Methermicoccaceae. All members of Methanosarcinaceae and Methermicoccaceae grow on methylated compounds along with H_2, acetate and CO_2 while Methanosaetaceae require formate as the electron donor. All Methanosarcinaceae members are halophilic or at least halotolerant, display a relatively broad pH range for growth (Maestrojuán *et al*., 1992), and difficult to isolate and maintain in pure culture. Almost all species in this are Gram negative except the Gram positive *Methanosarcina*. Their cellular morphologies (cocci, pseudosarcinae, flat polygons and sheathed rods) are diverse. The cocci vary in morphology, from regular to ellipsoidal spheres in chain or pair. Methermicoccaceae members are obligate anaerobes occurring as irregular single cocci or in clusters. They reduce methanol with H_2, but not H_2 plus CO_2 or methanol alone (Sprenger *et al.*, 2000).

Order Methanocellales

This order is closely related to orders Methanosarcinales and Methanomicrobiales as all the three belong to class Methanomicrobia. This order consists of only one family Methanocellaceae, one genus *Methanocella* and three culturable species, *M. arvoryzae*, *M. conradii* and *M. paludicola* (LØ and Lu, 2012). Methanocellales are phylogenetically close to Methanosarcinales. The order Methanocellales represents an evolutionary intermediate between the Methanomicrobiales (hydrogenotrophic) and multifaceted Methanosarcinales (hydrogenotrophic, aceticlastic and methylotrophic).

This order is the recently proposed euryarchaeotal phylum comprising of single genus *Methanocella*. These are mixotrophic Gram negative microbes that require acetate and CO_2 as primary carbon sources. Before the taxonomic description of *M. paludicola* and proposal of the order Methanocellales, this euryarchaeotal lineage had long been recognised as unculturable archaeal group Rice Cluster-I (RC-I) which would play a key role in CH_4 production in rice fields. It was noteworthy that most of the genetic components for tricarboxylic acid (TCA) cycle were absent in the *M. paludicola* genome. Methanocellales appears to have unexceptional aerotolerant abilities and all the three strains encode a substantial number of genes involved in antioxidant resistance (Erkel *et al.*, 2006). A robust antioxidant system would also need a strong O_2-tolerant energy source, as many reactions consume reducing equivalents (Imlay, 2008). Such energy source has so far escaped attention. The proposed Wolfe cycle is unlikely to be such a source as its activity is severely inhibited by O_2.

Order Methanopyrales

This order is represented by a single family Methanopyraceae with three species which reduce CO_2 with H_2 for methanogenesis. The most prominent species *Methanopyrus kandleri* is hydrogenotrophic, hyperthermophilic (84–110°C). It is rod-shaped, motile (*via* archaella), Gram positive archaeum and its cell wall contains archaeol lipids and a novel pseudomurein, covered by a detergent-sensitive protein layer. It inhabits marine hydrothermal system (Liu and Whitman, 2008, Garcia *et al.*, 2000).

Order Methanomassiliicoccales

This order was provisionally proposed as Methanoplasmatales which was consequently renamed according to Bacteriological Code as Methanomassilii coccales to avoid taxonomical confusion. The order follows the account of a new family Methanomassiliicoccaceae, and its type genus *Methanomassiliicoccus* belongs to class Thermoplasmata. These members of family Methanomassiliicoccaceae are obligate hydrogenotrophs and are defined on the basis of dendrogram generated by phylogenetic analysis of 16S rRNA gene sequence of a culturable representative, enriched culture and environmental clone sequences derived mainly from the alimentary canal, anaerobic digester, landfill leachate and rice field soil. The global contribution of Methanomassiliicoccales representatives to CH_4 emission could be large, considering that it constitutes one of the three dominant archaeal lineages in the rumen and termed as Rumen Cluster-III (RC-III) (Borrel *et al.* 2014). *M. luminyensis* genome contains various important genes which are specifically present in soil and sediment methanogens.

Methanomassiliicoccales contribute to the high diversity of methanogens in wetlands, along with Methanosarcinales, Methanobacteriales, Methanomicrobiales, Methanococcales and Methanocellales in temperate wetlands (Nercessian *et al.*, 1999; Upton *et al.*, 2000; Yavitt *et al.*, 2012; Andersen *et al.*, 2013), and Methanosarcinales, Methanobacteriales, Methanomicrobiales and Methanocellales in arctic wetlands (Hoj *et al.*, 2005; Tveit *et al.*, 2013), which suggests their ubiquitous distribution in wetlands. The methanogenesis pathway is a truncated one in this order (Borrel *et al.*, 2014), and both methylamines and methanol act as electron acceptors besides methanol (Poulsen *et al.*, 2013; Brugère *et al.*, 2014).

Morphophysiological properties of methanogens

Morphology and physiology has fundamentally been used for identification and characterisation of microbial populations of a community. Methanogens are morphologically very diverse with the size ranging from 0.4–1.7μm. Due to their low growth rates and fastidious nutritional and environmental requirements it is difficult to culture methanogens. Their shapes are traditionally not confined to rods or coccoid, although these are the predominant forms. They may occur as spirilla, filaments, sarcina and unusual flattened plate shapes with ultra-structural variations (Karakashev *et al.*, 2005). Sarcina-type cells proliferate as irregular-sized cells that tend to clump forming sand-like aggregates. Cocci-shaped cells vary in morphology from regular to ellipsoidal spheres arranged in pairs or chains. As for example, *Methanobacterium* sp. and *Methanopyrus* sp. are rods, *Methanococcus* sp. and *Methanosphaera* sp. are coccoids, *Methanoplanus* is a plate, *Methanospirillum* is a long thin spirillum, and *Methanosarcina* is a cluster of round cells (Sirohi *et al.*, 2010). *Methanoculleus* and *Methanogenium* are coccoid to irregular shapes, attributable to the loosely bound S-layers on the cell walls. Cell walls of most of the methanogens are very resistant to common (osmotic, ultrasonic, mechanical, enzymatic) cell-disruption strategies.

Factors like pH, temperature, nutrients and inhibitors concentrations influence physiological reactions. The basic microbial growth and physiological activity of methanogens lie in a mild acidic to mild alkaline pH. Most microbes grow best in neutral pH condition (the most pH sensitive are methanoarchaea) since the pH values may adversely affect metabolism by altering the chemical equilibrium of enzymatic reactions, or by actually destroying the enzymes. Low pH can cause the biological reactions chain in an ecological niche to breakdown. The buffering system provided by organic (*e.g.*, protein) and inorganic (*e.g.*, bicarbonate) compounds resist any drastic pH change.

Microbes exhibit optimal growth and metabolic rates within well-defined temperature ranges which is species or group specific, particularly at the upper limit which is defined by the thermostability of the protein molecules synthesised by each. As the temperature changes, methanogens are more stressed than others due to the faster growth rate of the other groups (such as the acetogens) which can achieve substantial catabolism in varying temperatures (Schmid and Lipper, 1969). Likewise, methanogens are most severely inhibited by slight nutrient deficiencies. Besides an organic carbon energy source, anaerobic bacteria appear to have relatively simple nutrient requirements that include N_2, P, Mg^{2+}, Na^+, Mn^{2+}, Ca^{2+}, Se, W and Co (Garcia *et al.*, 2000). Additional nutrients may be required for growth in digesting industrial waste or crop residue. The morphophysiological properties of the various methanogen orders are detailed in Table 2.

Unique characteristics of methanogens

Though methanogens have common characters within the phylum Euryarchaeota, they differ in characters specifically confined to each order. Methanobacteriales have doubled layered cell wall alongwith all conventional cell wall contents. Methanomicrobiales cells are special in morphology with pilus like appendages, form septum in asymmetrical cell division, and are non-motile. Methanosarcinales are the earliest amongst methanogens whose cells contain large vacuoles. Their cells contain methanochondroitin, a special and unique type of polysaccharide. The cells are devoid of thick rigid cell wall which is proteinaceous made up of S-layer proteins. They are capable of fermenting acetate as also *p*-aminobenzoate (PABA, folic acid precursor). Methanocellales, the only order which can tolerate atmospheric O_2, change their morphology during cell division. These encode a putative [NiFe] hydrogenase complex involved in oxidoreduction of hydrogen. The pseudomurein (pseudopeptidoglycan), the main cell wall component in archaea, consists of ornithine and lysine instead of NAG as in Methanopyrales. Unlike others, Methanomassiliicoccales contains granular materials in the cell wall. The genes

Table 2: The Order-wise morpho-physiological features of methanogens

Trait	Methanobacteriales	Methanococcales	Methanomicrobiales	Methanosarcinales	Methanocellales	Methanopyrales	Methanomassiliicoccales
Length (μm)	Dia 0.6-2.5 0.3-0.4×1.0-4.0	Dia 1.0-3.0	Dia 0.2-1.2 (*Methanocalculaceae*), 2.0-6.5×0.3, Dia. 0.2-0.8 (Methanoregulaceae)	Cell width 1.3±0.2 (*Methanomicrococcus*) Dia 0.9-1.0 (*Methanococcoides*)	1.8-2.4×0.3-0.6	0.5×2-14	Dia 0.850
Cell wall & composition	Pseudomurein, Cell wall is often double layered	Glycoprotein, C_{20} isopranyl glycerol ethers are abundant	Proteinaceous, *i.e.*, 70% amino acids, 11% lipids and 6.6% carbohydrates	Proteinaceous, has NAG & D-glucuronate (or D-galacturonate), C_{20} isopranyl glycerol ethers, methanochondroitin	(-)	Protein layer present outside the cell wall, Pseudomurein	Cell wall consists of one thin electron-dense layer and one thick transparent layer (140±20 nm)
Cellular (Core) lipids	Caldarchaeol, archaeol and hydroxyarchaeol (some spp.)	Caldarchaeol, archaeol, hydroxyarchaeol and Macrocyclic archaeol	Archaeol, Caldarchaeol	(-)	(-)	Archaeol, phytanyl diether	Tetraether lipids with butentriol or pentatnetriol, C_{20-25} and C_{25-25} isoprenoidal chain
Flagella (Archella) & Motility	Peritrichous, Most are nonmotile. Excp. *Methanothermus* sp.	Present, Motile (Excp. Non-motile *Methanotorris* sp)	Monotrichous, Highly Motile, (Excp. Non-motile *Methanogenium* sp., *Methanolacinia* sp.)	No flagellation, Non-motile, (Excp. *Methermicoccus* sp.)	Present, Motile (Excp. *M. paludicola*)	Polar tuft flagella, Motile	Absent, Non-motile

Contd.

Trait	Methanobacteriales	Methanococcales	Methanomicrobiales	Methanosarcinales	Methanocellales	Methanopyrales	Methanomassiliicoccales
Polar lipids and other chemicals	Glu., NAG, Myoin., Ethan., Ser. along with diether and tetraether	Glu, NAG, Ser, Ethan.	Polar lipids of diether and tetraether glycerols (*Methanospirillum* sp.)	(-)	(-)	(-)	Phosphatidyl glycerol butanetriol dibiphytanyl glycerol tetraether
Other special Characters	Forms flocs, (Methan othermaceae)	Cell wall composed of S-layer proteins	Often changes in morphology (rod to plate to coccoid) (*Methanoplanus* sp.) Cell division is asymmetrical, by septum formation, Blue auto-fluoroscence, *i.e.*, presence of (F_{420}) (*Methanocalculaceae*, *Methanosphaerula*, *Methanoregulaceae*) Filamentous cells in late-exponential phase	Cells are pseudosarcinae, Catabolise C_2H_6S Deazaflavine (F_{420}) present Often form cysts during exponential phase (*Methermicoccus*) PABA and sludge fluid are growth factors (*Methanosaeta*)	Change in morphology with duration (from rods to coccus) Cytoskeleton comprised of á-tubulin suppressor Encodes a putative (NiFe) hydrogenase complex Aerotolerance	Pseudomurein (contain ornithine and lysine) NAG is absent	Cell wall contained granular materials NAG is absent

NAG: N-acetylglucosamine; Excp: Exception; Glu: Glucose; Ser: Serine; Ethan: Ethanolamine; Myoin: Myoinositol; dia: Diameter; C_2H_6S: Dimethylsulfide; PABA: *p*-aminobenzoate; (-): No data given involved in the CO_2 reduction/methyl-oxidation pathways are absent in all methanogens.

Mechanisms of methanogenesis

Bacterial methane production

Strictly anaerobic methanogenic bacteria obtain their energy for growth and metabolism from the conversion of an array of substrates to CH_4. Methanogens occupy the terminal position in the anaerobic decomposition chain and depend on the degradation of complex organic compounds by chemoheterotrophic bacteria. In most instances, compounds serving as the substrates for methanogenesis are produced as end-products of eubacterial and eukaryotic fermentations and anaerobic oxidations of both complex and simple organic compounds (Jones *et al.*, 1987). The carbon compounds as substrates are majorly, simple CO_2 (with H_2 as reductant), formate and acetate along with methylamines (mono, di and trimethylamine, as well as tetra- methylammonium ion), dimethyl sulphide and simple secondary alcohol as methanol, iso-propanol etc. (Zinder, 1993). No other substances are used directly.

All Methanococcales, Methanobacteriales, Methanomicrobiales, Methanopyrales and Methanosarcinales members utilise H_2, CO_2 and formate as energy substrates with few exception (Rother and Krzycki, 2010). Only two genera (*Methanosarcina* sp. and *Methanosaeta* sp.) are capable to use acetate for methanogenesis (Zinder, 1993). Order Methanocellales and Methanomassiliicoccales use acetate and methanol along with H_2 as substrate for energy.

Fermentative and acetogenic bacteria in addition to methanogens collectively could convert coal (complex organic substrates) to CH_4. Reportedly, pyruvate is CH_4 production intermediate from coal (Isbister and Barik, 1993). Fermentative bacteria hydrolyse and ferment complex substrates to produce acetate, long chain fatty acids, CO_2, H_2, NH_4^+ and HS-. H_2-using acetogens produce acetate by consuming H_2 and CO_2. Additionally, they demethoxylate low-molecular-weight ligneous materials and ferment some hydroxylated aromatic compounds for acetate fermentation (McInerney and Bryant, 1981). The H_2, CO_2 and acetate produced by acetogens from fatty acids, alcohols, and some aromatic and amino acids are used by Methanogens. Often, acetate is the major precursor in an anaerobic environment, although it has not been demonstrated to serve as the sole electron donor for growth and methanogenesis in pure cultures (Zeikus *et al.*, 1980).

Conventionally, methanogenic bacteria are capable of producing energy by reducing CO_2 to CH_4 ($CO_2 + 4H_2 \rightarrow CH_4 + 2H_2O$). For others, carbon monoxide is a suitable substrate, while still for some others methyl alcohol or acetic acid

($CH_3COOH \rightarrow CH_4 + CO_2$). Order Methanosarcinales comprises the most versatile (some *Methanosarcina*) species capable of performing all these four methanogenic pathways, although some (*Methanosaeta*) are uniquely dependent on acetoclastic methanogenesis or H_2-dependent methylotrophic methanogenesis by using methanol, mono-, di-, tri-methylamine, and dimethyl sulphide (*Methanomicrococcus blatticola*) (Sprenger *et al.*, 2000).

The ability to use methanol *via* H_2-dependent methylotrophic methanogenesis is present in a few species belonging to Methanobacteriales. Among them, *Methanosphaera stadtmanae* displays obligate H_2-dependent methylotrophy from methanol and has lost the capability to reduce CO_2 to CH_4 or oxidise methanol to CO_2, although it has kept most of the corresponding enzymes that might be used in other metabolic pathways (Fricke *et al.*, 2006). Specialisation for H_2-dependent methylotrophic methanogenesis could appear as an adaptation to gut environments as both *M. stadtmanae* and *M. blatticola* were isolated from there (Borrel *et al.* 2013). A collection of different microbial species is a consortium and interdependencies such as interspecies H_2 transfer are common for methanogenic consortia (Zinder, 1993).

Biochemistry of Methanogenesis

The main carbon sources for methanogenesis in nature are the accumulating detritus carbon during autumn and early winter when the temperature is too low for high methanogenic activity, a new anoxic degradation of litter when the thawing of the peat begins, and carbon compound exudates from living plants (Saarnio *et al.*, 1997). Homoacetogenic bacteria compete against hydrogenotrophic methanogens to use the H_2 excreted by fermentative bacteria (Abram and Nedwell, 1978; Kotsyurbenko *et al.*, 2001). The methanoarchaea are distinguished by their ability to obtain all or most of their energy from methanogenesis. Methanogens that can grow without producing methane have also been identified.

Methanogenesis, an anaerobic respiration, requires the biosynthesis of six unusual coenzymes, a long, multistep pathway and a number of unique membrane-bound enzyme complexes for coupling to the proton motive force (Hedderich and Whitman, 2013). Basically, the process biochemistry is divided into four stages, hydrolysis, acidogenesis, acetogenesis, methanogenesis, carried out by four different microbial groups (Fig. 1). In nature, methanoarchaea depend strongly on other bacteria for essential nutrient like trace minerals, vitamins, acetate, amino acids or other growth factors.

Methanogens and methanotrophs are the two microbial communities involved in biogeochemical methane cycle in soil. Methanogens are anaerobes and are active in highly reduced, low E_h and flooded environments (Pazinato *et al.*, 2010).

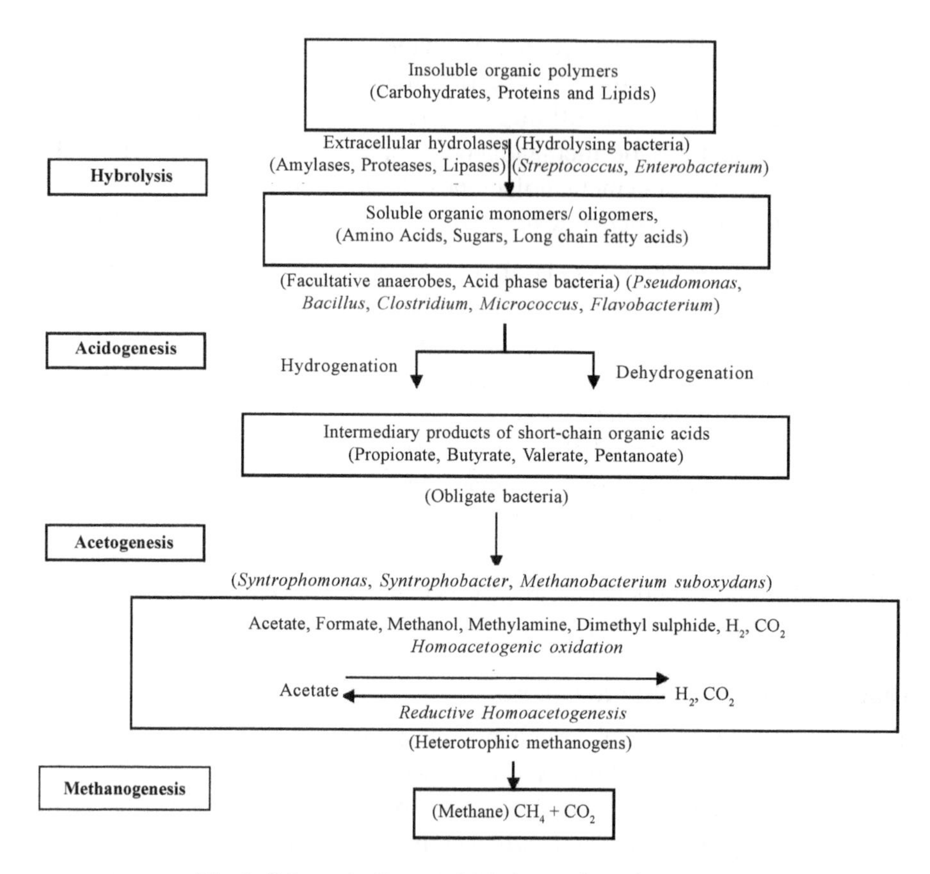

Fig. 1: Schematic diagram depicting methanation process

In contrast, methanotrophs account for CH_4 oxidisation before releasing it to the atmosphere. In contrast with methanogens, methanotrophs are mostly aerobic unicellular microbes active in the oxic soil (Fazli, 2013). Methanoarchaea of the orders Methanobacteriales, Methanococcales and Methanopyrales also possess an enzyme that could directly reduce methenyl-H_4MPT to methylene-H_4MPT using H_2 as the electron donor (Thauer *et al.*, 1996). Orders Methanobacteriales and Methanococcales are seen to contain an isoenzyme called methyl coenzyme-M reductase II (MCR-II) in addition to the MCR (Bult *et al.*, 1996; Lehmacher and Klenk, 1994; Pihl *et al.*, 1994).

Methanogens utilise numerous substrates with methyl coenzyme-M reductase (MCR) as their unique key enzyme. The MCR catalyses the reduction of methyl-coenzyme M accompanied by the release of CH_4 (Ellermann *et al.*, 1988). Some parts of the MCR operon are highly conserved, and all MCR operons appear to have evolved from a common ancestor (Springer *et al.*, 1995). Thus, this functional gene is therefore a helpful tool for taxonomical study.

The H_2 production by syntrophic bacteria decreases as H_2 production processes are less favoured at low temperatures. Thus, hydrogenotrophic methanogenesis is inhibited at low temperature as the substrates supply becomes insufficient. A similar effect of temperature is also observed in rice paddies. Relative contribution of hydrogenotrophic methanogenesis changes with depth or stratified layers in rivers and lakes.

Genomics of methanogens

The phylogenetic placement of methanogens in the archaeal dendrogram and the analysis of the core enzymes in methanogenesis as also previous analyses suggest that methanogenesis has a unique and early origin in the Euryarchaeota, likely after the divergence of Thermococcales. This has been followed by subsequent multiple independent losses in non-methanogenic euryarchaeal lineages (Brochier *et al.*, 2004; Gribaldo and Brochier-Armanet, 2006). Remnants of a methanogenic past are found in the Archaeoglobales genomes which still possess the enzymes for methanogenesis from H_2/CO_2, now involved in producing CO_2 from lactate oxidation (Mo¨ller-Zinkhan *et al.*, 1989). In 1996, the first complete genome sequence of a methanogen *M. jannaschii* was published (Bult *et al.*, 1996), and the first complete genome sequence of methanogen isolated from human sample *M. stadtmanae* DSM3091 was completed 10 years later (Fricke *et al.*, 2006).

All the five methanogenesis (Mcr) markers (A, B, C, D, and G) shared by seven methanogens orders are largely consistent with the ribosomal protein-based phylogeny, notably by recovering the monophyly of all orders. This strongly hints at the occurrence of horizontal gene transfer and vertical inheritance, and strengthens the likely methanogenesis emergence scenario after the Thermococcales emerged. It was subsequently lost multiple times independently during the Euryarchaeota evolutionary history. The origin of methylotrophy in Euryarchaeota, at least from methanol, seems parallel to that of methanogenesis from H_2/CO_2. The rRNA genes in most other methanogens are organised in an operon, but a separation of 5S, 16S, and 23S rRNA genes has been reported. Members of Methanomassiliicoccales have pseudogenized several tRNA genes and lack genes for lysine and selenosysteine (Lang *et al.*, 2015). Based on 16S rRNA clone libraries, all of the 1524 archaeal sequences examined from three healthy human faecal samples belonged to *M. smithii* (Eckburg *et al.*, 2005). Microbiome analysis from the 16S rRNA clone libraries revealed that *M. smithii* was the only archaeal species from healthy human faeces (Gill *et al.*, 2006). Using 16S rRNA and *mcrA* genes detection tools, it appeared that the faecal samples harboured only *M. smithii* although two of the six 16S rRNA amplicons and two of the eight mcrA amplicons matched to *M. stadtmanae* (Samuel *et al.*, 2007).

Future prospects

Methanogens are very diverse in terms of phylogeny and ecology. Their physiological diversity and the ability to anaerobically respire make them truly cosmopolitan in anaerobic environments, thriving in O_2-free extreme thermophillic environments to ruminant gut. The relative abundance of various methanogens is regulated by substrate availability, and other parameters like temperature, pH, and salinity. Methanogens are distinctive archaea responsible for the vast majority of metabolic CH_4 production on earth. Even increased human activity, such as increased farming, sewage processing plants and landfill sites, will tend to increase methanogenic CH_4 emissions. Methanogens use CO_2 to make a major flammable component of natural gas CH_4 which could be used as a renewable, carbon neutral energy substitute. Microbial CH_4 is more ecofriendly than other biofuels. Unlike production of conventional biofuels, methanogens do not compete with food production and other agricultural resources like land, irrigation and fertilisers. Additionally, by utilising microbial CH_4 as a fuel source, the uncontrolled release of CH_4 can be reduced, thereby reducing its greenhouse gas effect. Governments are now looking at the dual role of microbes as a source of fuel and biological wastes managers. Methanogens diversity is underestimated as indicated by various recent culture-independent studies. Further physiological and ecological investigations are needed to reveal their diversity and importance fully.

References

Abram JW, Nedwell DB, 1978. Inhibition of methanogenesis by sulfate reducing bacteria competing for transferred hydrogen. *Archives of Microbiology*, **117(1)**: 89–92.

Andersen R, Chapman SJ, Artz RRE, 2013. Microbial communities in natural and disturbed peatlands: a review. *Soil Biology and Biochemistry*, **57**: 979–94.

Apajalahti J, 2005. Comparative gut microflora, metabolic challenges, and potential opportunities. *The Journal of Applied Poultry Research*, **14(2)**: 444–53.

Balch WE, Wolfe RS, 1976. New approach to the cultivation of methanogenic bacteria: 2-mercaptoethanesulfonic acid (HS-CoM)-dependent growth of *Methanobacterium ruminantium* in a pressurized atmosphere. *Applied and Environmental Microbiology*, **32(6)**: 781–91.

Biavati B, Vasta M, Ferry JG, 1988. Isolation and characterization of *Methanosphaera cuniculi* sp. nov. *Applied and Environmental Microbiology*, **54(3)**: 768–71.

Boone DR, Whitman WB, Rouvière P, 1993. Diversity and taxonomy of methanogens. *In*: JG Ferry (Ed) *Methanogenesis*. Chapman & Hall Microbiology Series (Physiology/Ecology/Molecular Biology/Biotechnology), Springer, Boston, MA: 35–80.

Borrel G, O'Toole PW, Harris HM, Peyret P, Brugère JF, Gribaldo S, 2013. Phylogenomic data support a seventh order of methylotrophic methanogens and provide insights into the evolution of methanogenesis. *Genome Biology and Evolution*, **5(10)**: 1769–80.

Borrel G, Parisot N, Harris HMB, Peyretaillade E, Gaci N, Tottey W, Bardot O, Raymann K, Gribaldo S, Peyret P, O'Toole PW, Brugère JF, 2014. Comparative genomics highlights the unique biology of Methanomassiliicoccales, a Thermoplasmatales-related seventh order of methanogenicarchaea that encodes pyrrolysine. *BMC Genomics*, **15(1)**: 679.

Bousquet P, Ciais P, Miller JB, Dlugokencky EJ, Hauglustaine DA, Prigent C, van der Werf GR, Peylin P, Brunke E-G, Carouge C, Langenfields RL, Lathière J, Papa F, Ramonet M, Schmidt M, Steele LP, Tyler SC, White J, 2006. Contribution of anthropogenic and natural sources to atmospheric methane variability. *Nature*, **443(&110)**: 439–43.

Brochier C, Forterre P, Gribaldo S, 2004. Archaeal phylogeny based on proteins of the transcription and translation machineries: tackling the *Methanopyrus kandleri* paradox. *Genome Biology*, **5(3)**: R17.

Brauer SL, Cadillo-Quiroz H, Ward RJ, Yavitt JB, Zinder SH, 2011, Methanoregula boonei gen. nov., sp.nov., an acidiphilic methanogen isolated from an acidic peat bog. International *Journal of Systematic and Evolutionary Microbiology*, **61**(1):45-52.

Brugère JF, Borrel G, Gaci N, Tottey W, O'Toole PW, Malpuech-Brugère C, 2014. Archaebiotics: proposed therapeutic use of archaea to prevent trimethylaminuria and cardiovascular disease. *Gut Microbes*, **5(1)**: 5–10.

Bryant MP, 1972. Commentary on the Hungate technique for culture of anaerobic bacteria. *The American Journal of Clinical Nutrition*, **25(12)**: 1324–28.

Bult CJ, White O, Olsen GJ, Zhou L, Fleischmann RD, Sutton GG, Blake JA, FitzGerald LM, Clayton RA, Gocayne JD, Kerlavage AR, Dougherty BA, Tomb JF, Adams MD, Venter JC, 1996. Complete genome sequence of the methanogenic archaeon, *Methanococcus jannaschii*. *Science*, **273(5278)**: 1058–73.

Burggraf S, Fricke H, Neuner A, Kristjansson J, Rouviere P, Mandelco L, Woese CR, Stetter KO, 1990. *Methanococcus igneus* sp. nov., a novel hyperther-mophilic methanogen from a shallow submarine hydrothermal system. *Systematic and Appied Microbiology*, **13(3)**: 263–9.

Cadillo-Quiroz H, Yavitt JB, Zinder SH, 2009. *Methanosphaerula palustris* gen. nov., sp. nov., a hydrogenotrophic methanogen isolated from a minerotrophic fen peatland. *International Journal of Systematic and Evolutionary Microbiology*, **59(5)**: 928–35.

Capone DG, Kiene RP, 1988. Comparison of microbial dynamics in marine and freshwater sediments: Contrasts in anaerobic carbon catabolism. *Limnology and Oceanography*, **33(4part2)**: 725–49.

Casper P, Chan OC, Furtado AL, Adams DD, 2003. Methane in an acidic bog lake: the influence of peat in the catchment on the biogeochemistry of methane. *Aquatic Sciences*, **65(1)**: 36–46.

Chaudhary PP, Sirohi SK, 2009. Dominance of *Methanomicrobium* phylotype in methanogen population present in Murrah buffaloes (*Bubalus bubalis*). *Letters in Applied Microbiology*, **49(2)**: 274–7.

Chen YH, Prinn RG, 2005. Atmospheric modeling of high and low frequency methane observations: Importance of interannually varying transport. *Journal of Geophysical Research: Atmospheres*, **110(D10303)**: 1–27.

Cheng L., Qiu T-L., Yin X.B., Wu X-L., Hu G-Q, Dang Y, Zhang H. 2007. Methemicoccus shengliensis gen. nov., sp. Nov., a thermophilic methylotrophic methanogen isolated from oil-production water, and proposal of Methemicroccaceae fam.nov. *International Journal of Systematic and Evolutionary Microbiolgy*. **57**:2964-69.

Chin KJ, Lueders T, Friedrich MW, 2004. Archaeal community structure and pathway of methane formation on rice roots. *Microbial Ecology*, **47(1)**: 59–67.

Chin KJ, Lukow T, Conrad R, 1999. Effect of temperature on structure and function of the methanogenic archaeal community in an anoxic rice field soil. *Applied and Environmental Microbiology*, **65(6)**: 2341–9.

Conrad R, 1999. Contribution of hydrogen to methane production and control of hydrogen concentrations in methanogenic soils and sediments. *FEMS Microbiology Ecology*, **28(3)**: 193–202.

Conrad R, Phelps TJ, Zeikus JG, 1985. Gas metabolism evidence in support of the juxtaposition of hydrogen-producing and methanogenic bacteria in sewage sludge and lake sediments. *Applied and Environmental Microbiology*, **50(3)**: 595–601.

Dannenberg S, Conrad R, 1999. Effect of rice plants on methane production and rhizospheric metabolism in paddy soil. *Biogeochemistry*, **45(1)**: 53–71.

Deuser WG, Degens ET, Harvey GR, Rubin M, 1973. Methane in Lake Kivu: New data bearing on its origin. *Science*, **181(4094)**: 51–4.

Dridi B, Fardeau ML, Ollivier B, Raoult D, Drancourt M, 2012. *Methanomassiliicoccus luminyensis* gen. nov., sp. nov., a methanogenic archaeon isolated from human faeces. *International Journal of Systematic and Evolutionary Microbiology*, **62(8)**: 1902–7.

Eckburg PB, Bik EM, Bernstein CN, Purdom E, Dethlefsen L, Sargent M, Gill SR, Nelson KE, Relman DA, 2005. Diversity of the human intestinal microbial flora. *Science*, **308(5728)**: 1635–8.

Elberson MA, Sowers KR, 1997. Isolation of an aceticlastic strain of *Methanosarcina siciliae* from marine canyon sediments and emendation of the species description for *Methanosarcina siciliae*. *International Journal of Systematic and Evolutionary Microbiology*, **47(4)**: 1258–61.

Ellermann J, Hedderich R, Böcher R, Thauer RK, 1988. The final step in methane formation. Investigations with highly purified methyl-CoM reductase (component C) from *Methanobacterium thermoautotrophicum* (strain Marburg). *European Journal of Biochemistry*, **172(3)**: 669-77.

Erkel C, Kube M, Reinhardt R, Liesack W, 2006. Genome of Rice Cluster I Archaea — the key methane producers in the rice rhizosphere. *Science*, **313(5785)**: 370–2.

Falz KZ, Holliger C, Grosskopf R, Liesack W, Nozhevnikova AN, Müller B, Wehrli B, Hahn D, 1999. Vertical distribution of methanogens in the anoxic sediment of Rotsee (Switzerland). *Applied and Environmental Microbiology*, **65(6)**: 2402–8.

Fazli P, Man HC, Shah UKM, Idris A, 2013. Characteristics of methanogens and methanotrophs in rice fields: A review. *Asia-Pacific Journal of Molecular Biology and Biotechnology*, **21(1)**: 3–17.

Fricke WF, Seedorf H, Henne A, Krüer M, Liesegang H, Hedderich R, Gottschalk G, Thauer RK, 2006. The genome sequence of *Methanosphaera stadtmanae* reveals why this human intestinal archaeon is restricted to methanol and H_2 for methane formation and ATP synthesis. *Journal of Bacteriology*, **188(2):** 642–58.

Garcia JL, Patel BK, Ollivier B, 2000. Taxonomic, phylogenetic and ecological diversity of methanogen Archaea. *Anaerobe*, **6(4)**: 205–26.

Gill SR, Pop M, DeBoy RT, Eckburg PB, Turnbaugh PJ, Samuel BS, Gordon JI, Relman DA, Fraser-Liggett CM, Nelson KE, 2006. Metagenomic analysis of the human distal gut microbiome. *Science* **312(5778)**: 1355–9.

Gribaldo S, Brochier-Armanet C, 2006. The origin and evolution of Archaea: a state of the art. *Philosophical Transactions of the Royal Society of London B: Biological Sciences*, **361(1470)**: 1007–22.

Grotenhuis JT, Smit M, Plugge CM, Xu YS, Van Lammeren AA, Stams AJ, Zehnder AJ, 1991. Bacteriological composition and structure of granular sludge adapted to different substrates. *Applied and Environmental Microbiology*, **57(7)**: 1942–9.

Hackstein JH, Stumm CK, 1994. Methane production in terrestrial arthropods. *Proceedings of the National Academy of Sciences of the United States of America*, **91(12)**: 5441–5.

Hedderich R, Whitman WB, 2013. Physiology and biochemistry of the methane-producing Archaea. *In*: M Dworkin (Ed) *The Prokaryotes*. Springer, Berlin, Heidelberg: 635–662.

Høj L, Olsen RA, Torsvik VL, 2005. Archaeal communities in High Arctic wetlands at Spitsbergen, Norway (78° N) as characterized by 16S rRNA gene fingerprinting. *FEMS Microbiology Ecology*, **53(1)**: 89–101.

Holzapfel-Pschorn A, Conrad R, Seiler W, 1985. Production, oxidation and emission of methane in rice paddies. *FEMS Microbiology Letters*, **31(6)**: 343–51.

Hori T, Haruta S, Ueno Y, Ishii M, Igarashi Y, 2006. Dynamic transition of a methanogenic population in response to the concentration of volatile fatty acids in a thermophilic anaerobic digester. *Applied and Environmental Microbiology*, **72(2)**: 1623–30.

Imlay JA, 2008. Cellular defenses against superoxide and hydrogen peroxide. *Annual Review of Biochemistry*, **77**: 755–76.

Isbister JD, Barik S, 1993. Biogasification of low rank coals. *In*: DL Crawford (Ed) *Microbial Transformations of Low Rank Coals*. CRC Press, Boca Raton, Florida: 139–56.

Jones WJ, Nagel Jr. DP, Whitman WB, 1987. Methanogens and the diversity of archaebacteria. *Microbiology Reviews*, **51(1)**: 135–77.

Jones WJ, Leigh JA, Mayer F, Woese CR, Wolfe RS, 1983. *Methanococcus jannaschii* sp. nov., an extremely thermophilic methanogen from a submarine hydrothermal vent. *Archives of Microbiology*, **136(4)**: 254–61.

Karakashev D, Batstone DJ, Angelidaki I, 2005. Influence of environmental conditions on methanogenic compositions in anaerobic biogas reactors. *Applied and Environmental Microbiology*, **71(1)**: 331–8.

Kendall MM, Boone DR, 2006. Cultivation of methanogens from shallow marine sediments at Hydrate Ridge, Oregon. *Archaea*, **2(1)**: 31–8.

Kendall MM, Wardlaw GD, Tang CF, Bonin AS, Liu Y, Valentine DL, 2007. Diversity of Archaea in marine sediments from Skan Bay, Alaska, including cultivated methanogens, and description of *Methanogenium boonei* sp. nov. *Applied Environmental Microbiology*, **73(2)**: 407–14.

Kevbrin VV, Lysenko AM, Zhilina TN, 1997. Physiology of the alkaliphilic methanogen Z-7936, a new strain of *Methanosalsus zhilinaeae* isolated from Lake Magadi. *Microbiologia*, **66**: 261–6.

King GM, 1988. Methanogenesis from methylated amines in a hypersaline algal mat *Applied and Environmental Microbiology*, **54(1)**: 130–6.

Koga Y, Morii H, Akagawa-Matsushita M, Ohga I, 1998. Correlation of polar lipid composition with 16S rRNA phylogeny in methanogens: further analysis of lipid component parts. *Bioscience, Biotechnology, and Biochemistry*, **62(2)**: 230-6.

Kotelnikova S, Macario AJ, Pedersen K, 1998. *Methanobacterium subterraneum* sp. nov., a new alkaliphilic, eurythermic and halotolerant methanogen isolated from deep granitic groundwater. *International Journal of Systematic and Evolutionary Microbiology*, **48(2)**: 357–67.

Kotsyurbenko OR, Glagolev MV, Nozhevnikova AN, Conrad R, 2001. Competition between homoacetogenic bacteria and methanogenic archaea for hydrogen at low temperature. *FEMS Microbiology Ecology*, **38(2-3)**: 153–9.

Kristjansson JK, Schönheit P, Thauer RK, 1982. Different Ks values for hydrogen of methanogenic bacteria and sulfate reducing bacteria: an explanation for the apparent inhibition of methanogenesis by sulfate. *Archives of Microbiology*, **131(3)**: 278–82.

Krüger M, Frenzel P, Kemnitz D, Conrad R, 2005. Activity, structure and dynamics of the methanogenic archaeal community in a flooded Italian rice field. *FEMS Microbiology Ecology*, **51(3)**: 323–31.

Kuhner CH, Smith SS, Noll KM, Tanner RS, Wolfe RS, 1991. 7-mercaptoheptanoylthreonine phosphate substitutes for heat-stable factor (mobile factor) for growth of *Methanomicrobium mobile*. *Applied and Environmental Microbiology*, **57(10)**: 2891–5.

Kurr M, Huber R, König H, Jannasch JW, Fricke H, Trincone A, Kristjansson JK, Stetter KO, 1991. *Methanopyrus kandleri*, gen. and sp. nov. represents a novel group of hyperthennophilic methanogens, growing at 110°C. *Archives of Microbiology*, **156(4)**: 239–47.

Lang K, Schuldes J, Klingl A, Poehlein A, Daniel R, Brune A, 2015. New mode of energy metabolism in the seventh order of methanogens as revealed by comparative genome analysis of 'Candidatus Methanoplasma termitum'. *Applied and Environmental Microbiology*, **81(4)**: 1338–52.

Lauerer G, Kristjansson JK, Langworthy TA, König H, Stetter KO, 1986. *Methanothermus sociabilis* sp. nov., a second species within the Methanothermaceae growing at 97°C. *Systematic and Applied Microbiology*, **8(1-2)**: 100–5.

Leadbetter JR, Crosby LD, Breznak JA, 1998. *Methanobrevibacter filiformis* sp. nov., a filamentous methanogen from termite hindguts. *Archives of Microbiology*, **169(4)**: 287–92.

Lehmacher A, Klenk HP, 1994. Characterization and phylogeny of mcrII, a gene cluster encoding an isoenzyme of methyl coenzyme M reductase from hyperthermophilic *Methanothermus fervidus*. *Molecular and General Genetics MGG*, **243(2)**: 198–206.

Lin C, Raskin L, Stahl DA, 1997. Microbial community structure in gastrointestinal tracts of domestic animals: comparative analyses using rRNA-targeted oligonucleotide probes. *FEMS Microbiology Ecology*, **22(4)**: 281–94.

Liu Y, Boone DR, Choy C, 1990. *Methanohalophilus oregonense* sp. nov., a methylotrophic methanogen from an alkaline, saline aquifer. *International Journal of Systematic and Evolutionary Microbiology*, **40(2)**: 111–6.

Liu Y, Whitman WB, 2008. Metabolic, phylogenetic, and ecological diversity of the methanogenic Archaea. *Annals of the New York Academy of Sciences*, **1125(1)**: 171–89.

Lovley DR, 1985. Minimum threshold for hydrogen metabolism in methanogenic bacteria. *Applied and Environmental Microbiology*, **49(6)**: 1530–1.

Lu Y, Lueders T, Friedrich MW, Conrad R, 2005. Detecting active methanogenic populations on rice roots using stable isotope probing. *Environmental Microbiology*, **7(3)**: 326–36.

LØ Z, Lu Y, 2012. *Methanocella conradii* sp. nov., a thermophilic, obligate hydrogenotrophic methanogen, isolated from Chinese rice field soil. *PloS one*, **7(4)**: e35279.

Lueders T, Friedrich M, 2000. Archaeal population dynamics during sequential reduction processes in rice field soil. *Applied and Environmental Microbiology*, **66(7)**: 2732–42.

Lyimo TJ, Pol A, den Camp HO, Harhangi HR, Vogels GD, 2000. *Methanosarcina semesiae* sp. nov., a dimethylsulfide-utilizing methanogen from mangrove sediment. *International Journal of Systematic and Evolutionary Microbiology*, **50(1)**: 171–8.

Mackie RI, 2002. Mutualistic fermentative digestion in the gastrointestinal tract: diversity and evolution. *Integrative and Comparative Biology*, **42(2)**: 319–26.

Maestrojuan GM, Boone DR, Xun L, Mah RA, Zhang L (1990) Transfer of *Methanogenium bourgense*, *Methanogenium marisnigri*, *Methanogenium olentangyi*, and *Methanoculleus gen* nov., emendation of *Methanoculleus marisnigri* and Methanogenium, and description of new strains of *Methanoculleusbourgense* and *Methanoculleusmarisnigri*. *International Journal of Systematic Bacteriology*, **40**: 117–22.

Maestrojuán GM, Boone JE, Mah RA, Menaia JAGF, Sachs MS, Boone DR, 1992. Taxonomy and halotolerance of mesophilic *Methanosarcina* strains to species, and synonymy of *Methanosarcina mazei* and *Methanosarcina frisia*. *International Journal of Systematic and Evolutionary Microbiology*, **42(4)**: 561–7.

Martius C, Wassmann R, Thein U, Bandeira A, Rennenberg H, Junk W, Seiler W, 1993. Methane emission from wood-feeding termites in Amazonia. *Chemosphere*, **26(1-4)**: 623–32.

McHugh S, Carton M, Mahony T, O'Flaherty V, 2003. Methanogenic population structure in a variety of anaerobic bioreactors. *FEMS Microbiology Letters*, **219(2)**: 297–304.

McInerney MJ, Bryant MP, 1981. Review of methane fermentation fundamentals. *In*: DL Wise (Ed) [*Production of Fuel Gas from Organic Waste Products*]. *Fuel gas production from biomass*. CRC Press, Boca Raton, Florida: 19–46.

Meslé M, Dromart G, Oger P, 2013. Microbial methanogenesis in subsurface oil and coal. *Research in Microbiology*, **164(9)**: 959–72.

Miller TL, 1989. Genus II. *Methanobrevibacter*. *In*: JT Staley (Ed) *Bergey's Manual of Systematic Bacteriology*, Volume 3. Williams & Wilkins, Baltimore: 2178–83.

Miller TL, Wolin MJ, 1986. Methanogens in human and animal intestinal tracts. *Systematic and Applied Microbiology*, **7(2-3)**: 223–9.

Miller TL, 2001. Genus II. *Methanobrevibacter Balch* and Wolfe 1981, 216^{VP}. *In*: DR Boone, RW Castenholz, GM Garrity (Eds). Bergey's Manual of Systematic Bacteriology, Springer, New York: 218–26.

Möller-Zinkhan D, Borner G, Thauer R, 1989. Function of methanofuran, tetrahydromethanopterin, and coenzyme F420 in *Archaeoglobus fulgidus*. *Archives of Microbiology*, **152(4)**: 362–8.

Murray PA, Zinder SH, 1985. Nutritional requirements of *Methanosarcina* sp. strain TM-1. *Applied and Environmental Microbiology*, **50(1)**: 49–55.

Nercessian D, Upton M, Loyd D, Edwards C, 1999. Phylogenetic analysis of peat bog methanogen populations.*FEMS Microbiology Letters*, **173 (2)**: 425-9.

Newberry CJ, Webster G, Cragg BA, Parkes RJ, Weightman AJ, Fry JC, 2004. Diversity of prokaryotes and methanogenesis in deep subsurface sediments from the Nankai Trough, Ocean Drilling Program Leg 190. *Environmental Microbiology*, **6(3)**: 274–87.

Nozhevnikova AN, Chudina VI, 1984. Morphology of the thermophilic acetate methane bacterium *Methanothrix thermoacetophila* sp. nov. *Microbiology*, **53(5)**: 618–24.

Parkes RJ, Cragg BA, Banning N, Brock F, Webster G, Fry JC, Hornibrook E, Pancost RD, Kelly S, Knab N, Jørgensen BB, Rinna J, Weightman AJ, 2007. Biogeochemistry and biodiversity of methane cycling in subsurface marine sediments (Skagerrak, Denmark). *Environmental Microbiology*, **9(5)**: 1146-61.

Patel GB, Sprott GD, 1990. *Methanosaeta concilii* gen. sp. nov. ('*Methanothrix concilii*') and *Methanosaeta thermoacetophila* nom. rev., comb. nov. *International Journal of Systematic and Evolutionary* Microbiology, **40(1)**: 79–82.

Paterek JR, Smith PH, 1988. *Methanohalophilus mahii* gen. sp. nov., a methylotrophic halophilic methanogen. *International Journal of Systematic and Evolutionary Microbiology*, **38(1)**: 122–3.

Paul K, Nonoh JO, Mikulski L, Brune A, 2012. 'Methanoplasmatales': Thermoplasmatales-related Archaea in termite guts and other environments are the seventh order of methanogens. *Applied and Environmental Microbiology*, **78**: 8245-825.

Pazinato JM, Paulo EN, Mendes LW, Vazoller RF, Tsai, SM, 2010. Molecular characterization of the archaeal community in an Amazonian wetland Soil and culture-dependent isolation of methanogenic archaea. *Diversity*, **2(7)**: 1026–47.

Pihl TD, Sharma S, Reeve JN, 1994. Growth phase-dependent transcription of the genes that encode the two methyl coenzyme M reductase isoenzymes and N^5-methyltetrahydromethanopterin: coenzyme M methyltransferase in *Methanobacterium thermoautotrophicum* Delta H. *Journal of Bacteriology*, **176(20)**: 6384–91.

Poulsen M, Schwab C, Jensen BB, Engberg RM, Spang A, Canibe N, Hojberg O, Milinovich G, Fragner L, Schleper C, Weckwerth W, Lund P, Schramm A, Urich T, 2013. Methylotrophic methanogenic *Thermoplasmata* implicated in reduced methane emissions from bovine rumen. *Nature Communications*, **4**: 1428.

Rivard CJ, Henson JM, Thomas MV, Smith PH, 1983. Isolation and characterization of *Methanomicrobium paynteri* sp. nov., a mesophilic methanogen isolated from marine sediments. *Applied and Environmental Microbiology*, **46(2)**: 484–90.

Rother M, Krzycki JA, 2010. Selenocysteine, pyrrolysine, and the unique energy metabolism of methanogenic Archaea. *Archaea*, **2010**: 1–14.

Roy R, Kluber HD, Conrad R, 1997. Early initiation of methane production in anoxic rice soil despite the presence of oxidants. *FEMS Microbiology Ecology*, **24(4)**: 311–320.

Saarnio S, Alm J, Silvola J, Lohila A, Nykänen H, Martikainen PJ, 1997. Seasonal variation in CH_4 emissions and production, and oxidation potentials at microsites on an oligotrophic pine fen. *Oecologia*, **110(3)**: 414–22.

Sakai S, Ehan m, Tseng I-C Yamnag T, Braucer SL., Cadillo-Quiroz H, Zinder SH, Imachi H, 2012. Methanolinea mesophila sp. nov., a hydrogenotrophic methanogen isolated from ride field soil, and proposal of the archaeal family Methanoregulaceae fam. nov. within the order Methanomicrobiales. *International Journal of Systematic and Evolutionary Microbiology*, 62:1389-95.

Samuel BS, Hansen EE, Manchester JK, Coutinho PM, Henrissat B, Fulton R, Latreille P, Kim K, Wilson RK, Gordon JI, 2007. Genomic and metabolic adaptations of *Methanobrevibacter smithiito* to the human gut. *Proceedings of the National Academy of Sciences, USA*, **104(25)**: 10643–8.

Sandbeck KA, Ward DM, 1982. Temperature adaptations in the terminal processes of anaerobic decomposition of Yellowstone National Park and Icelandic hot spring microbial mats. *Applied and Environmental Microbiology*, **44(4)**: 844–51.

Savant DV, Shouche YS, Prakash S, Ranade DR, 2002. *Methanobrevibacter acididurans* sp. nov., a novel methanogen from a sour anaerobic digester. *International Journal of Systematic and Evolutionary Microbiology*, **52(4)**: 1081–7.

Schmid, LA, Lipper, RI, 1969. Swine wastes, characterization and anaerobic digestion. Proceedings Cornell University Conference on Agricultural Waste Management. Ithaca, NY.

Schulz S, Matsuyama H, Conrad R, 1997. Temperature dependence of methane production from different precursors in a profundal sediment (Lake Constance). *FEMS Microbiology Ecology*, **22(3)**: 207–13.

Schutz H, Seiler W, Conrad R, 1989. Processes involved in formation and emission of methane in rice paddies. *Biogeochemistry*, **7(1)**: 33–53.

Sekiguchi Y, Kamagata Y, Nakamura K, Ohashi A, Harada H, 1999. Fluorescence *in situ* hybridization using 16S rRNA-targeted oligonucleotides reveals localization of methanogens and selected uncultured bacteria in mesophilic and thermophilic sludge granules. *Applied and Environmental Microbiology*, **65(3)**: 1280–8.

Sirohi S, Pandey N, Singh B, Puniya A, 2010. Rumen methanogens: a review. *Indian Journal of Microbiology*, **50(3)**: 253–62.

Skillman LC, Evans PN, Naylor GE, Morvan B, Jarvis GN, Joblin KN, 2004. 16S ribosomal DNA-directed PCR primers for ruminal methanogens and identification of methanogens colonising young lambs. *Anaerobe*, **10(5)**: 277–85.

Smith PH, Hungate RE, 1958. Isolation and characterization of *Methanobacterium ruminantium* n. sp. *Journal of Bacteriology*, **75(6)**: 713–8.

Sorokin DY, Abbas B, Merkel AY, Irene W, Rijpstm C, Sinninghe Damte JS, sukhacheva MV, van Loosdrecht MCM, 2015. Methanosalsum natronophilum sp. nov., and Methanocalculus alkaliphilus sp. nov., haloalkaliphilic methanogens from hypersaline soda lake. *International Journal of Systematic and Evolutionary Microbiology*, **65**:3739-45.

Sowers KR, Ferry JG, 1983. Isolation and characterization of a methylotrophic marine methanogen *Methanococcoides methylutens* gen. nov. *Applied and Environmental Microbiology*, **45(2)**: 684-90.

Sowers KR, Ferry JG, 1985. Trace metal and vitamin requirements of *Methanococcoides methylutens* grown with trimethylamine. *Archives of Microbiology*, **142(2)**: 148–51.

Sowers KR, Gunsalus RP, 1988. Adaptation for growth at various saline concentrations by the archaebacterium *Methanosarcina thermophila*. *Journal of Bacteriology*, **170(2)**: 998–1002.

Sprenger WW, Van Belzen MC, Rosenberg J, Hackstein Johannes HP, Keltjens JT, 2000. *Methanomicrococcus blatticola* gen. nov., sp. nov., a methanol- and methylamine-reducing methanogen from the hindgut of the cockroach *Periplaneta americana. International Journal of Systematic and Evolutionary Microbiology*, **50(6)**: 1989–99.

Springer E, Sachs MS, Woese CR, Boone DR, 1995. Partial gene sequences for the A subunit of methyl-coenzyme M reductase (mcrI) as a phylogenetic tool for the family Methanosarcinaceae. *International Journal of Systematic and Evolutionary Microbiology*, **45(3)**: 554–9.

Stetter KO, Thomm M, Winter J, Wildgruber G, Huber H, Zillig W, Jane-Covic D, Konig H, Palm P, Wunderi S, 1981. *Methanothermus fervidus*, sp. nov., a novel extremely thermophilic methanogen isolated from an Icelandic hot spring. *Zentralblattfur Bakteriologie Mikrobiologie und Hygiene: I Abt Originale C: Allgemeine, angewandte und okologische Mikrobiologie*, **2(2)**: 166–78.

Stocker FT, Qin D, Plattner GK, Tiger MBM, Allen SK, Boschung Z, Nauels A, Xia Y, Bex V, Midgley PM, 2014. Climate change 2013: the physical science basis: Working Group I contribution to the Fifth assessment report of the Intergovernmental Panel on Climate Change. Cambridge University Press, The United States of America.

Sundset MA, Edwards JE, Cheng YF, Senosiain RS, Fraile MN, Northwood KS, Praestaneg KE, Glad T, Mathiesen SD, Wright AD, 2009. Rumen microbial diversity in Svalbard reindeer, with particular emphasis on methanogenic archaea. *FEMS Microbiology Ecology*, **70(3)**: 553–62.

Thauer RK, Klein AR, Hartmann GC, 1996. Reactions with molecular hydrogen in microorganisms: evidence for a purely organic hydrogenation catalyst. *Chemical Reviews*, **96(7)**: 3031–42.

Tveit A, Schwacke R, Svenning MM, Urich T, 2013. Organic carbon transformations in high-Arctic peat soils: key functions and microorganisms. *The ISME Journal*, **7(2)**: 299–311.

Upton M, Hill B, Edwards C, Saunders JR, Ritchie DA, Lloyd D, 2000. Combined molecular ecological and confocal laser scanning microscopic analysis of peat bog methanogen populations. *FEMS Microbiology Letters*, **193(2)**: 275–81.

Van Bruggen JJA, Zwart KB, Hermans JGF, Van Hove EM, Stumm CK, Vogels GD, 1986. Isolation and characterization of *Methanoplanusendo symbiosus* sp. nov., an endosymbiont of the marine sapropelic ciliate *Metopus contortus* Quennerstedt. *Archives of Microbiology*, **144(4)**: 367–74.

Von Klein D, Arab H, Volker H, Thomm M, 2002. *Methanosarcina baltica*, sp. nov., a novel methanogen isolated from the Gotland Deep of the Baltic Sea. *Extremophiles*, **6(2)**: 103–10.

Weimer PJ, Russell JB, Muck RE, 2009. Lessons from the cow: what the ruminant animal can teach us about consolidated bioprocessing of cellulosic biomass. *Bioresource Technology*, **100(2)**: 5323–31.

Whiticar MJ, 1999. Carbon and hydrogen isotope systematics of bacterial formation and oxidation of methane. *Chemical Geology*, **161(1-3)**: 291–314.

Whiticar MJ, Faber E, Schoell M, 1986. Biogenic methane formation in marine and freshwater environments: CO_2 reduction vs. acetate fermentation—Isotope evidence. *Geochimica et Cosmochimica Acta*, **50(5)**: 693–709.

Xun L, Boone DR, Mah RA, 1989. Deoxyribonucleic acid hybridization study of *Methanogenium* and *Methanocorpusculum* species, emendation of the genus *Methanocorpusculum*, and transfer of *Methanogenium aggregans* to the genus *Methanocorpusculum* as *Methanocorpusculum aggregans* comb. nov. *International Journal of Systematic and Evolutionary Microbiology*, **39(2)**: 109-11.

Yanagita K, Kamagata Y, Kawaharasaki M, Suzuki T, Nakamura Y, Minato H, 2000. Phylogenetic analysis of methanogens in sheep rumen ecosystem and detection of *Methanomicrobium mobile* by fluorescence *in situ* hybridization. *Bioscience, Biotechnology, and Biochemistry*, **64(8)**: 1737–42.

Yavitt JB, Yashiro E, Cadillo-Quiroz H, Stephen ZH, 2012. Methanogen diversity and community composition in peatlands of the central to northern Appalachian Mountain region, North America. *Biogeochemistry*, **109(1-3)**: 117–31.

Zabel HP, Konig H, Winter J, 1985. Emended description of *Methanogenium thermophilicum*, Rivard and Smith, and assignment of new isolates to this species. *Systematic and Applied Microbiology*, **6(1)**: 72–8.

Zeikus JG, Ben-Bassat A, Hegge PW, 1980. Microbiology of methanogenesis in thermal volcanic environments. *Journal of Bacteriology*, **143(1)**: 432–40.

Zellner G, Messner P, Kneifel H, Tindall BJ, Winter J, Stackebrandt E, 1989. *Methanolacinia* gen. nov., incorporating *Methanomicrobium paynteri* as *Methanolacinia paynteri* comb. nov. *The Journal of General and Applied Microbiology*, **35(3)**: 185–202.

Zhilina TN, Zavarzin GA, 1987. *Methanohalobium evestigatus* nov. gen. sp. - An extremely halophilic methane-forming archebacterium (*Methanohalobium evestigatus* n. gen., n. sp.- Ekstremal'no-galofil'naia metanoobrazuiushchaia arkhebakteriia). *Akademiia Nauk SSSR*, **293(2)**: 464–8.

Zinder SH, 1993. Physiological ecology of methanogens. *In*: JG Ferry (Ed) *Methanogenesis: Ecology, Physiology, Biochemistry and Genetics*. Chapman & Hall, New York: 128–206.

2

Biochemistry and Molecular Biology of Biomethanation

Snehasish Mishra[a,1], Sanjay K Ojha[b] and Tapan Kumar Adhya[c]

[a]*BDTC, Bioenergy Lab, School of Biotechnology, KIIT, Bhubaneswar Odisha – 751 024, India*
[b]*Department of Biophysics, All India Institute of Medical Sciences New Delhi – 110 024, India*
[c]*School of Biotechnology, KIIT, Bhubaneswar, Odisha, India – 751 024, India*

ABSTRACT

Methanogens are primarily anaerobic Archaebacteria representing the most primitive earthly creatures, alive since about 3.5 billion years. While oxygen inhibit their activity, they adapt well in a variety of ecological niche, viz., *the intestinal tracts of insects and animals including ruminants, sewage digesters, groundwater and deep soil/water, etc. They are involved in biomethanation that converts organic polymers, including the complex recalcitrant lignocelluloses, to CH_4 and CO_2. Enhancing the process by optimising the physical, chemical and molecular parameters is an interesting biotechnology. Molecular research is rapidly progressing on the genomics, gene and its regulation, and methanogenesis as it expresses. Identifying novel species expressing methanogenesis and manipulating the genetic make-up including through cloning in recent years will set goals and leads for future investigations. The biomethanation enzyme Methyl-CoM-reductase (Mcr) constitutes about 10% of the methanogen protein. Due to its abundance and significance, elucidating its structure and its synthesis and regulation mechanism has received much attention recently. Mcr-coding genes from methanogens have been cloned and sequenced. An approach at the molecular level to enhance biomethanation seems to lie in metagenomics. The microbiology, biochemistry, the roles of the participating microbes, their molecular biology and applications in methanogenesis are discussed in this chapter.*

Keywords: Methanogen, Methyl Co-M reductase, Metagenome, Archaea, Biogas

Abbreviations used

ATP : adenosine tri-phosphate
CoB : coenzyme B
CoM : coenzyme M
MB : methanogenic bacteria
Mcr : methyl coenzyme-M reductase
PCR : polymerase chain reaction
rRNA : ribosomal ribonucleic acid

Introduction

Renewable energy is a natural energy resource from both physical (*viz*., sunlight, wind, tides, geothermal heat) and biological (*e.g.*, bacterial-methanogenesis, algal-biodiesel, phytological-bioethanol) sources. Burning of coal, oil and natural gas is connected with GHG (including CO_2) emissions. Sourcing of energy from fossil sources is becoming more and more problematic essentially due to its non-renewable nature. Biomethane emits carbon-di-oxide at an intensity of 1:1 compared to 67.9, 95.8 and 96.7 for natural gas, diesel and gasoline respectively (Anon, 2005). In light of this, renewable energy is being promoted in various countries through their national programmes with long-term objectives of ensuring sustainable development through reduced greenhouse gas emissions and future energy supply. In 2008, renewable sources contributed approximately 19% of the global final energy consumption (Anon, 2011). Global survey hints at slow albeit steady decline in the dependency on nonrenewable energy sources.

Methane itself being a big contributor to global warming, understanding methanogenesis to its proper use in the process for human good while limiting its greenhouse gas effects is necessitated. Biogas is a mixture of gases (CH_4, CO_2) in defined proportions produced during anaerobiosis of organic substrates by specialised microbes (Ohmiya *et al.,* 2005), a class of bacteria known as methanogens. These bacteria transform 1- or 2-C compounds from the intermediates produced from the breakdown of the larger macromolecules to smaller forms. This context of biogas production from renewable resources or organic wastes has been of huge socioeconomic and environmental importance (Weiland, 2003; Sreekrishnan *et al.,* 2004).

Methanogens

Regarded among the oldest earthly creatures, the methanogens are extremophiles that are adept to thrive in harsh habitats. Their discovery established the kingdom *Archaeobacteria* that also includes some other extremophiles such as, halophiles, thermophiles, psychrophiles, radophiles, barophiles, etc. as also the sulphur-dependent organisms. Woese *et al.* (1990) proposed a separate kingdom for methanogens and other Archaeobacteria as Archaea. Currently, there is a domain Archaea that contains the two most prominent phyla Euryarchaeota and Crenarchaeota. Recently, a third phylum named Bathyarchaeota has also been proposed (He *et al.*, 2016). Methanogens are grouped under Euryarchaeota. Methanobacteria, Mathanomicrobia and Metanococcoi are the classes which comprises all the methanogens. *Methanobacterium formicicum* is the representative organism of the phylum Euryarchaeota.

History of Archaea and methanogens

Microbial fossils are found by looking for the membranes formed from isoprene chains unique to Archaea. These do not decompose at high temperatures and make good markers for the presence of ancient Archaea. These ancient life forms have also been found in the oldest known sediment (3.8 billion years old) on earth, in the Isua district of Greenland, which indicates that they appeared within one billion years of the earth's formation, in an atmosphere that was rich in ammonia and methane. They have also been found in Mesozoic, Paleozoic and Precambrian sediments and thus, it is thought that these were most likely the earliest inhabitants of the earth.

The activity of phylum Euryarchaeota in methanogenesis can be well studied on the basis of an essential enzyme, *i.e.*, Methyl CoenzymeM Reductase (Mcr). The Database when searched and filtered showed the following groups with their respective MCR active organism. Methanobacteriales (244), Methanococcales (101), Methanomicrobia (337), Methanopyrus kandleri (10) and Environmental samples (2094). Environmental samples representing the total number of organisms reported so far but yet to be cultured and classified at organismal level (http://www.uniprot.org/uniprot/?query=methyl+ coenzyme +reductase&by=taxonomy#131567,2157,28890).

Habitat of methanogens

Methanogens inhabit in some of the most extreme environments on earth, including the rumen of ruminants living on hydrogen and carbon-di-oxide produced by syntrophic microbes that help digest cellulose, as well as necessary for protein synthesis. They can be found in places like muck of swamps and marshes, hydrothermal vents, porous rock, sewage sludge, termite-gut and oil-contaminated groundwater at underground oil storage facilities (Watanabe *et al.*, 2002). Based on their natural habitat, some are thermophiles, the methanogens found in volcanic hot springs, mud volcanoes and solfataras, where temperatures span from 40–100°C and in marine environments in undersea hydrothermal vents where the temperatures can reach up to 350°C due to high pressure. Psychrophily is rare among methanogens with only a few species being identified till now (Nozhevnikova *et al.*, 2003). Also, there are halophiles (Riffat and Krongthamchat, 2006), acidophiles (Zhou and Nanqi, 2007), alkaliphiles (Thakker and Ranade, 2002), radophiles, and so on. Malakahmed *et al.* (2009) reported a 93% bacterial, 5% protozoan and 2% fungal population in a 50l anaerobic bioreactor (ABR), using 75% kitchen waste and 25% sewage sludge as substrate respectively. They showed that, fast-growing bacteria which are robust enough to grow on high substrate concentration and reduced (acidic) pH were dominant in the acidification zone of the ABR, *i.e.*, the front compartment of reactor. The terminal part of this ABR exhibited slower scavenging bacteria that grow excellently at high (alkaline) pH.

Morphology of methanogens

Archaea exhibit a wide variety of shapes, sizes, and ultrastructural variations, unlike bacterial cells. Two shapes, *i.e.*, rods and coccoid, seem to dominate though. Examples of rods are *Methanobacterium* spp. and *Methanopyrus kandleri*, and coccoids include *Methanococcus* and *Methanosphaera*. *Methanoculleus* and *Methanogenium* exhibit coccoid to irregular shapes, possibly due to the loosely bound S-layers on the walls. Methanogens are not just limited to these shapes, but also include a plate (*Methanoplanus*), long thin spiral (*Methanospirillum*), and cluster of round (*Methanosarcina*) cells. Methanogens are known to lack murein, though some may contain pseudomurein, which can only be distinguished from its bacterial counterpart through chemical analysis (Sprott and Beveridge, 1994; König, 1988). Methanogens that do not possess pseudomurein have at least one paracrystalline array (S-layer), the proteins that fit together in an array like jigsaw pieces that do not covalently bind to one another, in contrast to a cell wall that is one giant covalent bond. The S-layer proteins of some methanogens (*e.g.*, *Methanococcus* spp.) are glycosylated thereby facilitating stability.

Identification of methanogens – culture-based techniques

Methanogens requiring very low redox potential for the growth are perhaps the strictest anaerobes. Many workers have defined ways for its growth but a modified Hungate culture technique has been the most appropriate one. Use of Freter-type anaerobic glove box with an inner ultra low oxygen chamber has been described by Edwards and McBride (1975) to isolate and grow methanogens. The inner chamber is specially modified to maintain the redox potential and pressure necessary to grow methanogens. For this, the chamber is periodically flushed with H_2 and CO_2 (80:20). Cultures are plated in the outer anaerobic glove box and immediately placed in the inner chamber. Though this method is relatively expensive than Hungate procedures, it offers unique advantages like low skill, manual dexterity, and allows routine testing procedures.

Biology of methanogens – culture-independent

It is long recognised that standard culture methods fail to adequately represent the enormous microbial diversity that exists in nature. To avoid reliance on cultivation, development of a variety of culture-independent methods, many of them coupled with high-throughput DNA sequencing, has allowed microbial diversity to be explored in even greater detail (Handelsman, 2005; Harris *et al.*, 2004; Moreira and Lopez-Garcia, 2002; Rappe and Giovannoni, 2003). These include screening of expression libraries with immune serum, nucleic acid subtractive methods, small molecule detection with mass spectroscopy and many more (Relman, 2002). Sequence-based methods are more in application now-a-days because of their general applicability and the continued expansion of high-throughput, low cost, sequencing capacity.

The basis of culture-independent identification of Archaeal species is sequence analysis of the sufficiently well-conserved (across species) rRNA genes that can be readily amplified using random PCR primers based on highly conserved sequences, yet are sufficiently diverse to differentiate archaeal species. To maximise the utility of 16S rRNA gene analysis for species determination, the entire 16S rRNA gene is amplified and sequenced in its entirety through bi-directional sequencing of cloned 16S amplicons (Hugenholtz, 2002). After sequencing, 16S sequences are clustered into groups and a threshold of sequence similarity is established (usually, 98 or 99%) to distinguish genus and species. This approach has been applied to biogas-producing microbial communities as well (Huang *et al.*, 2002, 2005; McHugh *et al.*, 2003; Mladenovska *et al.*, 2003; Shigematsu *et al.*, 2006; Tang *et al.*, 2007; Klocke *et al.*, 2007,2008). Using 16S rRNA and RFLP, Joulion *et al.* (1998) characterised four major groups of methanogens from rice field soil. While PCR amplification of 16S rRNA sequences has been of enormous value, there are some loopholes to this approach and corrected by hybridisation-based methods such as *in situ* hybridisation with species or strain-specific 16S oligonucleotides applied to the original (or similar) sample (Amann *et al.*, 1995; Bosshard *et al.*, 2000).

In addition to the 16S-rDNA target, other marker genes such as *mcrA* encoding the α-subunit of methyl coenzyme-M reductase have been used to elucidate the composition of methanogenic consortia (Lueders *et al.*, 2001; Luton *et al.*,

2002; Friedrich, 2005; Juottonen *et al*., 2006; Rastogi *et al*., 2008). Mcr catalyses the last step of methanogenesis and is conserved among all methanogens. Phylogenetic inference with *mcrA* sequences is similar to that obtained with 16S rRNA gene sequences, suggesting no lateral transfer (Bapteste *et al*., 2005). Moreover, Mcr is absent in all non-methanogens, with the exception of the anaerobic methane-oxidising *Archaea*, which are closely related to the methanogens (Hallam *et al*., 2003). Due to the fact that methanogens may be examined exclusively in an environment, *mcrA* has been increasingly used for phylogenetic analysis coupled with, or independent of, 16S rRNA genes.

Metagenomics

The metagenome of a biogas-producing microbial community from a production-scale biogas plant fed with renewable primary products has been analysed by applying the ultrafast 454-pyrosequencing technology. Community structure analysis of the fermentation sample revealed that *Clostridia* from the phylum Firmicutes is the most prevalent taxonomic class, whereas, species of the order *Methanomicrobiales* are dominant among methanogenic *Archaea* (Krause *et al*., 2008a). Many sequence reads could be allocated to the genome sequence of the Archaeal methanogen *Methanoculleus marisnigri* JR1. This result indicated that species related to those of the genus *Methanoculleus* play a dominant role in hydrogenotrophic methanogenesis in the analysed fermentation sample (Schlüter *et al*., 2008).

Short-read-length libraries are generally not preferred for metagenomic characterisation of microbial communities (Wommack *et al*., 2008). On the other hand, other authors describe the phylogenetic classification of short environmental DNA fragments obtained by high-throughput sequencing technologies (Krause *et al*., 2008b; Manichanh *et al*., 2008). Franke-Whittle *et al.* (2009) developed Anaerochip (a molecular tool), oligonucleotide probes targeting the 16S rRNA gene of methanogens. It allows screening for the presence or absence of most lineages of mesophilic and thermophilic methanogens within complex anaerobic samples in a single test. Application of this microarray to complex samples should result in a greater knowledge of the methanogenic communities. The study showed the dominance of *Methanoculleus* in a suboptimally operating acidified anaerobic biowaste digester. As per NCBI database, sequencing of 62 organisms of phylum Euryarchaeota have been completed and other 69 are in progress (Table 1; Fig. 1). Researchers suggest that higher GC content provides stability to the genome, thermo-stability to the DNA, and conservation through generations. The table has thus been arranged as per the total GC contents in a descending order. The total genome accounts for the chromosome and the plasmid, and the ones with an asterisk mention the estimated size.

Table 1: The Metabolic classification of Methanogenic Bacteria (MB)

Sl.No.	Group	Substrate	Representative species	Equation
1.	MB I	Acetate	*Methanosaeta* spp.	$CH_3COOH \rightarrow CO_2 + CH_4$
2.	MB II	Hydrogen and Formate	*Methanobrevibater* spp. *Methanogenium* spp.	$CO_2 + H_2O \rightarrow CH_4 + 2H_2O$
3.	MB III	Methylated Compounds	*Methanolobus* spp. *Methanococcus*	$CH_3OH \rightarrow CH_4 + CO_2 + 2H_2O$ $4CH_3NH_2 + 2H_2O \rightarrow 3CH_4 + CO_2 + 4NH_4^+$
4.	MB IV	Acetate, H_2 & Methylated Compounds	*Metahanosarcina* spp.	Combinations of all

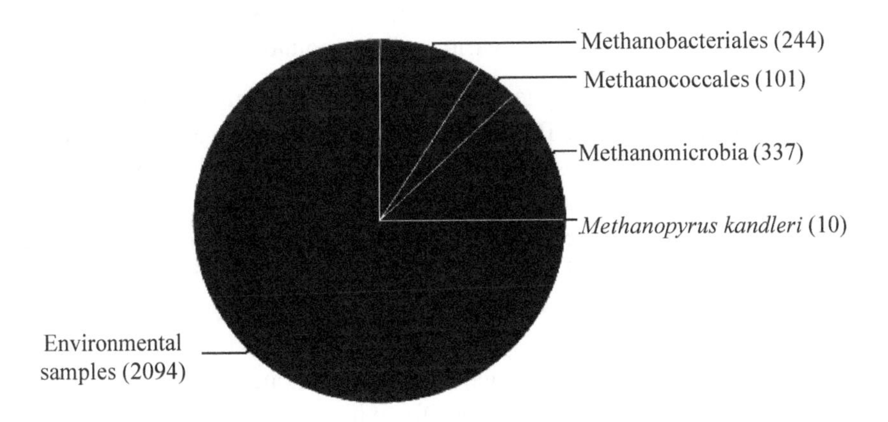

Fig. 1: A pie diagram showing the break-up of various methanogens from the NCBI database

Methanogenesis

Biogas is produced from agricultural, municipal, and industrial waste biomass with the help of syntrophic bacteria and methanogens. These are physiologically united as a consortium of methane producers in anaerobic digestion. Though the main substrates are acetate, H_2 and CO_2, methylamines, CO, formate and methanol are also converted to CH_4. Methanogenic metabolism is unusual as H_2, CO_2, formate, methylated C1 compounds and acetate are used as energy and carbon sources.

Substrate for methanogenesis

In contrast to their huge phylogenetic diversity, methanogens can only use a few simple substrates, most of them being C1 compounds, like CO_2, formate, methanol and methylamines. In fact, the metabolic profile is restricted to only one or two of the above substrates, exceptions being *Methanosarcina* and *Methanolacina*.

Carbon-di-oxide reduction by molecular hydrogen, followed by formate utilisation is the common energy-yielding reaction in methanogens. Acetate is a substrate for *Methanosarcina* and *Methanosaeta*, while methylotrophic genera (*e.g.*, most members of the *Methanosarcinaceae*) utilise methanol, several methylamines or methylsulphide. Furthermore, some species grow on primary and secondary short-chain alcohols. Many species are dependent on special growth factors like vitamins and amino acids. All methanogens can use ammonium as a nitrogen source. A few species (*e.g.*, *Methanosarcina barkeri* and *Methanococcus thermolithotrophicus*) fix molecular nitrogen too. Methanogenic bacteria (MB) can be categorised into four groups based on the substrate use (Table 2). MB I, II and III are the groups which exclusively use acetate, formate and methylated compounds, respectively. MB IV, a comprehensive group, can use a variety of compounds as substrate.

All catabolic processes finally lead to the formation of a mixed disulphide from coenzyme M and coenzyme B that functions as an electron acceptor of certain anaerobic respiratory chains. Molecular hydrogen, reduced coenzyme F_{420} or reduced ferredoxin is used as electron donors. The redox reactions are coupled to proton translocation across the cytoplasmic membrane. The resulting electrochemical proton gradient is the driving force for ATP synthesis as catalysed by an A_1A_0-type ATP synthase. Other energy-transducing enzymes involved are the membrane-integral methyltransferase and the formylmethanofuran dehydrogenase complex. The former enzyme is a unique, reversible sodium ion pump that couples methyl-group transfer with the Na^+ transport across the membrane. The formylmethanofuran dehydrogenase is a reversible ion-pump that catalyses formylation and deformylation of methanofuran.

Stages of methanogenesis

Methanogenesis is a three-stage process (Fig. 2). Stage I microbes (fermentative bacteria) initiate by hydrolysing and fermenting complex insoluble organics to simple acids, alcohol and others.

In stage II, the intermediate products are transformed into acetic acids, and H_2 and CO_2 through acetogenesis. Methanogens come into action in stage III of the whole process. They utilise the products thus formed in stages I and II thereby producing methane.

Biochemistry of methanogenesis

Methanogenesis is an anaerobic respiration, and oxygen inhibits methanogens. Terminal electron acceptor here is the carbon of low molecular weight compounds CO_2 and acetic acid:

Table 2: List of the methanogens whose gene sequences have been annexed in the NCBI database

Sl.No.	Organism / strain	Family/Class	*Size	GC	Ref Seq	Center
01	*Methanoculleus marisnigri JR1*	Methanomicrobiaceae / Methanomicrobia	2.4781	62.1	NC_009051.1	DOE Joint Genome Institute
02	*Methanopyrus kandleri AV19*	Methanopyraceae / Methanopyri	1.69497	61.2	NC_003551.1	Fidelity Systems
03	*Methanosphaerula palustris E1-9c*	Unclassified / Methanomicrobiales	2.92292	55.4	NC_011832.1	DOE Joint Genome Institute
04	*Methanocella paludicola SANAE*	Methanocellaceae/ Methanomicrobia	*3	54.9	NC_013665.1	NITE
05	*Methanoregula boonei 6A8*	Methanomicrobiaceae/ Methanomicrobia	2.54294	54.5	NC_009712.1	DOE Joint Genome Institute
06	*Methanosaeta thermophila PT*	Methanosaetaceae/ Methanomicrobia	1.87947	53.5	NC_008553.1	DOE Joint Genome Institute
07	*Methanocorpusculum labreanum Z*	Methanocorpusculaceae / Methanomicrobia	1.8	50.0	NC_008942.1	DOE Joint Genome Institute
08	*Methanoplanus petrolearius DSM 11571*	Methanomicrobiaceae / Methanomicrobia	*2.8	50	NC_014507.1	DOE Joint Genome Institute
09	*Methanothermobacter thermautotrophicus str. Delta H*	Methanobacteriaceae / Methanobacteria	1.75138	49.5	NC_000916.1	Oscient Pharmaceuticals Corporation
10	*Methanospirillum hungatei JF-1*	Methanospirillaceae / Methanobacteria	3.54474	45.1	NC_007796.1	DOE Joint Genome Institute
11	*Methanosarcina acetivorans C2A*	Methanosarcinaceae / Methanomicrobia	5.75149	42.7	NC_003552.1	Broad Institute
12	*Methanohalophilus mahii DSM 5219*	Methanosarcinaceae / Methanomicrobia	2	42.6	NC_014002.1	DOE Joint Genome Institute
13	*Methanosarcina mazei Go1*	Methanosarcinaceae / Methanomicrobia	4.1	41.5	NC_003901.1	University of Goettingen

Contd.

Sl.No.	Organism / strain	Family/Class	*Size	GC	Ref Seq	Center
14	*Methanococcoides burtonii DSM 6242*	Methanosarcinaceae / Methanomicrobia	2.57503	40.8	NC_007955.1	DOE Joint Genome Institute
15	*Methanosarcina barkeri str. Fusaro*	Methanosarcinaceae / Methanomicrobia	4.87341	39.2	NC_007355.1	DOE Joint Genome Institute
16	*Methanocaldococcus infernus ME*	Methanocaldococcaceae / Methanococci.	*1.3	33.5	NC_014122.1	DOE Joint Genome Institute.
17	*Methanococcus maripaludis C6*	Methanococcaceae / Methanococci	1.74419	33.4	NC_009975.1	DOE Joint Genome Institute
18	*Methanococcus maripaludis C7*	Methanococcaceae / Methanococci	1.77269	33.3	NC_009637.1	DOE Joint Genome Institute
19	*Methanococcus maripaludis S2*	Methanococcaceae / Methanococci	1.66114	33.1	NC_005791.1	University of Washington
20	*Methanococcus maripaludis C5*	Methanococcaceae / Methanococci	1.8083	33.0	NC_009135.1	DOE Joint Genome Institute
21	*Methanobrevibacter ruminantium M1*	Curculionoidea / Methanobacteria	*2.9	32.6	NC_013790.1	AgResearch
22	*Methanocaldococcus fervens AG86*	Methanocaldococcaceae / Methanococci	1.522	32.2	NC_013156.1	DOE Joint Genome Institute
23	*Methanocaldococcus sp. FS406-22*	Methanocaldococcaceae / Methanococci	*1.812	32.0	NC_013887.1	DOE Joint Genome Institute.
24	*Methanocaldococcus vulcanius M7*	Methanocaldococcaceae / Methanococci	*1.7157	31.6	NC_013407.1	DOE Joint Genome Institute.
25	*Methanocaldococcus jannaschii DSM 2661*	Methanocaldococcaceae / Methanococci	1.73997	31.3	NC_000909.1	TIGR
26	*Methanococcus vannielii SB*	Methanococcaceae / Methanococci	1.72005	31.3	NC_009634.1	DOE Joint Genome Institute
27	*Methanobrevibacter smithii ATCC* 35061	Methanobacteriaceae / Methanobateria	1.85316	31.0	NC_009515.1	Genome Sequencing Center, Washington Univ.

Contd.

Sl.No.	Organism / strain	Family/Class	*Size	GC	Ref Seq	Center
28	*Methanococcus aeolicus Nankai*-3	Methanococcaceae / Methanococci	1.5695	30.0	NC_009635.1	DOE Joint Genome Institute
29	*Methanococcus voltae A*3	Methanococcaceae / Methanococci	*1.9	28.6	NC_014222.1	DOE Joint Genome Institute
30	*Methanosphaera stadtmanae* DSM 3091	Methanobacteriaceae / Methanobacteria	1.7674	27.6	NC_007681.1	University of Goettingen
31	*Methanohalobium evestigatum* Z-7303	Methanosarcinaceae / Methanomicrobia	*2.36	NA	NC_014253.1	DOE Joint Genome Institute
32	*Methanothermobacter marburgensis str. Marburg*	Curculionoidea / Methanobacteria	*1.6044	NA	NC_014408.1	Georg-August-University Goettingen.
33	Methanothermus fervidus DSM 2088 Methanobacteria	Methanothermaceae /	*1.2	NA	NC_014658.1	DOE Joint Genome Institute

Source: http://www.ncbi.nlm.nih.gov/genomes/lproks.cgi; *Size values provided are only estimated, otherwise genome size is calculated based on existing sequences

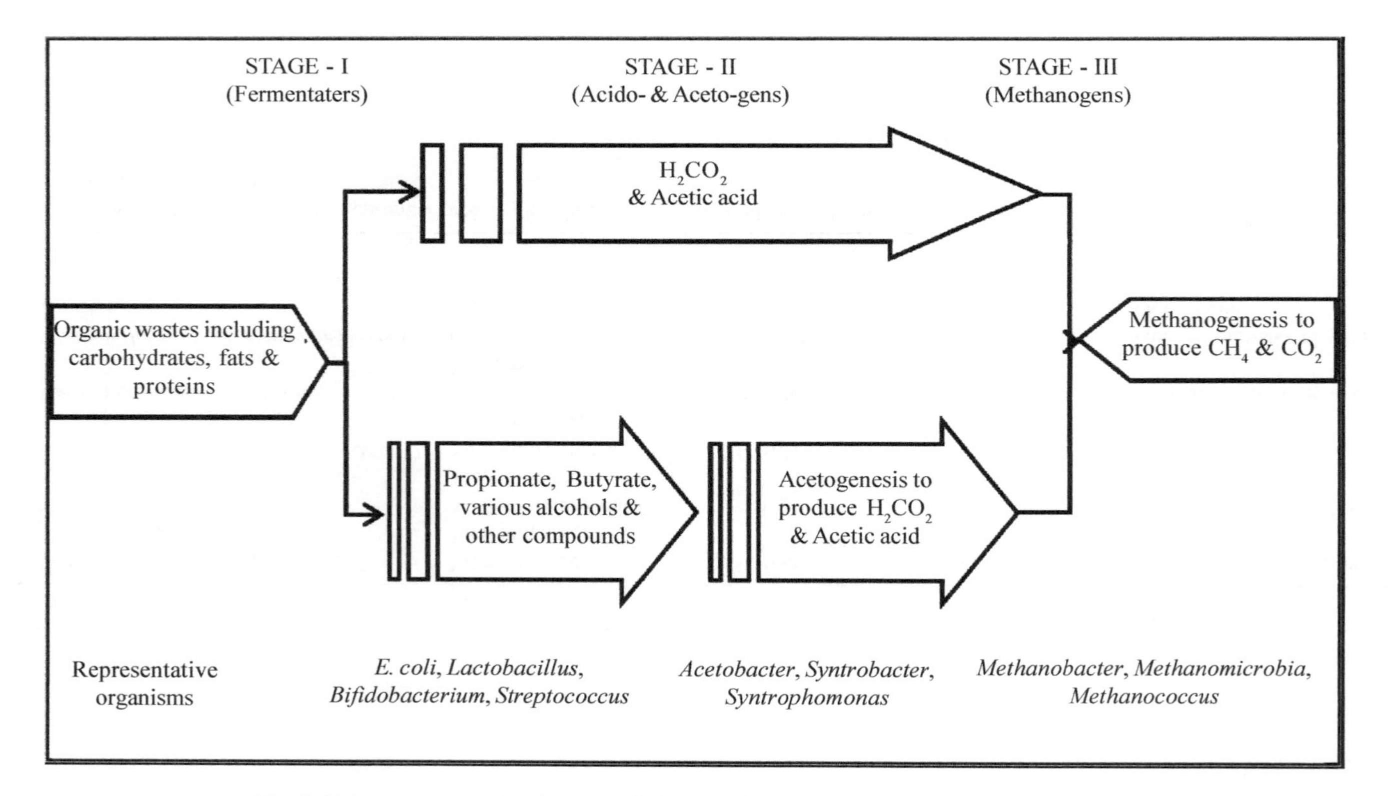

Fig. 2: Schema representing the stages in biotransformation of organic material to methane

$$CO_2 + 4H_2 \rightarrow CH_4 + 2H_2O \quad (1)$$

and

$$CH_3COOH \rightarrow CH_4 + CO_2 \quad (2)$$

Methanogenic pathway utilising CO_2 and H_2 involves methanogenic specific enzymes that catalyses unique reactions using novel coenzymes (Fig. 2). Methanofuran, the first C1 carrier found only in methanogenic and sulphur-reducing Archaea, is reduced to Formylmethanofuran as CO_2 binds to it. The reaction is catalysed by Formylmethanofuran dehydrogenase (Fmd) (Thauer *et al.*, 1993). This reaction is the only endergonic (" $G^{æ\%}$ = +16kJ/mol) reaction in the whole process. The required energy is sourced from sodium ion membrane potential (Kaesler and Schonheit, 1989a; Kaesler and Schonheit, 1989b). The second C1 carrier is Tetrahydromethanopterin (H_4MPT) and the formyl group is now transferred to H_4MPT, catalysed by Formylmethanofuran-H_4MPT formyltransferase (Ftr) (Donnelly and Wolfe, 1986; Breitung and Thauer, 1990). The Formyl H_4MPT is then changed to N^5N^{10}-Methenyl-H_4MPT and the reaction is catalysed by Methenyl-H_4MPT cyclohydrolase (Mch) (Breitung *et al.*, 1991) (ΔG = -4.6kJ/mol). The N^5N^{10}-Methenyl-H_4MPT is now reduced to Methylene-H_4MPT in two ways, either F_{420}-dependent or independent.

The independent reduction is catalysed by H_2-forming methylene-H_4MPT dehydrogenase (Hmd) (Thauer *et al.*, 1996) and the dependent pathway which also requires F_{420}-reducing hydrogenase (Frh) (not shown in the figure) for F_{420} reduction is catalysed by F_{420}-dependent methylene-H_4MPT dehydrogenase (Mtd) (Thauer *et al.*, 1993). Reduction of Methylene-H_4MPT to Methyl-H_4MPT is now F_{420}-dependent and is catalysed by Methylene-H_4MPT cyclohydrolase (Mer). Then, the methyl group from Methyl-H_4MPT is transferred to a third C1 carrier, *i.e.*, Coenzyme M. The reaction is catalysed by Methyl-H_4MPT-coenzyme M methyltransferase (Mtr), (DiMarco *et al.*, 1990; Gottschalk and Thauer, 2001).

Mtr is an integral membrane protein complex of 670kDa. The negative free energy change of this reaction (-30kJ/mol) is conserved by sodium ion membrane potential. This is a typical methyltransferase that is coupled with ion transport and energy conservation. The sodium ion membrane potential that is formed by Mtr reaction is mainly used as a driving force for the first reaction (Gottschalk and Thauer, 2001). Methyl-coenzyme M is finally reduced to methane by Methyl-coenzyme M reductase (Mcr) (Thauer, 1998). The reductant here is Coenzyme B, which is the archaeal methanogen characteristic. Coenzyme B and Coenzyme M are oxidised to the corresponding heterodisulphide (Grabarse *et al.*, 2001). The heterodisulphide is an important intermediate of the energy metabolism in methanogens since it is substrate of an energy conservation reaction catalysed by heterodisulphide reductase/hydrogenase (Hdr) system. In this reaction heterodisulphide is reduced to Coenzyme M and Coenzyme B (Hedderich *et al.*, 1994). The reaction steps along with the free-energy equivalents and the various catalytic factors involved in the process are shown in Fig. 3.

Methyl-coenzyme M reductase

Methyl-Coenzyme M Reductase (MCR) is Archaea enzyme that catalyses methane formation by combining the hydrogen donor coenzyme B and the methyl donor coenzyme M. It has two active sites, each occupied by the nickel-containing F_{430} cofactor (Thauer, 1998). The conversion is presented as CH_3-S-CoM + HS-CoB $\rightarrow$ CH_4 + CoB-S-S-CoM.

All known methanogens express the enzyme Methyl-Coenzyme M Reductase (MCR), which catalyses the terminal step in biogenic methane production (Ferry, 1999; Reeve *et al.*, 1997; Thauer, 1998). The presence of MCR is considered a diagnostic indicator of methanogenesis (Ferry, 1992; Lueders *et al.*, 2001; Luton *et al.*, 2002, Reeve *et al.*, 1997, Thauer, 1998). The genomes of all methanogenic archaea encode at least one copy of the *mcrA* operon (Reeve *et al.*, 1997; Thauer, 1998).

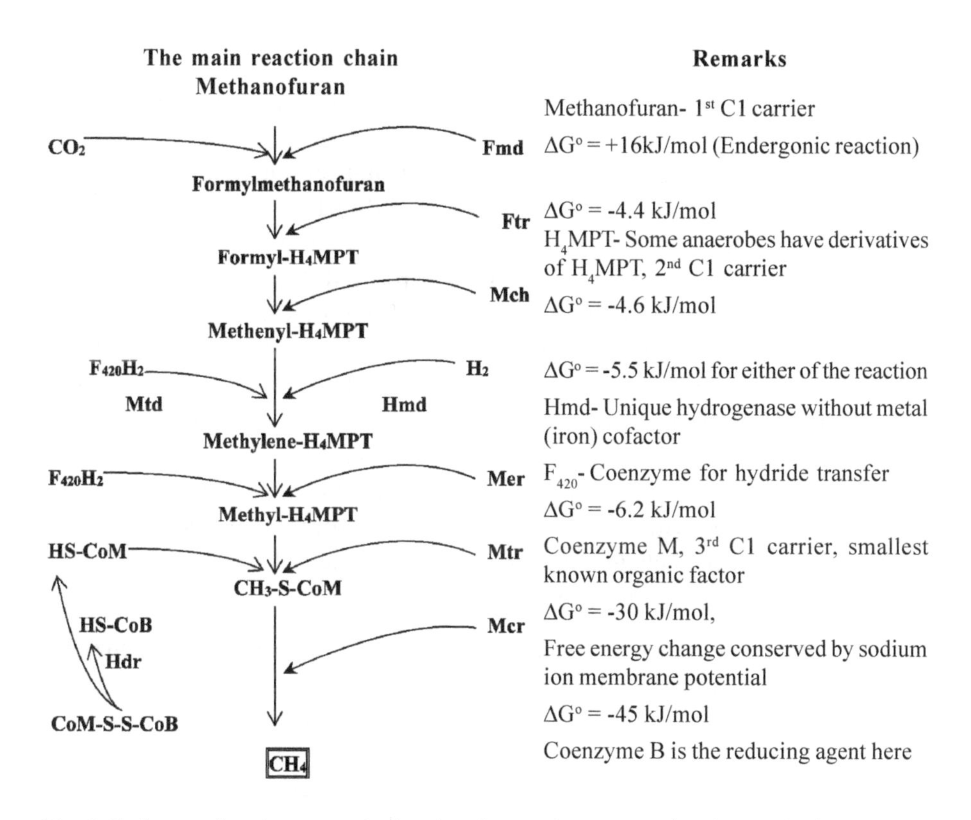

Fig. 3: Pathway of methanogenesis showing the reaction steps and major catalysis
Note: All "Gæ% data are referenced from Thauer (1998). Abbreviations/acronyms:
Fmd: Formylmethanofuran dehydrogenase; **Ftr**: Formylmethanofuran-H_4MPT formyltransferase; **Mch**: Methenyl-H_4MPT cyclohydrolase; **H_4MPT**: Tetrahydromethanopterin (H_4MPT); **Hmd**: H_2-forming methylene-H_4MPT dehydrogenase; **Mer**: Methylene-H_4 MPT reductase; **Mtd**: F_{420} dependent methylene-H_4MPT dehydrogenase; **HS-B**: Heterodisulphide of coenzyme B; **Mtr**: Methyl-H_4MPT coenzyme M methyltransferase; **Mcr**: Methyl-coenzyme M reductase; **Hdr**: Heterodisulphide reductase/hydrogenase; **Frh**: F_{420}-reducing hydrogenase; **CoM-S-S-CoB**: Heterodisulphide of coenzyme M and coenzyme B

Composed of two alpha (*mcrA*), beta (*mcrB*) and gamma (*mcrG*) subunits, the *mcrA* holoenzyme catalyses heterodisulphide formation between coenzyme M and coenzyme B from methyl-coenzyme M and coenzyme B and the subsequent release of methane (Ellermann *et al.*, 1998). Functional constraints on its catalytic activity have resulted in a high degree of MCR amino acid sequence conservation, even between phylogenetically distant methanogenic lineages (Luton *et al.*, 2002; Reeve *et al.*, 1997). This conserved primary structure is used to develop degenerate PCR primers for recovering naturally occurring *mcrA* fragments from a variety of environments (Lueders *et al.*, 2001; Luton *et al.*, 2002).

Bioenergetics

In whole of the pathway, two of the reactions are coupled to the formation of chemical gradients that drive ATP synthesis, the membrane-bound N^5-methyltetrahydromethanopterin coenzyme M methyltransferase in the CO_2 reduction and acetate fermentation, and reduction of CoM-S-S-CoB. Methyltransferase is an integral membrane-bound complex that generates a sodium ion gradient across the membrane during methyl transfer. The complex contains factor III of which the Co^+ atom functions as a super-reduced nucleophile accepting the methyl group from CH_3-H_4MPT producing CH_3-Co^{3+} in the first of the two partial reactions catalysed by the enzyme. The second partial reaction involves transfer of the methyl group from CH_3-Co^+ to CoM, producing CH_3-S-CoM and regenerating the activated Co^+ form of the corrinoid. It is proposed that sodium ion translocation is accomplished by a permease associated with MtrA and that the energy for translocation is derived from a conformational change in MtrA during the methylation-demethylation cycle of Co^+/CH_3-Co^{3+}(Harms and Thauer, 1996).

The second energy-generating step is the demethylation of methyl-coenzyme M and reduction of the heterodisulphide CoM-S-S-CoB catalysed by methyl-coenzyme M and heterodisulphide reductases. In cell extracts, the methyl-coenzyme M reductase is generally inactive and experiments suggest that activation occurs by reduction of the protein-bound coenzyme F_{430} to the Ni(I) state (Ferry, 2002). The electron donor for activation of methyl-coenzyme M reductase is ferredoxin. A membrane-bound electron transport chain delivers electrons to the heterodisulphide, generating a proton gradient that drives ATP synthesis. The relative positions of CoM, CoB and F_{430} in the crystal structure of the methyl-CoM reductase is consistent with a nucleophilic attack of Ni(I) on CH_3-S-CoM and formation of a $[_{F430}]$Ni(III)-CH_3. In the next step Ni(III) oxidises HSCoM, producing CS-CoM thiyl radical and $[F_{430}]$Ni(II)-CH_3. Finally, protonolysis releases CH_4 and the thiyl radical is coupled to 2 S-CoB to form CoB-S-S-CoM with the excess electron transferred to Ni(II) forming Ni(I) (Ermler *et al.*, 1997).

The energy transfer values (in kJoules) as the basic energetic reaction are:

$$CO_2 + 4H_2 \rightarrow CH_4 + 2H_2O \qquad \Delta G^{0'} = -130.4 \text{ kJ/mol } CH_4 \tag{3}$$

and

$$4HCOO^- + 4H^+ \rightarrow CH_4 + 3CO_2 + 2H_2O \quad \Delta G^{0'} = -119.5 \text{ kJ/mol } CH_4 \tag{4}$$

Conclusion

Use of fossil fuel as an energy source is integral part of our daily life, but this is an unsustainable resource owing to their finite reserves and negative environmental effects. Recycling and minimising waste are two main and major objectives of waste management strategies globally. Biogasification seems to have a solution to both of these ever-growing global menaces. In a full-scale system, several environmental conditions will be varying constantly owing to the complexity and variability of organic wastes. It is therefore important to predict environmental conditions having the largest impact. Further, the molecular approach can help to identify the ecology-, abundance- and/or activity-wise relevant microbes. These microbes can then be the subject of detailed studies as a target for directed cultivation.

Majority of prokaryotes living in natural environments are rather inconspicuous. Several molecular techniques have been developed in order to overcome the lack of information about the bacterial function by cultivation-independent methods. Despite the progress made in linking the identification of distinct microbes with their functions *in situ*, it may still be necessary to isolate or enrich novel bacteria to reveal their metabolic potential under various environmental conditions. The results of molecular ecology research has established that experimental strategies based on the combination of molecular techniques with traditional cultivation-dependent methods have great potential in revealing some of the hidden complexity of natural microbial ecosystems. The opportunities for the discovery of new organisms and the development of resources based on microbial diversity are greater than ever before. DNA sequences have finally given the microbiologists a way to define microbial phylogeny. The sequences are the bases of tools that will allow microbiologists to explore the distribution and function of environmental microbes.

Metagenomics, a new lens to screen the methanogens, has revolutionised the understanding of the entire living world. In Metagenomics, the power of genomic analysis is applied to entire communities of microbes, bypassing the need to isolate and culture individual microbial species. This new approach will bring to light the many abilities of the methanogens. A combined approach of high throughput metagenomics and massive environmental data monitoring is necessary to find correlations between the environment and community (Knights *et al.*, 2010). In addition, ecological principles can aid in selecting for superior communities that, for example, are rich in parallel metabolic pathways (Hashsham *et al.*, 2000), have high evenness (Wittebolle *et al.*, 2009) and are either resistant, resilient, or redundant (Allison and Martiny, 2008) to sustain a stable bioprocess.

Methanogens have been studied since long but still a lot await discovery. A lab-scale feasible technology needs scaling-up to commercial level with proper dissemination programmes for the rural and urban society. Beside molecular and biochemical aspects, there are other many means that help to understand and enhance biogas production, *e.g.*, physical, physiochemical, nature of substrate and many more. Based on this knowledge, an engineer makes decisions on the designing, inoculation, and operation of the full-scale system to obtain the sufficient kinetic rates and yields for process viability. Breakthroughs like better processing technique for methane to be used as source of energy is also envisaged.

Acknowledgements

The authors are thankful to the MNRE, Govt. of India for kindly providing the grants under the BDTC to carry out research on the subject. Thanks are also due to KIIT for the infrastructure and logistic support.

References

Allison SD, Martiny JBH, 2008. Resistance, resilience, and redundancy in microbial communities. *Proceedings of the National Academy of Sciences USA*, **105(Suppl. 1)**: 11512–9.

Alpana S, Vishwakarma P, Adhya TK, Inubushi K, Dubey SK, 2017. Molecular ecological perspective of methanogenic archaeal community in rice agroecosystem. *Science of the Total Environment*, **596**: 136–46.

Amann RI, Ludwig W, Schleifer KH, 1995. Phylogenetic identification and *in situ* detection of individual microbial cells without cultivation. *Microbiology Reviews*, **59(1)**: 143–69.

Anon, 2005. http://www.compressednaturalgas.net/, www.biomethane.com, Renewable Energy Institute, Texas, US, March 11, 2011.

Anon, 2011. Renewable energy, http://en.wikipedia.org/wiki/Renewable_energy, Wikimedia Foundation, Inc., March 11, 2011.

Bapteste E, Brochier C, Boucher Y, 2005. Higher-level classification of the archaea: evolution of methanogenesis and methanogens. *Archaea*, **1(5)**: 353–63.

Bosshard PP, Santini Y, Grüter D, Stettler R, Bachofen R, 2000. Bacterial diversity and community composition in the chemocline of the meromictic alpine Lake Cadagno as revealed by 16S rDNA analysis. *FEMS Microbiology Ecology*, **31(2)**: 173–82.

Breitung J, Schmitz RA, Stetter KO, Thauer RK, 1991. N^5,N^{10}-Methenyltetrahydro-methanopterin cyclohydrolase from the extreme thermophile *Methanopyrus kandleri*: increase of catalytic efficiency (K_{car}/K_m) and thermostability in the presence of salts. *Archives of Microbiology*, **156(6)**: 517–24.

Breitung J, Thauer RK, 1990. Formylmethanofuran: Tetrahydromethanopterin formyltransferase from *Methanosarcina barkeri* Identification of N^5-formyltetrahydromethanopterin as the product. *FEBS Letters*, **275(1-2)**: 226–30.

DiMarco AA, Sment KA, Konisky J, Wolfe RS, 1990. The formylmethanofuran: tetrahydro methanopterin formyltransferase from *Methanobacterium thermoautotrophicum* "H. Nucleotide sequence and functional expression of the cloned gene. *Journal of Biological Chemistry*, **265(1)**: 472–6.

Edwards T, McBride BC, 1975. New method for the isolation and identification of methanogenic bacteria. *Applied Microbiology*, **29(4)**: 540–5.

Ellermann J, Hedderich R, Böcher R, Thauer RK, 1988. The final step in methane formation: Investigations with highly purified methyl-CoM reductase (component C) from *Methanobacterium thermoautotrophicum* (strain Marburg). *European Journal of Biochemistry*, **172(3)**: 669–77.

Ermler U, Grabarse W, Shima S, Goubeaud M, Thauer RK, 1997. Crystal structure of methyl-coenzyme M reductase: the key enzyme of biological methane formation. *Science*, **278(5342)**: 1457–62.

Ferry JG, 1992. Biochemistry of methanogenesis. *Critical Reviews in Biochemistry and Molecular Biology*, **27(6)**: 473–503.

Ferry JG, 1999. Enzymology of one-carbon metabolism in methanogenic pathways. *FEMS Microbiology Review*, **23(1)**: 13–38.

Ferry JG, 2002. Methanogenesis Biochemistry. *In*: Yixian Zheng (Ed) Encyclopedia of Life Sciences, Macmillan Publishers Ltd, Nature Publishing Group, United States: 1–9.

Franke-Whittle IH, Goberna M, Pfister V, Insam H, 2009. Design and development of the ANAEROCHIP microarray for investigation of methanogenic communities. *J. Microbiological Methods*, **79(3)**: 279–88.

Friedrich MW, 2005. Methyl-coenzyme M reductase genes: unique functional markers for methanogenic and anaerobic methane-oxidizing Archaea. *Methods in Enzymology*, **397**: 428–42.

Gottschalk G, Thauer RK, 2001. The Na^+-translocating methyltransferase complex from methanogenic Archaea. *Biochimica et Biophysica Acta (BBA)-Bioenergetics*, **1505(1)**: 28–36.

Grabarse W, Mahlert F, Duin EC, Goubeaud M, Shima S, Thauer RK, Lamzin V, Ermler U, 2001. On the mechanism of biological methane formation: structural evidence for conformational changes in methyl-coenzyme M Reductase upon substrate binding. *Journal of Molecular Biology*, **309(1)**: 315–30.

Hales BA, Edwards C, Ritchie DA, Hall G, Pickup RW, Saunders JR, 1996. Isolation and identification of methanogen-speciûc DNA from blanket bog peat by PCR ampliûcation and sequence analysis. *Applied and Environmental Microbiology*, **62(2)**: 668–75.

Hallam SJ, Girguis PR, Preston CM, Richardson PM, DeLong EF, 2003. Identification of methyl coenzyme M reductase A (mcrA) genes associated with methane-oxidizing archaea. *Applied Environmental Microbiology*, **69(9)**: 5483–91.

Handelsman J, 2005. Metagenomics: application of genomics to uncultured microorganisms. *Microbiology and Molecular Biology Reviews*, **69(1)**: 195.

Harms U, Thauer RK, 1996. The corrinoid-containing 23-kDa subunit MtrA of the energy-conserving N^5-methyl-tetrahydromethanopterin: coenzyme M methyltransferase complex from *Methanobacterium thermoautotrophicurn*. EPR spectroscopic evidence for a histidine residue as a cobalt ligand of the cobamide. *European Journal of Biochemistry*, **241(1)**: 149–54.

Harris JK, Kelley ST, Pace NR, 2004. New perspective on uncultured bacterial phylogenetic division OP11. *Applied Environmental Microbiology*, **70(2)**: 845–9.

Hashsham SA, Fernandez AS, Dollhopf SL, Dazzo FB, Hickey RF, Tiedje JM, Criddle CS, 2000. Parallel processing of substrate correlates with greater functional stability in methanogenic bioreactor communities perturbed by glucose. *Applied Environmental Microbiology*, **66(9)**: 4050–7.

He Y, Li M, Perumal V, Feng X, Fang J, Xie J, Sievert SM, Wang F, 2016. Genomic and enzymatic evidence for acetogenesis among multiple lineages of the archaeal phylum Bathyarchaeota widespread in marine sediments. *Nature Microbiology*, **1(6)**: 16035.

Hedderich R, Koch J, Linder D, Thauer RK, 1994. The heterodisulfide reductase from *Methanobacterium thermoautotrophicum* contains sequence motifs characteristic of pyridine-nucleotide-dependent thioredoxin reductases. *European Journal of Biochemistry*, **225(1)**: 253–61.

Huang LN, Zhou H, Chen YQ, Luo S, Lan CY, Qu LH, 2002. Diversity and structure of the archaeal community in the leachate of a full-scale recirculating landfill as examined by direct 16S rRNA gene sequence retrieval. *FEMS Microbiology Letters*, **214(2)**: 235–40.

Huang LN, Zhu S, Zhou H, Qu LH, 2005. Molecular phylogenetic diversity of bacteria associated with the leachate of a closed municipal solid waste landûll. *FEMS Microbiology Letters*, **242(2)**: 297–303.

Hugenholtz P, 2002. Exploring prokaryotic diversity in the genomic era. *Genome Biology*, **3(2)**: 0003.1-0003.8.

Hugenholtz P, Goebel BM, Pace NR, 1998. Impact of culture-independent studies on the emerging phylogenetic view of bacterial diversity. *Journal of Bacteriology*, **180(18)**: 4765–74.

Joulian C, Ollivier B, Patel BKC, Roger PA, 1998. Phenotypic and phylogenetic characterization of dominant culturable methanogens isolated from ricefield soils. *FEMS Microbiology Ecology*, **25(2)**: 135–45.

Juottonen H, Galand PE, Yrjälä K, 2006. Detection of methanogenic Archaea in peat: comparison of PCR primers targeting the mcrA gene. *Research in Microbiology*, **157(10)**: 914–21.

Kaesler B, Schönheit P, 1989a. The role of sodium ions in methanogenesis: Formaldehyde oxidation to CO_2 and $2H_2$ in methanogenic bacteria is coupled with primary electrochemical potential of Na^+ translocation at a stoichiometry of 2–3 Na^+/CO_2. *European Journal of Biochemistry*, **184(1)**: 223–32.

Kaesler B, Schönheit P, 1989b. The sodium cycle in methanogenesis: CO_2 reduction to the formaldehyde level in methanogenic bacteria is driven by a primary electrochemical potential of Na^+ generated by formaldehyde reduction to CH_4. *European Journal of Biochemistry*, **186(1-2)**: 309–16.

Klocke M, Mähnert P, Mundt K, Souidi K, Linke B, 2007. Microbial community analysis of a biogas-producing completely stirred tank reactor fed continuously with fodder beet silage as mono-substrate. *Systematic and Applied Microbiology*, **30(2)**: 139–51.

Klocke M, Nettmann E, Bergmann I, Mundt K, Souidi K, Mumme J, Linke B, 2008. Characterization of the methanogenic Archaea within two-phase biogas reactor systems operated with plant biomass. *Systematic and Applied Microbiology*, **31(3)**: 190–205.

Knights D, Costello EK, Knight R, 2010. Supervised classification of human microbiota. *FEMS Microbiology Reviews*, **35(2)**: 343–59.

König H, 1988. Archaeobacterial cell envelopes. *Canadian Journal of Microbiology*, **34(4)**: 395–406.

Krause L, Diaz NN, Edwards RA, Gartemann KH, Krömeke H, Neuweger H, Pühler A, Runte KJ, Schlüter A, Stoye J, Szczepanowski R, Tauch A, Goesmann A, 2008a. Taxonomic composition and gene content of a methane-producing microbial community isolated from a biogas reactor. *Journal of Biotechnology*, **136(1-2)**: 91–101.

Krause L, Diaz NN, Goesmann A, Kelley S, Nattkemper TW, Rohwer F, Edwards RA, Stoye J, 2008b. Phylogenetic classification of short environmental DNA fragments. *Nucleic Acids Research*, **36(7)**: 2230–9.

Kroes I, Lepp PW, Relman DA, 1999. Bacterial diversity within the human subgingival crevice. *Proceedings of National Academy of Science*, **96(25)**: 14547–52.

Loy A, Lehner A, Lee N, Adamczyk J, Meier H, Ernst J, Schleifer KH, Wagner M, 2002. Oligonucleotide microarray for 16S rRNA gene-based detection of all recognized lineages of sulfate-reducing prokaryotes in the environment. *Applied and Environmental Microbiology*, **68(10)**: 5064–81.

Loy A, Schulz C, Lücker S, Schöpfer-Wendels A, Stoecker K, Baranyi C, Lehner A, Wagner M, 2005. 16S rRNA gene-based oligonucleotide microarray for environmental monitoring of the betaproteobacterial order "Rhodocyclales". *Applied and Environmental Microbiology*, 71(3): 1373–86.

Lueders T, Chin KJ, Conrad R, Friedrich M, 2001. Molecular analyses of methyl-coenzyme M reductase α-subunit (mcrA) genes in rice field soil and enrichment cultures reveal the methanogenic phenotype of a novel archaeal lineage. *Environmental Microbiology*, **3(3)**: 194–204.

Luton PE, Wayne JM, Sharp RJ, Riley PW, 2002. The mcrA gene as an alternative to 16s rRNA in the phylogenetic analysis of methanogen populations in landfill. *Microbiology*, **148(11)**: 3521–30.

Malakahmad A, Zain SM, Basri NA, Kutty SM, Isa MH, 2009. Identification of anaerobic microorganisms for converting kitchen waste to biogas. *World Academy of Science, Engineering and Technology*, **3**: 12–20.

Manichanh C, Chapple CE, Frangeul L, Gloux K, Guigo R, Dore J, 2008. A comparison of random sequence reads versus 16S rDNA sequences for estimating the biodiversity of a metagenomic library. *Nucleic Acids Research*, **36(16)**: 5180–8.

McHugh S, Carton M, Mahony T, O'Flaherty V, 2003. Methanogenic population structure in a variety of anaerobic bioreactors. *FEMS Microbiology Letters*, **219(2)**: 297–304.

Mladenovska Z, Dabrowski S, Ahring BK, 2003. Anaerobic digestion of manure and mixture of manure with lipids: biogas reactor performance and microbial community analysis. *Water Science and Technology*, **48(6)**: 271–8.

Moreira D, López-García P, 2002. The molecular ecology of microbial eukaryotes unveils a hidden world. *Trends in Microbiology*, **10(1)**: 31–8.

Nozhevnikova AN, Zepp K, Vazquez F, Zehnder AJ, Holliger C, 2003. Evidence for the Existence of psychrophilic methanogenic communities in anoxic sediments of deep lakes. *Applied and Environmental Microbiology*, **69(3)**: 1832–5.

Ohmiya K, Sakka K, Kimura T, 2005. Anaerobic bacterial degradation for the effective utilization of biomass. *Biotechnology and Bioprocess Engineering*, **10(6)**: 482–93.

Rappé MS, Giovannoni SJ, 2003. The uncultured microbial majority. *Annual Reviews in Microbiology*, **57(1)**: 369–94.

Rastogi G, Ranade DR, Yeole TY, Patole MS, Shouche YS, 2008. Investigation of methanogen population structure in biogas reactor by molecular characterization of methyl-coenzyme M reductase A (mcrA) genes. *Bioresource Technology*, **99(13)**: 5317–26.

Reeve JN, Nölling J, Morgan RM, Smith DR, 1997. Methanogenesis: genes, genomes, and who's on ûrst? *Journal of Bacteriology*, **179(19)**: 5975.

Relman DA, 2002. New technologies, human-microbe interactions, and the search for previously unrecognized pathogens. *The Journal of Infectious Diseases*, **186(Suppl. 2)**: S254–8.

Riffat R, Krongthamchat K, 2006. Specific methanogenic activity of halophilic and mixed cultures in saline wastewater. *International journal of Environmental Science & Technology*, **2(4)**: 291–9.

Rudi K, Skulberg OM, Skulberg R, Jakobsen KS, 2000. Application of sequence-specific labeled 16S rRNA gene oligonucleotide probes for genetic profiling of cyanobacterial abundance and diversity by array hybridization. *Applied Environmental Microbiology*, **66(9)**: 4004–11.

Schlüter A, Bekel T, Diaz NN, Dondrup M, Eichenlaub R, Gartemann KH, Krahn I, Krause L, Krömeke H, Kruse O, Mussgnug JH, Neuweger H, Niehaus K, Pühler A, Runte KJ, Szczepanowski R, Tauch A, Tilker A, Viehöver P, Goesmann A, 2008. The metagenome of a biogas-producing microbial community of a production-scale biogas plant fermenter analysed by the 454-pyrosequencing technology. *Journal of Biotechnology*, **136(1-2)**: 77–90.

Schmidt TM, DeLong EF, Pace NR, 1991. Analysis of a marine picoplankton community by 16S rRNA gene cloning and sequencing. *Journal of Bacteriology*, **173(14)**: 4371–8.

Shigematsu T, Era S, Mizuno Y, Ninomiya K, Kamegawa Y, Morimura S, Kida K, 2006. Microbial community of a mesophilic propionate-degrading methanogenic consortium in chemostat cultivation analyzed based on 16S rRNA and acetate kinase genes. *Applied Microbiology and Biotechnology*, **72(2)**: 401–15.

Small J, Call DR, Brockman FJ, Straub TM, Chandler DP, 2001. Direct detection of 16S rRNA in soil extracts by using oligonucleotide microarrays. *Applied and Environmental Microbiology*, **67(10)**: 4708–16.

Springer E, Sachs MS, Woese CR, Boone DR, 1995. Partial gene-sequences for the A subunit of methyl-coenzyme M reductase (mcrI) as a phylogenetic tool for the family Methanosarcinaceae. *International Journal of Systematic and Environmental Microbiology*, **45(3)**: 554–9.

Sprott GD, Beveridge JT, 1994. Microscopy. *In*: JG Jerry (Ed) Metanogenesis: Ecology, physiology, biochemistry and genetics, Springer: 85–7.

Sreekrishnan TR, Kohli S, Rana V, 2004. Enhancement of biogas production from solid substrates using different techniques—a review. *Bioresource Technology*, **95(1)**: 1–10.

Stahl DA, Lane DJ, Olsen GJ, Pace NR, 1984. Analysis of hydrothermal vent-associated symbionts by ribosomal RNA sequences. *Science*, **224(4647)**: 409–11.

Stahl DA, Lane DJ, Olsen GJ, Pace NR, 1985. Characterization of a Yellowstone hot spring microbial community by 5S rRNA sequences. *Applied and Environmental Microbiology*, **49(6)**: 1379–84.

Tang YQ, Fujimura Y, Shigematsu T, Morimura S, Kida K, 2007. Anaerobic treatment performance and microbial population of thermophilic upûow anaerobic filter reactor treating awamori distillery wastewater. *Journal of Bioscience and Bioengineering*, **104(4)**: 281–7.

Thakker CD, Ranade DR, 2002. An alkalophilic *Methanosarcina* isolated from Lonar crater. *Current Science*, **82(4)**: 455–8.

Thauer RK, 1998. Biochemistry of methanogenesis: a tribute to Marjory Stephenson. 1998 Marjory Stephenson prize lecture. *Microbiology*, **144(9)**: 2377–2406.

Thauer RK, Hedderich R, Fischer R, 1993. Biochemistry: Reactions and enzymes involved in methanogenesis from CO_2 and H_2. *In*: Ferry JG (Ed) Methanogenesis: Ecology, physiology, biochemistry & genetics. Chapman & Hall. Inc., New York & London: 209–52.

Thauer RK, Klein AR, Hartmann GC, 1996. Reactions with molecular hydrogen in microorganisms: Evidence for a purely organic hydrogenation catalyst. *Chemical Reviews*, **96(7)**: 3031–42.

Ward DM, Weller R, Bateson MM, 1990. 16S rRNA sequences reveal numerous uncultured microorganisms in a natural community. *Nature*, **345(6270)**: 63.

Watanabe K, Kodoma Y, Hamamura N, Kaku N, 2002. Diversity, abundance, and activity of archaeal populations in oil-contaminated groundwater accumulated at the bottom of an underground crude oil storage cavity. *Applied and Environmental Microbiology*, **68(8)**: 3899–907.

Weiland P, 2003. Production and energetic use of biogas from energy crops and wastes in Germany. *Applied Biochemistry and Biotechnology*, **109(1-3)**: 263–74.

Whitman WB, Coleman DC, Wiebe WJ, 1998. Prokaryotes: the unseen majority. *Proceedings of National Academy of Sciences USA*, **95(12)**: 6578–83.

Wittebolle L, Marzorati M, Clement L, Balloi A, Daffonchio D, Heylen K, De Vos P, Verstraete W, Boon N, 2009. Initial species evenness favours functionality under selective stress. *Nature*, **458(7238)**: 623–6.

Woese CR, 1982. Archaebacteria and cellular origins: an overview. *Zentralblatt für Bakteriologie Mikrobiologie und Hygiene: I. Abt. Originale C: Allgemeine, angewandte und ökologische Mikrobiologie*, **3(1)**: 1–17.

Woese CR, Fox GE, 1977. Phylogenetic structure of the prokaryotic domain: the primary kingdoms. *Proceedings of the National Academy of Sciences USA*, **74(11)**: 5088–90.

Woese CR, Kandler O, Wheels ML, 1990. Towards a natural system of organisms: proposal for the domains Archaea, Bacteria, and Eucarya. *Proceedings of the National Academy of Sciences USA*, **87(12)**: 4576–9.

Wommack KE, Bhavsar J, Ravel J, 2008. Metagenomics: read length matters. *Applied and Environmental Microbiology*, **74(5)**: 1453–63.

Zhou X, Nanqi R, 2007. Acid resistance of methanogenic bacteria in a two-stage anaerobic process treating high concentration methanol wastewater. *Frontiers of Environmental Science & Engineering in China*, **1(1)**: 53–56.

Zhou Z, Pan J, Wang F, Gu JD, Li M, 2018. Bathyarchaeota: globally distributed metabolic generalists in anoxic environments. *FEMS Microbiology Reviews*, **42(5)**: 639–55.

3

Methane Driven Biogeochemical Processes in Terrestrial Ecosystem

K. Bharati, Neha Ahirwar and S.R. Mohanty[1]

Indian Institute of Soil Science, Berasia Road, Nabibagh, Bhopal – 462 038 Madhya Pradesh, India
[1]Corresponding Author: mohantywisc@gmail.com

ABSTRACT

Methanogenesis is an important biogeochemical process takes place in anaerobic ecological sites. It is the last step of terminal electron accepting process of anaerobic respiration. Methanogens synthesise CH_4 using acetate and CO_2 as precursor molecules. Former types of methanogens are referred as acetoclastic methanogens and the later ones as hydrogenotrophic methanogens. After biogenesis of CH_4, it is emitted to atmosphere and causes global warming. In aerobic ecosystem CH_4 is oxidised by methanotrophs to CO_2 a less global warming molecule. However, in anoxic environment CH_4 acts as an electron donor and interacts with various electron acceptors. Methane modulates denitrification, Fe reduction, SO_4 reduction and organic matter decomposition. Manuscript elucidates methanogenesis and highlights the pathways of interaction between methanogenesis and cycling of electron acceptors in various anaerobic ecosystems in relation to global C and N balance.

Keywords: Anaerobiosis, Climate change, Elemental cycle, Methanogenesis, Redox process

Abbreviations used

ANME : Anaerobic methanotroph
DAMO : Denitrifying anaerobic methane oxidation
SRB : sulphate-reducing bacteria
VFA : Volatile fatty acid; XANES: 'X-ray absorption near edge structure'.

Introduction

In terrestrial ecosystem, wetlands and flooded rice soils are the large source of CH_4. Methane production through methanogenesis also occurs substantially in animal's enteric regions, anaerobic digesters including biogas plants, composting pits etc. Methane production is the final product of anaerobic respiration (methanogenic). The production of CH_4 by anaerobic methanogens includes reduction of methanol, CO_2 and cleavage of acetate, as well as biosynthesis of methylated compounds. Methanogens are two types based on the carbon substrate they use: acetoclastic and hydrogenotrophic. These two groups of methanogens play a important role in most of the biogenically produced CH_4. Acetate is used as carbon source for CH_4 production by the acetoclastic methanogenic archaea. These groups are represented as Methanosarcinaceae and Methanosaetaceae. Methanogens belonging to Methanosaetaceae are abundant in paddy soil when acetate concentration is low, while Methanosarcinaceae are dominant during higher acetate concentration. Hydrogenotrophic methanogens use H_2 and CO_2 for CH_4 production. The hydrogenotrophic methanogens are Methanocellales, Methanomicrobiales, Methanosarcinales, and Methanobacteriales. A major fraction of CH_4 produced from flooded rice fields is oxidised at aerobic soil compartments.

Microbial mediated methane (CH_4) oxidation contributes significantly in the mitigation of global atmospheric CH_4 budget. It is estimated that about 10–40 Tg of atmospheric CH_4 is oxidised (consumed) annually by the methane oxidisers (Reeburgh, 2003). Microbial CH_4 oxidation occurs in many natural ecosystems. Soils act as a major sink for atmospheric CH_4 (Suwanwaree and Robertson, 2005). CH_4 is produced from flooded rice field under anaerobic condition and the emitted CH_4 is oxidised under aerobic condition. In wet soils, CH_4 oxidation decreases with high soil moisture (Mohanty *et al.*, 2014), but at low moistures CH_4 oxidation do not exhibit high correlation with moisture content. Under flooded anoxic condition anaerobes are predominant and reduce the aerobic microbial processes. However, flooded soil does not always result into the formation of uniform reduced profile. A thin, oxidised surface horizon occurs over the deep reduced soil horizon. This occurs due to the diffusion of dissolved oxygen from the overlying flood water into the surface water-soil interface. In flooded rice paddies the rhizosphere remains oxidised because of the delivery of O_2 from roots into the soil (Bodelier and Frenzel, 1999). In submerged soil, anaerobic microbial redox metabolism proceeds by the sequential reduction of inorganic electron acceptors including O_2, NO_3^-, Fe^{3+}, SO_4^{2-}, and CO_2.

The sequence of a reduction of terminal electron acceptors is based on the thermodynamics principle. This reduction, proceeds with the reduction of available electron acceptors which have the most positive redox potential

(Zehnder and Stumm, 1988). Previously, studies have been undertaken to investigate the impact of oxidised electron acceptors on methanogens in flooded rice soil (Bond and Lovley, 2002). Anaerobic bacteria like denitrifiers, dissimilatory iron reducers, sulphate reducers, and methanogens remain active in presence of high labile organic material. They often compete for common reduced carbon sources (Carucci *et al*., 1999). In flooded soil ecosystem aerobic CH_4 oxidation is inhibited due to O_2 limitation and due to the predominance of reduced species like S^{2-} (Van Bodegom *et al*., 2001). In anerobic soil ecosystem CH_4 oxidation occurs at less reduced site through reduction of NO_3^-, Fe^{3+} and SO_4^{2-} reduction (Murase and Kimura, 1996). So far the anaerobic CH_4 oxidation is poorly understood because the microorganisms for anaerobic CH_4 oxidation have not been characterised. The current manuscript elucidates mechanism of methanogenesis and the interaction of CH_4 with Fe, SO_4 and organic matter under anoxic environment in the perspective of global C and N balance.

Methanogenesis

Methanogenesis is an important ecological process results CH_4 production from the organic matter decomposition. The anaerobic decomposition of organic matter includes three stages: hydrolysis-acidogenesis, acetogenesis and methanogenesis. Generally, the complex substrate (organic matter) is primarily polymers of carbohydrates and proteins (Fig. 1). These substrates are first converted to soluble organic compounds. Subsequently, the products are further broken down to soluble monomers.

Soluble monomers undergo fermentation to various intermediate products such as volatile fatty acids (VFAs). These VFAs are converted into acetic acid, CO_2 and H_2 by acetogenic bacteria. Anerobically the methanogenic archaea produce CH_4 from acetic acid and H_2. Although acetate is well known precursor of CH_4 production by the methanogens, H_2 is also an important substrate for CH_4 production. Methane production through H_2 and CO_2 is carried out by hydrogenotrophic methanogens, In this context, some methanogens have evolved a mechanisms for syntrophic acetate oxidation. In this process first the acetate oxidise to H_2 and CO_2. Subsequently, CH_4 is produced through hydrogenotrophic methanogenesis. Such CH_4 production pathway generally occurs in thermophilic methanogenic reactors. Therefore, CH_4 production through acceleration of hydrogenotrophic methanogens has gained importance to increasing the proportion of methane in the biogas. Hydrogen is the intermediate product and precursor for methanogenesis. It is produced from organic wastes by biological fermentation. However, the yield of H_2 is limited. Therefore, to enhance the CH_4 production, supplement of moderate amount of H_2 is an efficient method for biomethanation.

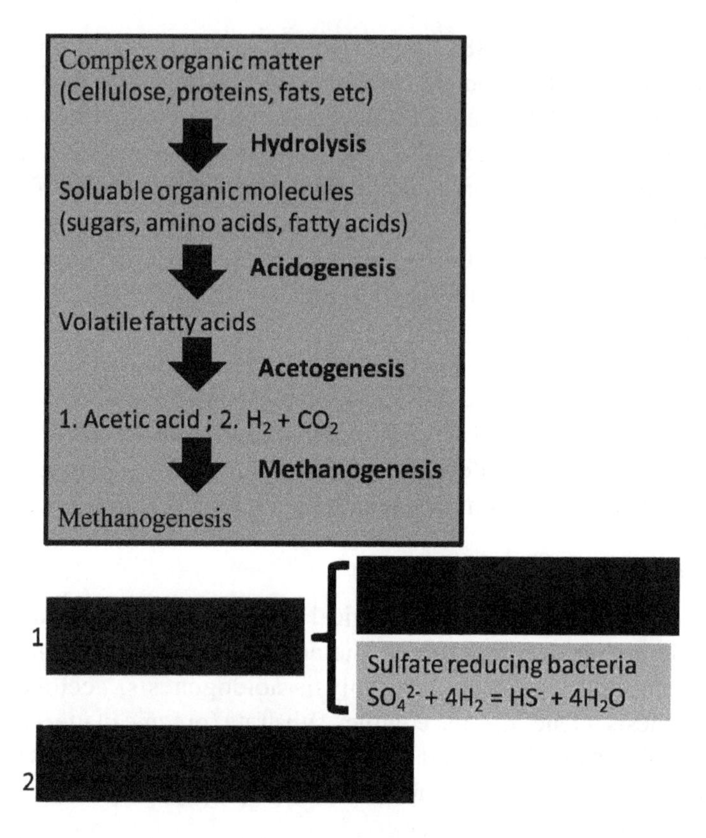

Fig. 1: Methanogenesis in terrestrial ecosystem: Methane is produced either from acetate (acetoclastic methanogens) or CO_2 (hydrogenotrophic methanogens) from decomposing complex substrates

Recently, bio-electrochemical systems have been suggested as a promising method for production of methane and hydrogen from wastewater and waste activated sludge. However, the detailed mechanism underlying biological methanogenesis mostly remains unclear. One proposed mechanism of bioelectrochemical pathway highlights methanogenesis by utilising the abiotically produced H_2. The H_2 molecules can be produced through electrochemical reaction particularly from the cathode terminal. The released H_2 is used to reduce CO_2 to CH_4. The later (CO_2) can be produced from acetate oxidation. In addition, elemental metals also serve as electron source for CH_4 production through depolarisation of cathode. Among different metals, zero-valence iron (Fe^0) found to the very efficient to enhance CH_4 production in anaerobic biological waste treatment plants (*e.g.*, wastewater and waste activated sludge). Therefore, addition of zero-valence iron to activated sludge can enhance CH_4 production by about 40-45%. Although zero-valence iron acts as an electron donor, but it can also enhance acidogenesis and acetogenesis for CH_4 production.

Methane coupled anaerobic ammonia oxidation

The anaerobic ammonium oxidation is commonly known as 'anammox'. Anaerobic CH_4 oxidation coupled to anammox is an important microbial process. It is also referred as denitrifying methane oxidation (DAMO). The denitrifying anaerobic methane oxidation processes plays important role in the global nitrogen cycle (Fig. 2).

The anammox was first evaluated in a wastewater treatment plant. In this experiment ammonium and nitrite were consumed to generate gaseous N_2 and NO_3^- (Isaka *et al.*, 2017). Recently, anammox has been reported to be ubiquitous in many natural environments (Hu *et al.*, 2011). In anoxic marine systems up to 50% of nitrogen gas (N_2) is produced through anammox (Van De Vossenberg *et al.*, 2008). The microbial species carry out DAMO were first enriched from freshwater canal sediments (Lu *et al.*, 2015). These microbial groups involved in DAMO are candidatus *Methylomirabilis oxyfera* (~80%), and anaerobic methanotrophic archaea (ANME-2d) (~10%). Another study showed that *M. oxyfera* can reduce nitrite (NO_2^-) to nitric oxide (NO). Thus the evolved oxygen can be used for methane oxidation using nitric oxide (Ettwig *et al.*, 2010). It was shown that the DAMO culture is generally dominated by ANME-2d and *M. Oxyfera* (Hu *et al.*, 2015). The ANME-2d, known as *Candidatus Methanoperedens nitroreducens*, oxidises CH_4 to CO_2 through reverse

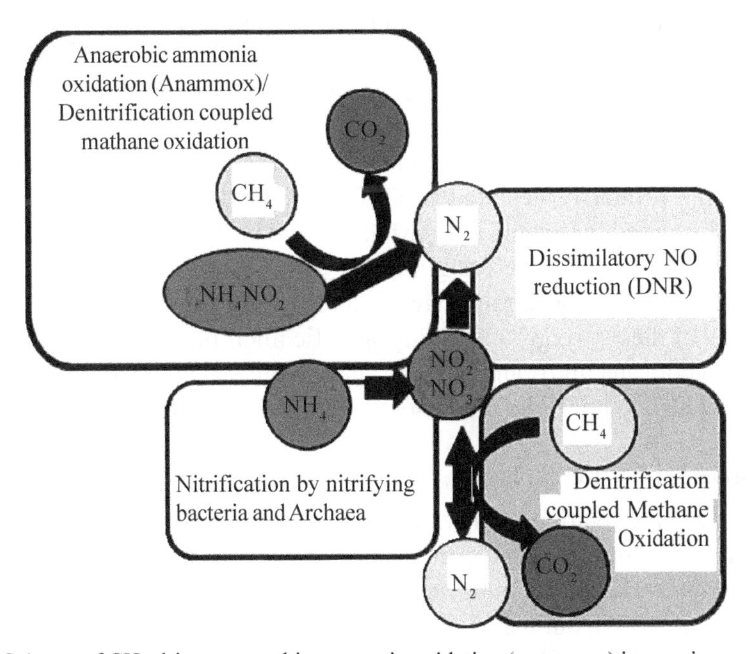

Fig. 2: Schema of CH_4-driven anaerobic ammonia oxidation (anammox) in anoxic ecosystem: denitrifying CH_4 oxidation for N-removal from wastewaters and for global N- and C-cycle

methanogenesis, and reduction of NO3 to NO_2^- (Ding *et al*., 2017). Till today the microbial groups involved in DAMO are known to be *M. nitroreducens*. These are nitrate-driven DAMO archaea and *M. oxyfera*, a nitrite-driven DAMO bacteria.

The most speculated ecology of anammox coupled DAMO was the oxic–anoxic interface in sediments or water columns predominated by methane and nitrogenous compounds. In these ecological niches organisms belonging to anammox and DAMO coexist. These organisms interact through cross-feed or substrate competition. For example, *M. nitroreducens* carry out denitrification (reduction of NO_3^- to NO_2^-), which serves as substrate for anammox and for *M. oxyfera*. On the other hand, anammox bacteria produce NO_3^-, a substrate for *M. nitroreducens*. An enriched coculture of anammox bacteria and *M. oxyfera* has been observed in a sequential batch reactor (Luesken *et al*., 2011). The reactor was fed with nitrite, ammonium and methane. Reactor was also inoculated with *M. nitroreducens* and anammox bacteria.

The temporal (seasonal) variation of anammox and DAMO bacterial community structures and their abundance estimated was explored in a sewage sludge treatment plants (Xu *et al*., 2017). Data indicated that bacterial species modulating anammox and DAMO coexisted in the sewage sludge in different seasons. The bacterial abundance positively correlated with each other ($P<0.05$). Bacterial abundance of anammox and DAMO was high in autumn and winter. This reflected that these seasons were favourable for the growth of anammox and DAMO bacteria. The community structure of anammox and DAMO bacteria shifted by seasonal changes. The *Candidatus Brocadia* genus of anammox bacteria was dominant in spring and summer, while an unknown cluster was primarily detected in autumn and winter. Similarly, bacterial community varied seasonally in DAMO; group B bacteria of DAMO dominated in spring and summer, whereas group A was dominant in autumn and winter. Slurry pH and NO_3 concentration significantly ($P<0.01$) affected the community structures of these two groups. Using specific microbial groups depending on the season both anammox and DAMO can be regulated in wastewater treatment plants and also to tailor the processes for nitrogen removal from wastewaters.

Methane-driven Fe cycling

CH_4 has high activation energy, so few specialised groups of archaea and bacteria are capable of using it. Among these microbial groups, aerobic methane-oxidising bacteria are the most commonly studied organisms (Trotsenko and Khmelenina, 2005), but recently anaerobic CH_4 oxidation has been identified (Knittel *et al*., 2005). The organisms mediating anaerobic CH_4 oxidation has been found to belong to distinct classes within methanogens (phylum

Euryarchaeota). These functional microbial groups normally produce methane under anoxic conditions in deep sea. The anaerobic methanotrophic (ANME) archaea have vast potential to regulate atmospheric CH_4 concentration. Because they prevent the emission of the large quantities of methane from the oceans' sediments and gas hydrates (Knittel *et al.*, 2005).

Until a decade ago, sulphate was identified as a known electron acceptor for anaerobic methane oxidation (Milucka *et al.*, 2012). However, a variety of other predominant environmentally oxidised molecules like Fe^{3+}, Mn^{4+}, NO_3^- (Haroon *et al.*, 2013), and NO_2^- (Ettwig *et al.*, 2010) favour electron acceptors than sulphate. However, there is lack of information on the anaerobic methanotrophs which metabolise CH_4 oxidation by coupling with metal oxide reduction is less known. Anaerobic CH4 oxidation coupled through nitrogenous electron acceptors, like NO_3/NO_2, and the microbial species carrying out these processes have been identified. The methanotrophic bacteria *Methylomirabilis oxyfera*, of the NC10 phylum, produces oxygen from NO_2, and uses this oxygen for methane oxidation. Thus they are analogous to aerobic methanotrophs (Ettwig *et al.*, 2010). In addition, archaea belonging to *Methanosarcinales* (referred to as ANME-2d), is coenriched with *M. oxyfera* (Raghoebarsing *et al.*, 2005). These are found to couple the CH_4 oxidation to the reduction of NO_2 or NO_3 (Haroon *et al.*, 2013).

Weathering of rock in one way provides significant amounts of electron acceptors like iron, manganese, and other metal oxides to freshwater bodies (Martin and Meybeck, 1979), and ultimately oceans. However, evidence of anaerobic CH_4 oxidation coupled to the reduction of iron and manganese is scarce. By isotope tracer studies, the coupling of methane oxidation to reduction of the electron acceptors has been demonstrated. Based on these methodology it has been studied that anaerobic CH_4 oxidation takes place in aquifer (Amos *et al.*, 2012), lake water (Crowe *et al.*, 2011), lake sediments (Thamdrup and Schubert, 2013), marine sediments (Egger *et al.*, 2014), and mud volcano (Chang *et al.*, 2012).

Iron plays an important role in CH_4 cycling. Oxidised Fe (Fe^{3+}) can oxidise sulphide (S^{2-}) to form sulphate (SO_4). This resulting SO_4 mediates anaerobic CH_4 oxidation. Such microbial activity has been observed (Hansel *et al.*, 2015). Few studies have identified anoxic methane oxidation coupled to Fe^{3+} or Mn^{4+} reduction in naturally enriched samples, where ANME-2 archaea abundance increased. In marine samples, ANME-2a, ANME-2c, and other archaea were found to oxidise methane through Fe^{3+} and AQDS reduction (Scheller *et al.* 2016). In natural ecosystem metals occur in different forms, which are more difficult to access. Thus, the microorganisms are capable to couple the oxidation of methane with the reduction of metal oxides dominan in the environment, which is less understood.

Anaerobic methane oxidation was estimated in the sediment of Twentekanaal, a Dutch canal (Raghoebarsing *et al.*, 2006). Sediment was enriched with a continuous supply of methane, nitrite, and nitrate. Using archaeal methanotrophs Fe^{3+} and Mn^{4+} dependent methane oxidation was evaluated. Use of nitrate as the only electron acceptor in the mineral medium caused archaea to proliferate. The archaeal community was constituted of sequences homologous to cluster AAA, *Methanoperedens nitroreducens*, *M. nitroreducens* MPEBLZ, and *M. oxyfera.* The enrichment culture coupled CH_4 oxidation (amount of $^{13}CO_2$ produced to against the reduction of nitrate) followed a theoretical stoichiometry: $5CH_4+8NO_3+8H^+ = 5CO_2+4N_2+14H_2O$. Nitrate reduction in the enrichment culture was largely (>95%) carried out by nitrate depending methanotrophs. All genes necessary for nitrate reduction (NO_3 reduction to NO_2 and reduction of NO_2 to NH_4) are present in the genome of archaeal groups belonging to cluster AAA (Arshad *et al.*, 2015).

Methane-driven S cycling

Production of CH_4 in anaerobic ecosystem plays important role in the S cycle. Methane is anaerobically oxidised by coupling with the sulphate reduction. This SO_4 driven CH_4 oxidation is influenced by the partial pressure of CH_4. The SO_4 reduction rate of the sediment from Hydrate Ridge was significantly stimulated at an increased partial pressure of CH_4 (Nauhaus *et al.*, 2005). At a concentration of 0 and 0.15 MPa of CH_4, there was a positive linear correlation between the CH_4 partial pressure and the anaerobic oxidation of methane (AOM) and sulphate reduction (SR) rates of an anaerobic methanotrophic enrichment obtained from Eckernförde Bay sediment (Meulepas *et al.*, 2009). In a study, the SO_4 reduction rate by anaerobic CH_4 oxidation in microbial mats of the Black Sea increased 10–15 fold when the partial pressure of methane increased from 0.2–10.0 MPa (Deusner *et al.*, 2010). Similarly, the affinity constant (K_m) for methane was about 37mM (equivalent to 3MPa CH_4) in the anaerobic sediment of the Gulf of Cádiz (Zhang *et al.*, 2010). The rate of anaerobic CH_4 oxidation increases with CH_4 partial pressure due to the more negative Gibbs free energy change (ΔG). The growth of anaerobic methanotrophs was faster when the CH_4 partial pressure is increased. Bioreactor studies have been carried out with high methane partial pressure (Zhang *et al.*, 2010). However, it is not clear how the CH_4 concentration stimulates anaerobic methane-oxidising archaea (ANME) and sulphate-reducing bacteria. This is important to develop microbial consortia to mitigate CH_4 production from the anaerobic ecosystems.

Both of the anaerobic methane-oxidising archaea (ANME) and sulphate reducing bacteria (SRB) are slow growers. This limits the ability of the researchers to study functionality of these microbial groups. A study was carried out to investigate the effect of the CH_4 partial pressure on methane production and methane oxidation in Eckernförde Bay sediment of the Baltic Sea (Timmers *et al.*, 2015). Results imply that the syntrophic association can be flexible. Both the CH_4 partial pressure and S^{2-} concentration influenced the growth of ANME-SRB microbial consortia. The subtype ANME-2a/b was abundant when the CH_4 partial pressure was 10.1MPa or 0.11mPa. However, the subtype ANME-2c was enriched only at 10.1MPa of CH_4. The sulphate reducing bacteria growing at 10.1MPa belonged mainly to the SEEP-SRB2 and Eel-1 groups and the Desulfuromonadales. The typical SRB belong to SEEP-SRB1 group. An increase in one group also stimulated the other, like the increase of ANME-2a/b stimulated SEEP-SRB2. At low CH_4 concentration SRB was found to be associated only with ANME-2c.

Methane and organic matter cycling

To mitigate CH_4 emission it is important to know how it is produced from the organic matter in natural ecosystem. Therefore, finding key process that aids to control biosynthesis as well as decomposition of CH_4 is important. Anaerobic oxidation of methane coupled with the reduction of electron acceptors plays significant role in mitigating atmospheric CH_4 budget. Wetlands contain very high organic matter, and the wetlands contribute significantly to the atmospheric methane. Nevertheless, microbial processes regulating CH_4 emissions in these environments are less understood. Organic matter also contains redox functional groups which can mediate AOM. In a study, using tracer analysis with $^{13}CH_4$ and spectroscopic evidence it was revealed that anaerobic methane oxidation is linked to the organic matter reduction. This process is dominant in tropical wetlands. Study suggests that microbial reduction of natural organic matter may largely contribute to the decline of CH_4 emissions from tropical wetlands. Study provided a novel conceptual pathway in the carbon cycle in which the slowly decaying natural organic matter like humic fraction in organotrophic ecosystem can act as terminal electron acceptor to enhance anaerobic methane oxidation.

The anaerobic decomposition of peat soils occurs through several microbial processes resulting production of CO_2 and CH_4. Thus, in peat soils, a high ratio of CO_2:CH_4 is due to the involvement of organic matter as electron acceptor. A study was conducted to test if organic matter (OM) suppresses CH_4 production and increases anaerobic CO_2 production (Yu *et al.*, 2015). This was linked with the organic matter acting as electron acceptor and the sulphur cycle that maintained the CO_2 production through bacterial sulphate reduction (BSR). Peat samples were fortified with sulphate or sulphide and incubated anaerobically for six weeks at 30°C. And the CO_2 and CH_4 were measured from the peat samples. CO_2:CH_4 production was at a ratio of 3:2. The electron accepting capacity of organic matter available in peat was 2.36μeq/cm³/d. Sulphate addition to peat samples significantly increased the production of CO_2 while significantly suppressed CH_4 production. After deducting the electron accepting capacity of the added SO_4 (0.97~2.81μeq/cm³/d), the values of electron accepting capacity of the organic matter reached 3.88–4.85μeq/cm³/d. The contribution of organic bound S moieties was further evaluated by XANES spectroscopy by measuring natural abundance of δ34S. Results demonstrated that SO_4 reduction involved both addition of H_2S and sulphate to organic matter lead to formation of reduced and partially oxidised organic sulphur species. The original peat contained 70.5% reduced organic compounds. The reduced organic compounds are generally R-S-H, R-S-R, R-S-S-R, and the oxidised organic compounds were in the form of R-SO_3, R-SO_2-R, R-SO_4-R. The organic matter added with H_2S or sulphate comprised of 75.7–81.1% reduced organic S, and only 21.1–18.9 % oxidised S. Study suggest that organic matter contributes to anaerobic decomposition of CH_4 through SO_4 reduction.

To define the linkage between natural organic matter and anaerobic oxidation of methane, experiments were performed by incubating freshly collected wet sediments from a tropical wetland (Valenzuela *et al.*, 2017), Sediment was amended with ^{13}C-CH_4 (0.67 atm). This was done to test the efficiency of the microbial community to perform anaerobic oxidation of methane through the reduction of the humic fraction of natural organic matter (NOM). Study demonstrates that electron-accepting functional groups of natural organic matter (e.g., quinones) stimulated anaerobic oxidation of methane by serving as a terminal electron acceptor. It was found that sulphate reduction contributed up to 42.5% of the anaerobic oxidation of methane (AOM). The microbial reduction of NOM concomitantly occurred. When the wetland sediment was supplemented with external NOM, it improved the electron-accepting capacity of the sediment. In this condition the AOM increased significantly accounting to ~100nmol $^{13}CH_4$ oxidised/cm³/day. Spectroscopic analyses of the sediment showed that quinone fraction of NOM were heterogeneously distributed in the wetland sediment. Reduction of these quinine moieties occurred during the course of AOM. Moreover, the stoichometry of enrichment of natural sediment and AOM was related to the reduction of the anthraquinone-2,6-disulphonate (AQDS) an analogue of humic substance. Microbial populations involved in AOM coupled to NOM reduction were dominated by diverse microbial groups from putative AOM-associated archaea.

Conclusions and perspective

Methanogenesis is a key microbial process that modulates both aerobic and anaerobic biogeochemical processes. The strict anaerobic processes can be managed to enhance CH_4 production through different approaches. For example, strategies like addition of H_2 to stimulate hydrogenotrophic methanogenesis can significantly increase CH_4 production. Recently, electrobiochemical process has been found to be a potential approach to enhance CH_4 production. This method potentially stimulates hydrogenotrophic methanogenesis. CH_4 can be explored to enhance denitrification or anammox. Thus, anoxic CH_4 oxidation coupled to anaerobic CH_4 oxidation can be a potential approach to remove NO_3 from wastewater. Anoxic CH_4 oxidation coupled to SO_4 reduction found to be the most general means of anoxic CH_4 oxidation. However, the microbial species involved in the SO_4 dependent anaerobic CH_4 oxidation is not well studied. Further studies in these lines may help improve wastewater and other natural ecosystem management strategies through these microbial groups.

References

Amos RT, Bekins BA, Cozzarelli IM, Voytek MA, Kirshtein JD, Jones EJ, Blowes DW, 2012. Evidence for iron mediated anaerobic methane oxidation in a crude oil contaminated aquifer. *Geobiology*, **10(6)**: 506–17.

Arshad A, Speth DR, de Graaf RM, Op den Camp HG, Jetten MS, Walte CU, 2015. A metagenomics-based metabolic model of nitrate-dependent anaerobic oxidation of methane by methanoperedens-like archaea. *Frontiers in Microbiology*, **6**: 1–14.

Bodelier PLE, Frenzel P, 1999. Contribution of methanotrophic and nitrifying bacteria to CH_4 and NH_4^+ oxidation in the rhizosphere of rice plants as determined by new methods of discrimination. *Applied and Environmental Microbiology*, **65(5)**: 1826–33.

Bond DR, Lovley DR, 2002. Reduction of Fe(III) oxide by methanogens in the presence and absence of extracellular quinones. *Environmental Microbiology*, **4(2)**: 115–24.

Carucci A, Kühni M, Brun R, Carucci G, Koch G, Majone M, Siegrist H,1999. Microbial competition for the organic substrates and its impact on EBPR systems under conditions of changing carbon feed. *Water Science and Technology*, **39(1)**: 75–85.

Chang YH, Cheng TW, Lai WJ, Tsai WI, Sun CH, Lin LH, Wang PL, 2012. Microbial methane cycling in a terrestrial mud volcano in eastern Taiwan. *Environmental Microbiology*, **14(4)**: 895–908.

Crowe SA, Katsev S, Leslie K, Strum A, Magen C, Nomosatryo S, Pack MA, Kessler JD, Reeburgh WS, Roberts JA, Gonzalez L, Douglas HG, Mucci A, Sundby B, Fowle DA, 2011. The methane cycle in ferruginous Lake Matano. *Geobiology*, **9(1)**: 61–78.

Deusner C, Meyer V, Ferdelman TG, 2010. High pressure systems for gas phase free continuous incubation of enriched marine microbial communities performing anaerobic oxidation of methane. *Biotechnology and Bioengineering*, **105(3)**: 524–33.

Ding ZW, Lu YZ, Fu L, Ding J, Zeng RJ, 2017. Simultaneous enrichment of denitrifying anaerobic methane-oxidising microorganisms and anammox bacteria in a hollow-fiber membrane biofilm reactor. *Applied Microbiology & Biotechnology*, **10(1)**: 437–46.

Egger M, Rasigraf O, Sapart CJ, Jilbert T, Jetten MSM, Rockmann T, vander Veen C, Banda N, Kartal, B, Etwig KF, Slomp CP, 2014. Iron-mediated anaerobic oxidation of methane in brackish coastal sediments. *Environmental Science and Technology*, **49(1)**: 277–83.

Ettwig KF, Butler MK, Le Paslier DL, Pelletier E, Mangenot S, Kuypers MMM, Schreiber F, Dutilh BE, Zedelius J, de Beer D, Gloerich J, Wessels HJCT, van Alen T, Luesken F, Wu ML, van de Pas-Schoonen KT, Op den Camp HJM, Janssen-Megens EM, Francoijs KJ, Stunnenberg H, Weissenbach J, Jetten MSM, Strous M, 2010. Nitrite-driven anaerobic methane oxidation by oxygenic bacteria. *Nature*, **464(7288)**: 543–8.

Hansel CM, Lentini CJ, Tang Y, Johnston DT, Wankel SD, Jardine PM, 2015. Dominance of sulfur-fueled iron oxide reduction in low-sulfate freshwater sediments. *The ISME Journal*, **9(11)**: 2400–12.

Haroon MF, Hu S, Shi Y, Imelfort, M, Keller J, Hugenholtz P, Yuan Z, Tyson GW, 2013. Anaerobic oxidation of methane coupled to nitrate reduction in a novel archaeal lineage. *Nature*, **500(7464)**: 567–70.

Hu B, Shen L, Xu X, Zheng P, 2011. Anaerobic ammonium oxidation (anammox) in different natural ecosystems. *Biochemical Society Transactions*, **39(6)**: 1811–6.

Hu S, Zeng RJ, Haroon MF, Keller, J., Lant PA, Tyson GW, Yuan, Z, 2015. A laboratory investigation of interactions between denitrifying anaerobic methane oxidation (DAMO) and anammox processes in anoxic environments. *Scientific Reports*, **5**: 8706.

Isaka K, Kimura Y, Matsuura M, Osaka T, Tsuneda S, 2017. First full-scale nitritation-anammox plant using gel entrapment technology for ammonia plant effluent. *Biochemical Engineering Journal*, **122**: 115–22.

Knittel K, Losekann T, Boetius A, Kort R, Amann R, 2005. Diversity and distribution of methanotrophic Archaea at Cold Seeps. *Applied and Environmental Microbiology*, **71(1)**: 467–79.

Lu YZ, Ding ZW, Ding J, 2015. Design and evaluation of universal 16S rRNA gene primers for high-throughput sequencing to simultaneously detect DAMO microbes and anammox bacteria. *Water Research*, **87**: 385–94.

Luesken FA, van Alen TA, van Der BE, Frijters C, Toonen G, Kampman C, Hendrickx TLG, Zeeman G, Temmink H, Strous M, Op den Camp HJM, Jetten MSM, 2011. Diversity and enrichment of nitrite-dependent anaerobic methane oxidising bacteria from wastewater sludge. *Applied Microbiology Biotechnology*, **92(4)**: 845–54.

Martin JM, Meybeck M, 1979. Elemental mass-balance of material carried by major world rivers. *Marine Chemistry*, **7(2)**: 173–206.

Meulepas RJ, Jagersma CG, Khadem AF, Buisman CJN, Stams AJM, Lens PNL, 2009. Effect of environmental conditions on sulfate reduction with methane as electron donor by an Eckernforde Bay enrichment. *Environmental Science and Technology*, **43(17)**: 6553–9.

Milucka J, Ferdelman TG, Polerecky L, Franzke D, Webgener G, Schmid M, Liberwith I, Wagner M, Widdel F, Kuypers MM, 2012. Zero-valent sulphur is a key intermediate in marine methane oxidation. *Nature*, **491(7425)**: 541–6.

Mohanty SR, Kollah B, Sharma VK, Singh AB, Singh S, Rao AS, 2014. Methane oxidation and methane driven redox process during sequential reduction of a flooded soil ecosystem. *Annals of Microbiology*, **64(1)**: 65–74.

Murase J, Kimura M, 1996. Methane production and its fate in paddy fields IX. Methane flux distribution and decomposition of methane in the subsoil during the growth period of rice plants. *Soil Science and Plant Nutrition*, **42(1)**: 187–90.

Nauhaus K, Treude T, Boetius A, Krüger M, 2005. Environmental regulation of the anaerobic oxidation of methane: a comparison of ANME-I and ANME-II communities. *Environmental Microbiology*, **7(1)**: 98–106.

Raghoebarsing AA, Pol A, van de Pas-Schoonen KT, Smolders AJP, Ettwig KF, Rijpstra WIC, Schouten S, Damste JSS, Op den Camp HJM, 2006. A microbial consortium couples anaerobic methane oxidation to denitrification. *Nature*, **440(7086)**: 918–21.

Raghoebarsing AA, Smolders AJP, Schmid MC, 2005. Methanotrophic symbionts provide carbon for photosynthesis in peat bogs. *Nature*, **436(7086)**: 1153–6.

Reeburgh WS, 2003. Global methane biogeochemistry. *Treatise on Geochemistry*, **4**: 347.

Scheller S, Yu H, Chadwick GL, McGlynn SE, Orphan VJ, 2016. Artificial electron acceptors decouple archaeal methane oxidation from sulfate reduction. *Science*, **351(6274)**: 703–7.

Suwanwaree P, Robertson GP, 2005 Methane oxidation in forest, successional, and no-till agricultural ecosystems. *Soil Science Society of American Journal*, **69(6)**: 1722–9.

Thamdrup B, Schubert CJ, 2013. Anaerobic oxidation of methane in an iron rich Danish freshwater lake sediment. *Limnology and Oceanography*, **58(2)**: 546–54.

Timmers PHA, Gieteling J, Widjaja-Greefkes HCA, Plugge CM, Stams AJM, Lens PNL, Meulepas RJW, 2015. Growth of anaerobic methane-oxidizing archaea and sulfate-reducing bacteria in a high-pressure membrane capsule bioreactor. *Applied Environmental Microbiology*, **81(4)**: 1286–96.

Trotsenko YA, Khmelenina VN, 2005. Aerobic methanotrophic bacteria of cold ecosystems. *FEMS Microbiology Ecology*, **53(1)**: 15–26.

Valenzuela EI, Prieto-Davo A, Lopez-Lozano NE, Hernández-Eligio A, Vega-Alvarado L, Juárez K, García-González AS, Lopez MG, Francisco JC, 2017. Anaerobic methane oxidation driven by microbial reduction of natural organic matter in a tropical wetland. *Applied Environmental Microbiology*, **83(11)**: 1-15.

Van Bodegom P, Goudriaan J, Leffelaar P, 2001. A mechanistic model on methane oxidation in a rice rhizosphere. *Biogeochemistry*, **55(2)**: 145–77.

van de Vossenberg J, Rattray JE, Geerts W, Kartal B, van Niftrik L, van Donselaar EG, Sinninghe DJS, Strous M, Jetten MS, 2008. Enrichment and characterisation of marine anammox bacteria associated with global nitrogen gas production. *Environmental Microbiology*, **10(11)**: 3120–9.

Xu S, Lu W, Mustafa MF, Caicedo LM, Guo H, Fu X, Wang H, 2017. Co-existence of anaerobic ammonium oxidation bacteria and denitrifying anaerobic methane oxidation bacteria in sewage sludge: Community diversity and seasonal dynamics. *Microbial Ecology*, **74(4)**: 832–40.

Yu Z, Peiffer S, Gottlicher J, Knorr KH, 2015. Contributions of organic matter and organic sulfur redox processes to electron flow in anoxic incubations of peat. Geophysical Research Abstracts, EGU2015-10963, 2015, EGU General Assembly, Vol. 17.

Zehnder AJ., Stumm W, 1988. Geochemistry and biogeochemistry of anaerobic habitats. *In*: AJ Zehnder (Ed) *Biology of Anaerobic Microorganisms*, John Wiley & Sons, New York: 1–38.

Zhang Y, Henriet JP, Bursens J, Boon N, 2010. Stimulation of *in vitro* anaerobic oxidation of methane rate in a continuous high-pressure bioreactor. *Bioresource Technology*, **101(9)**: 3132–8.

4

Bioenergetics and Evolutionary Relationship in Biomethanation with Respect to MCR

Haragobinda Srichandan[a,b,1], Puneet K. Singh[c], Ritesh Pattnaik[d] and Snehasish Mishra[c]

[a]*Mineral Resources Research Division, Korean Institute of Geoscience and Mineral Resources, Yuseong-gu, Deajeon, South Korea*
[b]*Department of Material Science and Engineering, Chungnam National University Daejeon, 300-764, South Korea*
[c]*BDTC, Bioenergy Division, School of Biotechnology, KIIT University Bhubaneswar-24, India*
[d]*School of Biotechnology, KIIT, Bhubaneswar–24, Odisha, India*
[1]*Corresponding Author: hara.biol@gmail.com*

ABSTRACT

Biomethanation comprises of four interdependent stages in series, hydrolysis, acidogenesis, acetogenesis and methanogenesis. Methanogenesis, the final stage of biomethanation, is accompanied with the reduction of substrates that include organic acids such as CH_3COOH, and others such as CO_2 and CH_3OH to CH_4 by methanogenic and methanotrophic archaea, usually under anaerobic conditions. Strictly anaerobic methanogenic archaea are found in the ruminant intestine, sewage digester, groundwater and deep soil/water etc. In methanogenesis, H_2 acts as the electron donor while substrates CO_2, CH_3OH and probably CoM-S-S-CoB (for CH_3COOH) act as terminal electron acceptors and reduce to methane, a process known as anaerobic respiration. This conversion is associated with the electrogenic proton motive force along with sodium motive force to drive the ATP synthesis by A_0A_1 ATP synthase, known as respiratory phosphorylation. The mechanism of conversion of methane from CO_2 by methanogens with cytochromes is same as those without cytochromes except that the reduction of heterodisulphide CoM-S-S-CoB is performed by the membrane-bound VhoACG-HdrDE complex for former while by cytoplasmic MvhADG–HdrABC complex for the later.

*The final step of methanogenesis for all the substrates is catalysed by Methyl CoM reductase (*Mcr*). Methyl CoM reductase (*Mcr*) is a hexamer composed of two α-, β- and γ-subunits along with two* F_{430} *cofactors (each containing Ni). The final step of methanogenesis as carried out by* Mcr *is proposed to be proceeding by three mechanisms, each of which differs by the type of intermediates. Mechanism I results in methyl-N(III) intermediate, while mechanism II and III result in Ni(II)-thiolate and Ni(III)-S-CoM intermediates respectively.*

Keywords: Methanogens, Methanogenesis, Methyl CoM reductase, Anaerobic, Oxidation, Reduction, Membrane bound complex, Cytoplasmic Complex, ATP synthesis

Abbreviations used

ATP : adenosine triphosphate
CoB : coenzyme B
CoM : coenzyme M
Mcr : methyl CoM reductase
Fd : ferredoxin

Introduction

Fossil fuels as energy sources deplete with time. Also, their burning is associated with the emission of greenhouse gases like CO_2, CO, NO_2 and SO_2 etc. Due to these, energy from renewable sources including biomass is gaining universal popularity. The contribution of renewable energy sources like wind, solar radiation, geothermal heat and oceanic tide etc. is 12.9% to global energy need. Many nations are promoting such renewable energies through national schemes with long-term goal to reduce greenhouse gas (GHG) emission. Decreasing dependency on the non-renewable energy sources in recent time is a positive step forward. Biogas emits naturally from anaerobic ecosystems. Being a combination of gases like CH_4 and CO_2, biogas may contribute to global warming as both CH_4 and CO_2 are potential GHG. Organic wastes like kitchen refuse, agricultural residues, industries, municipal sludge, sewage sludge, etc. can be artificially converted to biogas in controlled conditions by imitating the natural mechanism. Instead of uncontrolled discharge into the environment, the generated biogas can be collected, stored and used as a fuel for cooking, electricity generation, etc.

Biogas production or biogasification is essentially an anaerobic process, its major useful component being CH_4. CH_4 in the biogas is the energy source which, upon burning, releases CO_2 to environment that is further taken up by plants to generate biomass thereby resulting in no net CO_2 emission. A glaring case of carbon sequestration, the process is thus carbon neutral. Hence, biogas is considered a strong candidate as fossil fuel alternative. Biogas generation (biomethanation) from organic wastes comprises of four interdependent sequential stages, hydrolysis, acidogenesis, acetogenesis and methanogenesis. The first stage can be both aerobic and anaerobic while the next three proceed under anaerobic (or at least partially anaerobic) conditions. The beauty of the process, therefore, is that a closed biogasification system automatically carries out all four stages successfully, the available O_2 in the system reaching a subminimal level in the first step thereby setting the stage for the next three steps.

Hydrolysis

The first step of biogasification is hydrolysis leading to the breakdown of macromolecules like polysaccharides, lipids and proteins into their simpler molecules or monomers like sugars, fatty acids and amino acids that are further used as energy and carbon sources by later group of microbes (Adekunle *et al.*, 2015). Microbes secrete various extracellular enzymes to breakdown macromolecules into simpler units (monomers). For instance, the microbes that degrade sugars are known as saccharolytic and those degrading proteins are proteolytic (Adekunle *et al.* 2015). Table 1 lists a few extracellular enzymes. This is an important step in biomethanation as it makes substrate use easy for the succeeding groups of microbes.

Table 1: Some important hydrolytic enzymes and their function

Enzyme	Substrates	Breakdown Products
Proteinase	Protein	Aminoacids
Cellulase	Cellulose	Cellobiose, glucose
Hemicellulase	Hemicellulose	Sugars like glucose, xylose, mannose and arabinose
Amylase	Starch	Glucose
Pectinase	Pectin	Sugars like galactose, arabinose and polygalacticuronic acid
Lipase	Fats	Fatty acids and glycerol

The microbes capable of producing hydrolytic enzymes include both aerobes and anaerobes, from mesophiles to thermophiles. *Clostridium thermocellum* and *Clostridium stercorarium* demonstrated appreciable hydrolysing activity at a temperature range of 55–60°C (Zverlov *et al.* 2010), although majority of them are mesophiles. Alongwith bacteria, these also include aerobic and anaerobic fungi (Doi, 2008). Table 2 lists a few mesophilic aerobic and anaerobic cellulolytic bacteria and Table 3 lists a few mesophilic aerobic and anaerobic cellulolytic fungi, alongwith their habitat. Production of free extracellular hydrolytic microbial enzymes is common, and some microbes are capable of producing extracellular enzyme complex containing various enzymes. An instance

Table 2: A list of active mesophilic aerobic and anaerobic bacteria producing cellulolytic enzymes

Species	Habitat	Species	Habitat
		Aerobic	
Acetivibrio cellulolyticus	Sewage sludge	*Bacillus megaterium*	Soil
Bacteroides cellulosolvens	Sewage sludge	*Bacillus pumilus*	Soil, Dead plant
Butyrivibrio fibrisolvens	Bovine rumen	*Cellulomonas fimi*	Soil
Clostridium acetobutylicum	Soil	*Cellulomonas flavigena*	Soil, Leaf litter
Clostridium aldrichii	Wood digester	*Cellulomonas gelida*	Soil
Clostridium cellobioparum	Soil	*Cellulomonas iranensis*	Forest humus soil
Clostridium cellulofermentans	Dairy farm soil	*Cellulomonas persica*	Forest humus soil
Clostridium cellulolyticum	Decayed grass	*Cellulomonas uda*	Sugar cane field
Clostridium cellulovorans	Wood chips	*Cellvibrio gilvus*	Bovine faeces
Clostridium herbivorans	Pig intestine	*Cellvibrio mixtus*	Soil
Clostridium hungatei	Soil	*Pseudomonas fluorescens*	Soil, water
Clostridium josui	Compost	*Streptomyces antibioticus*	Soil
Clostridium papyrosolvens	Paper mill	*Streptomyces cellulolyticus*	Soil
Fibrobacter succinogenes	Rumen	*Streptomyces lividans*	Soil
Ruminococcus albus	Rumen	*Streptomyces reticuli*	Soil
Ruminococcus flavefaciens	Rumen		
		Anaerobic	
Acetivibrio cellulolyticus	Sewage sludge	*Clostridium cellulolyticum*	Decayed grass
Bacterioides cellulosolvens	Sewage sludge	*Clostridium josui*	Compost pile
Butyrivibrio fibrisolvens	Bovine rumen	*Clostridium papyrosolvens*	Paper mill
Clostridium acetobutylicum	Soil	*Ruminococcus albus*	Rumen
Clostridium cellulovorans	Wood-chip pile	*Ruminococcus flavefaciens*	Rumen
Clostridium cellobioparum	Bovine rumen		

of one such complex is cellulosome made up of a scaffolding protein, coupled with a variety of cellulolytic and hemicellulolytic enzymes. It includes a range of cellulases, pectinases and hemicellulases (xylanases, mannanases, easterases, debranching glycosidases, lichenases).

Made up of scaffolding protein, cellulosome adheres to enzyme by cohesins with the help of dockerins (Fig. 1a,b). The scaffold also contains cellulose-binding domain (CBD) through which the cellulosome binds with substrate firmly and guide the hydrolysing enzymes to the site. Collective action of the complex breaks the difficult molecules efficiently (Doi, 2008).

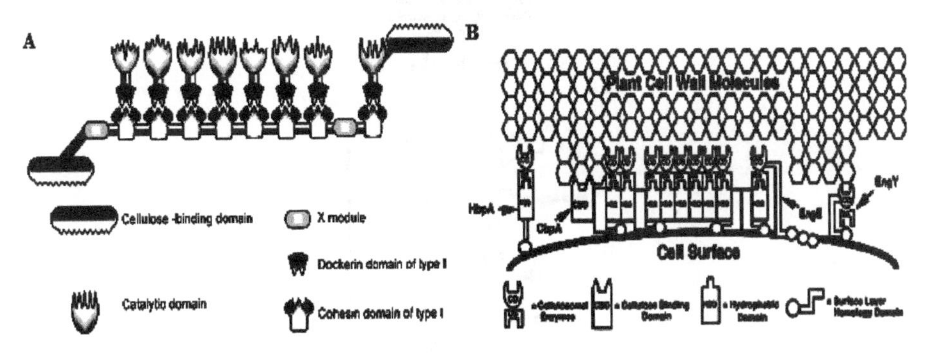

Fig. 1. A: Model of the scaffolding protein CipC of *Clostridium cellulolyticum*; B: Model of a cellulosome attached to substrate and cell surface (Reproduced from Desavaux, 2005 and Doi, 2008)

Cellulosomes in *Priomyces*, *Clostridium thermocellum*, *Clostridium stercorarium* have been studied systematically. Few anaerobic mesophilic cellulosome producers are listed in Table 3.

Table 3: A list of active mesophilic aerobic and anaerobic fungi producing cellulolytic enzymes

Aerobic fungus	Habitat	Anaerobic fungus	Habitat
Aspergillus niger	Soil, decomposing wood, agricultural waste	*Anaeromyces mucronatus*	Rumen and intestinaltract of large herbivorous animals
Phanerochaete chrysosporium		*Caecomyces communis*	
Piptoporus betulinus		*Cyllamyces aberensis*	
Pycnoporus cinnabarinus		*Neocallimastix frontalis*	
Rhizopus stolonifer		*Orpinomyces* sp.	
Serpula lacrymans		*Piromyces* sp.	
Sporotrichum pulverulentum		*Piromyces equi*	
Trichoderma reesei (Hypocrea jecorina)		*Piromyces* sp. *strain*	

Being soluble, the major product of microbial cellulolysis, cellobiose is efficiently utilised for energy (Belaich *et al*., 1997; Petitdemange *et al*., 1984; Giallo *et al*., 1983). Depolymerisation of cellulose by cellulosome liberates soluble sugars like glucose and soluble cellodextrins, *i.e.*, from cellobiose (disaccharide) to celloheptose (heptosaccharide) (Pereira *et al*., 1988). After incorporation into the cell, cellobiose converts to its corresponding monosaccharaide, *i.e.*, glucose units. The growth of *Clostridium cellulolyticum* on cellobiose is shown in Fig. 2.

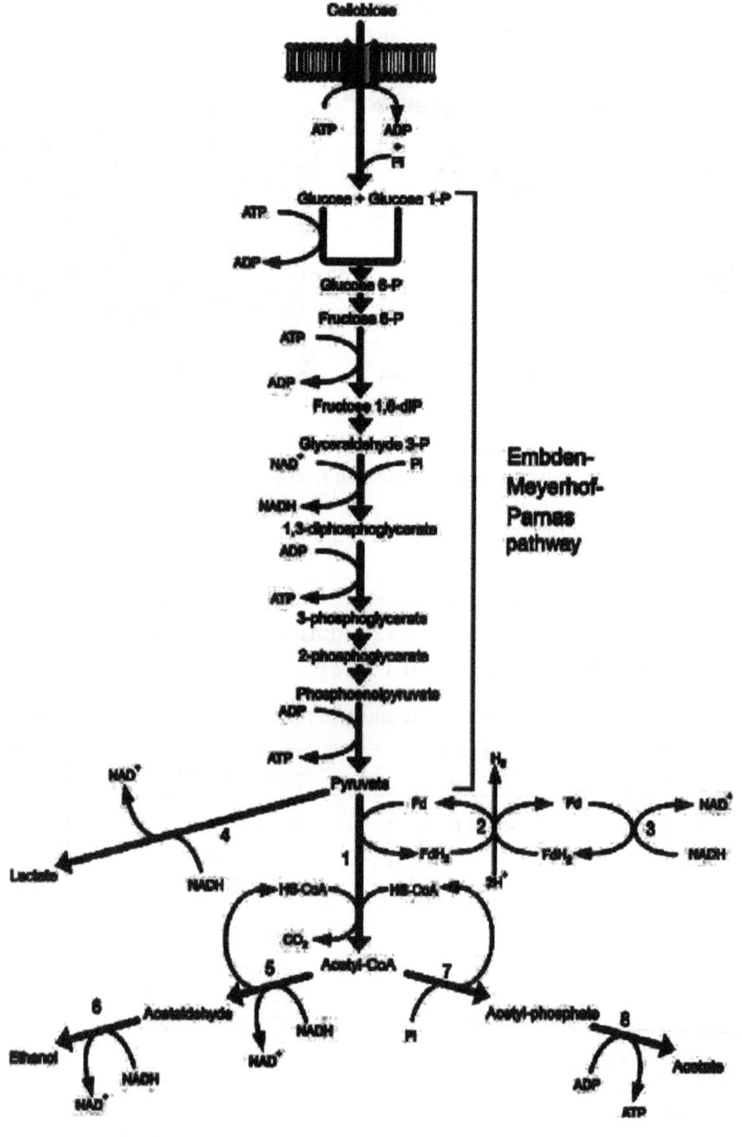

Fig. 2: Catabolic pathway of cellobiose in *C. cellulolyticum* (Reproduced from Desvaux, 2005)

Fig. 3 presents the adherence pattern of *Clostridium thermocellum* and *C. cellulolyticum* to cellulose fibre. *C. cellulolyticum* remains in close contact with cellulose, and the bacterial cells release is interrelated with available accessible cellulose (Desvaux, 2005; Gelhaye *et al.*, 1993). The ecological aspects of cellulolytic microbes are mentioned in Tables 2 and 3. Aerobic cellulolytic bacteria are usually found in humus soil, water, animal faeces, sugarcane field, plant substances and leaf litter, and the cellulolytic anaerobic bacteria are largely found in soil, decomposing plant materials, sewage sludge, rumen, and termite gut (Table 2).

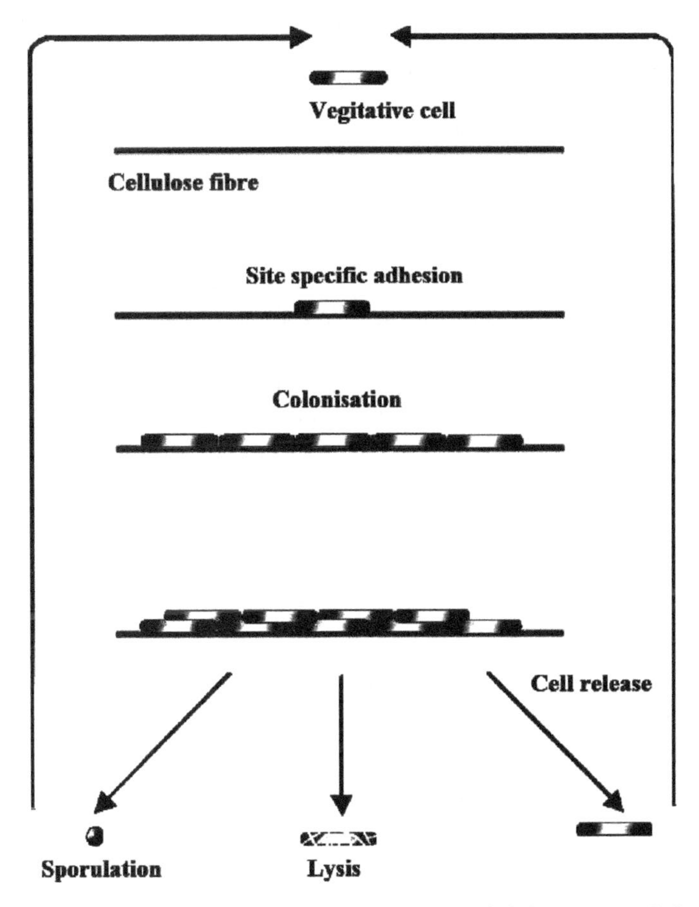

Fig. 3: The model showing the colonisation by *Clostridium cellulolyticum* on cellulose

Plant substances are also degraded by aerobic fungi found in decomposing plant materials, soil, wood and agricultural wastes (Table 3). An important cellulolytic fungi in degradation is *Trichoderma reesei. Anaeromyces*, *Cyllamyces*, *Caeomyces*, *Orpinomyces*, *Pironmyces* and *Neocallimastix* are the six popular genera of anaerobic fungi exhibiting cellulolytic activity.

Acidogenesis

An array of facultative and obligate anaerobes are capable of converting simpler molecules from the hydrolysed macromolecules to shorter chain organic (*viz.*, butyric, acetic, lactic, propionic) acids, and hydrogen, carbon-di-oxide and alcohols (Adekunle and Okolie, 2015; Kangle *et al.*, 2012). Such microbial conversion to obtain energy (ATP) by phosphorylations is collectively termed as fermentation, and the stoichiometry is (Kangle *et al.*, 2012):

$$C_6H_{12}O_6 + 2H_2O \rightarrow 2CH_3COOH + 2CO_2 + 4H_2 \quad (1a)$$

The same equation with Gibbs free energy can also be presented as:

$$C_6H_{12}O_6 + 4H_2O \rightarrow 2CH_3COOH + 2H_2CO_3 + 2CO_2 + 4H_2,$$

$$\Delta G^{o} = -206.1 \text{kJ/mole gulucose} \quad (1b)$$

$$C_6H_{12}O_6 \rightarrow 2CH_3CH_2OH + 2CO_2, \Delta G^{o} = -235 \text{ kJ/mole gulucose} \quad (2)$$

$$C_6H_{12}O_6 + 2H_2 \rightarrow 2CH_3CH_2COOH + 2H_2O, \Delta G^{o}=-235 \text{ kJ/mole gulucose} \quad (3)$$

The fermentation pathway of glucose to lactate, acetate and ethanol is shown in Fig. 2. The catabolic pathway in *C. tyrobutaricum* is shown in Fig. 4. In general, glucose breaks down to pyruvate *via* Embden-Meyerhof-Parnas (EMP) pathway followed by its conversion to Acetyl-CoA. Acetyl-CoA is the important metabolic intermediate that proceeds for either acetate or butyrate formation. Phosphotransacetylase (PTA) converts Acetyl-CoA to acetyl phosphate followed by the conversion of acetyl phosphate to acetic acid by acetate kinase (AK).

On the other hand in the butyrate forming pathway, phosphotransbutyrylase (PTB) converts Butyryl-CoA to butyryl phosphate, and then butyrate kinase (BK) converts butyryl phosphate to butyric acid. Fig. 4b shows the fermentation of glucose to propionic acid, carried out by the propionic acid bacteria (PAB). Glucose is converted to phospoenol pyruvate and then to pyruvate *via* EMP pathway. In the next step, pyruvate and methyl malonyl CoA react to form Propionyl-CoA and oxaloacetate. Oxaloacetate is converted to malate which then converts to fumarate. Fumarate converts to succinate which accepts the CoA from Propionyl-CoA and converts to Succinyl-CoA. Propionyl-CoA converts to propionate. Succinyl-CoA then converts to Methylmalonyl-CoA.

Following equation is a theoretical representation of propionate fermentation from glucose forming a total of six ATPs from 1.5 molecules of glucose (Suwannakham, 2005):

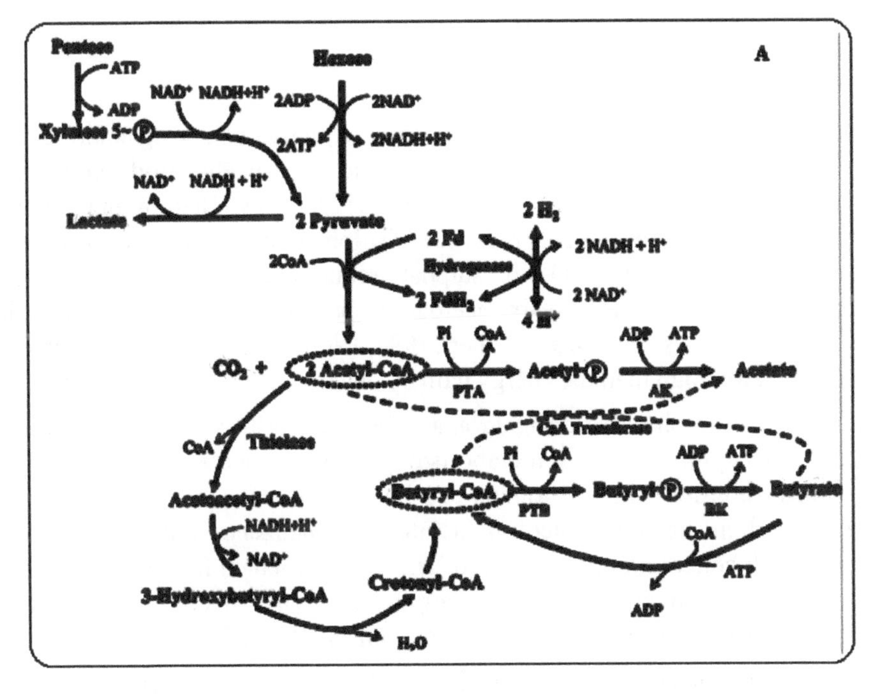

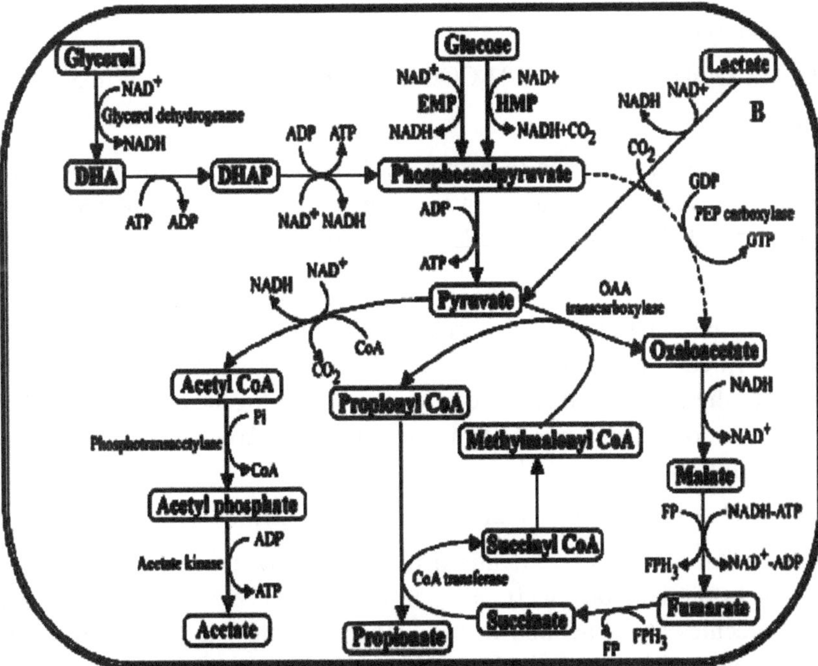

Fig. 4. A: Catabolism of glucose to butyrate in *Clostridium tyrobutaricum* (Reproduced from Liu *et al.*, 2006); B: Fermentation pathway of propionate from glucose by EMP and HMP pathways

$$1.5\ C_6H_{12}O_6 \rightarrow 2\ C_3H_6O_2 + CH_3COOH + CO_2 + H_2O + 6\ ATP \quad (4)$$

Acetogenesis

The process is driven by the partial H_2 pressure generated during acidogensis. The conversion process to acetic acid and CO_2 is thermodynamically unfavourable until the partial pressure of H_2 is too low (Beguin and Aubert, 1994), naturally achieved in the presence of sulphate reducer. Hence, the excess H_2 must be transferred to SO_4^{2-} by sulphate reducing bacteria to produce H_2S. H_2 is also transferred to CO_2 by acetogenic bacteria to produce acetic acid during acetogenesis (Beguin and Aubert, 1994). Normally, energy gain in aerobes is significantly high than the energy gain by anaerobes (Muller, 2001).

Acetogenic microbes are obligate anaerobes that utilise CO_2 as a terminal electron acceptor in energy metabolism to produce acetate (Fuchs, 1986). Certain proton-reducing acetogens use H^+ as terminal electron acceptor to oxidise certain substrates (alcohol, lactate, monobenzenes and fatty acids) to acetate with concomitant formation of H_2. The long-chain fatty acids, ethanol and lactate are converted to acetate. Fatty (organic) acid is converted to acetic acid by H^+-reducing (H_2-producing) acetogens as (Eqns 5 and 6):

$$CH_3CH_2CH_2COOH + 2H_2O \rightarrow 2CH_3COOH + 2H_2,\ \Delta G^{o'} = 48.1\ \text{kJ/mole butyrate} \quad (5)$$

$$CH_3CH_2COOH + 3H_2O \rightarrow CH_3COOH + H_2CO_3 + 3H_2,\ \Delta G^{o'}=76.5\ \text{kJ/mole} \quad (6)$$

Acetogens convert hexose to acetic acid by oxidising to pyruvate *via* the Embden-Meyerhoff-Parnas pathway and then to acetate by acetyl-CoA, producing ATP through substrate-level phosphorylation (Eqn. 7):

$$C_6H_{12}O_6 + 2H_2O + 4ADP + 4\ Pi \rightarrow 2CH_3COOH + 2CO_2 + 4ATP + 8H^+ \quad (7)$$

This fermentation initiates during acidogensis although major amount of acetic acid is formed during acetogenesis. The CO_2 generated during the process is utilised as the terminal electron acceptor during acetogenesis for which H_2 acts the electron donor reducing CO_2 to acetic acid *via* the Wood-Ljungdahl pathway (Eqn. 8; Muller, 2003):

$$2CO_2 + 4H_2 \rightarrow CH_3COOH + 2H_2O\ \text{Ä}G^{o'} = -95\ \text{kJ/mol} \quad (8)$$

The two CO_2 molecules generated from glucose fermentation to acetic acid are used to produce one acetic acid molecule. Hence, a total of three acetic acid molecules are generated from glucose. During this formation, formate dehydrogenase reduces CO_2 by H_2 to formate. The conversion of CO2 molecules to acetic acid (e.q. 8) is initiated by formate dehydrogenase which reduces CO_2 by H_2 to formate.

The formate is then activated and bound to tetrahydrofolate (H_4F), the resulting formyl-H_4F reaction being catalysed by formyltetrahydrofolate synthetase (Fig. 5a). Formyl-H_4F is converted to methenyl-H_4F by formyltetrahydrofolate cyclohydrolase. Methenyl-H_4F is then reduced to methylene-H_4F by H_2 being added, catalysed by methylenetetrahydrofolate dehydrogenase. The methylene-H_4F is reduced to CH_3-H_4F as another H_2 molecule is added, catalysed by methylenetetrahydrofolate reductase. The CH_3^- of CH_3-H_4F is then taken up by the CFeSP (a protein composed of iron sulphur clusters and a corrinoid cofactor) to form methyl-CFeSP, the reaction being catalysed by methyl transferase. CODH/ACS (aka CO dehydrogenase/acetyl-CoA synthase) reduces CO_2 to CO by H_2 molecule (Doukav, *et al.*, 2002; Ragsdale and Kumar, 1996; Muller, 2003). The methyl group reacts with CO on CODH/ACS with the addition of CoA as a result of which acetyl-CoA generates (Fig. 5a). Phosphotransacetylase converts acetyl-CoA to acetyl phosphate by adding phosphate group with concomitant release of CoA. In the next step, acetyl phosphate is converted to acetate by transferring phosphate group to ADP resulting in ATP catalysed by acetate kinase. During this process, one mole ATP is used to form HCO-H_4F and one mole ATP is formed from substrate level phosphorylation. Hence, the balance ATP generated is zero at substrate level phosphorylation, the net ATP generation resulting from the electron transport chain coupled with [H] motive force and/or ion-mediated ATP synthesis. Bioenergetics opinion of understanding put acetogens into two groups (Ljungdahl, 1994): i) the Na-dependent acetogens, *e.g.*, *Acetobacterium woodii* (Muller and Gottschalk, 1994); and ii) the H-dependent acetogens, *e.g.*, *Moorella thermoacetica* (earlier, *Clostridium thermoaceticum*). The later mentioned group comprises cytochromes that result in H^+ motive electron transport chain. The former group constitutes no cytochromes although it contains membrane-bound coronoids associated with Na^+ motive force (Muller, 2003). The [H^+] motive force and Na^+ motive force in case of *Acetobacterium woodii* and *Moorella thermoacetica* respectively are associated with ATP by ATP synthase.

Methanogens: cell morphology and abundance

Methanogens belong to the domain Archaea, phylum Euryarchaeota. Their origin is identified from the molecular fossils that showed the presence of the signature archaeal isoprene chains. Studied on the CH_4-rich 3.8 billion years old sediments in the Isua district of Greanland suggest the presence of CH_4 since that time. The active phylum Euryarchaeota is well studied based on the Methyl CoM reductase, an important enzyme for methanogenesis. Methanogens lack peptidoglycan layer in cell wall and exhibit great diversity in their cell wall composition. The cell wall organisation may range from a simple and non-rigid surface layer containing protein and glycoprotein subunits to hard ‘psudomurein’ analogues to eubacterial murein (Sirohi *et al.*, 2010). *Methanobrevibacter* and *Methanobacterium* have psudomurein, *Methanosarcina* has heteropolysaccharide, while *Methanomicrobium* has protein in their cell walls (Hook *et al.*, 2010).

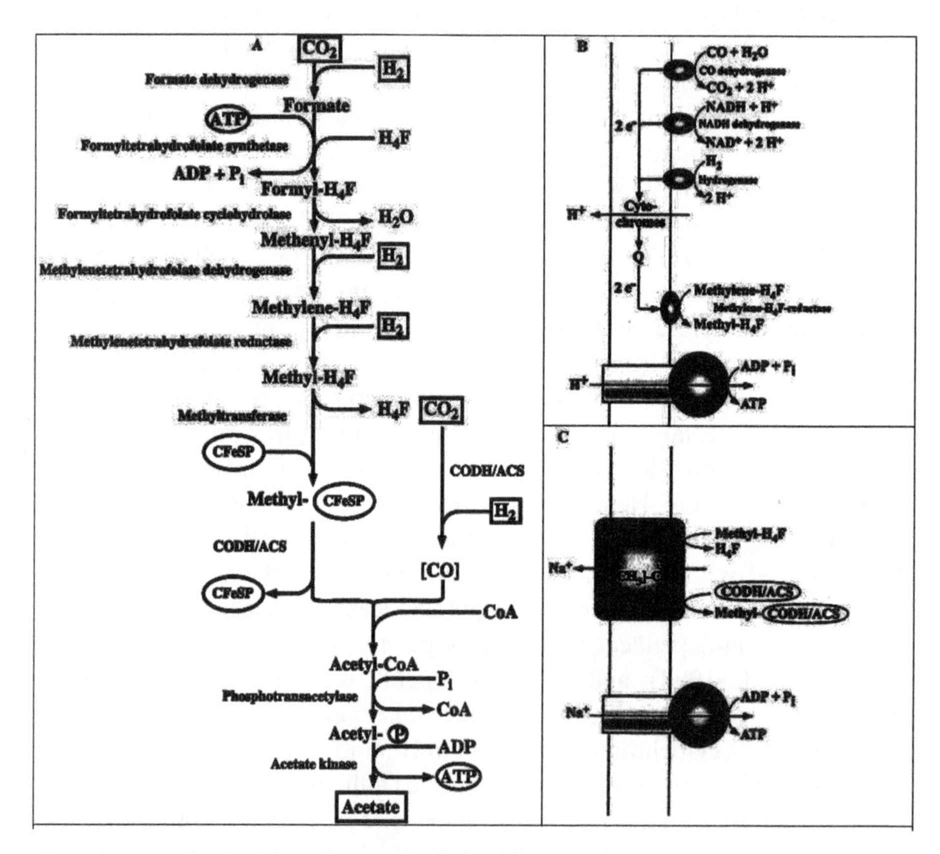

Fig. 5. a: Reduction of CO_2 by H_2 to produce CH_3COOH; Electron transport chain mediated ATP synthesis **b:** H^+-dependant; c: Na^+-dependent (Reproduced from Muller, 2003)

Cell characteristics can vary greatly among methanogens. Shape-wise, *Methanobacteriales* may vary from short lance shaped cocci to long filamentous rods (Sirohi *et al.*, 2010). *Methanobacterium* comprises of curled to linear non-spore-forming rods that are bigger in size with long filaments. *Methanobeacteruim* may be non-motile in nature or sometimes motile due to the presence of fimbre. *Methanobrevibacter* comprises of short less motile rods or cocci, occurring in pairs but usually in chains upto 20 or more cells, the width ranging from 0.5–1.0µm (Sirohi *et al.*, 2010). Methanococcales comprises of Gram negative irregular cocci, while *Methanococcus* is regular to irregular single or paired highly motile cocci (Sirohi *et al.*, 2010). *Methanomicrobiales* are Gram-variable rods or cocci. *Methanomicrobiacae* is Gram-negative coccus or somewhat curved or linear rod, and *Methanomicrobium* are small, straight or weakly bent rods having rounded ends (Sirohi *et al*., 2010). *Methanospirillum* is composed of motile small rods forming continual spiral filament. *Methanosarcinaceae* are large, round to pleomorphic Gram-positive cells of 1.5–2.5μm dia (Sirohi *et al*., 2010).

Methanogens are the glariest examples of the various forms of extremophiles. Based on their natural habitat preference, thermophilic methanogens grow at high (40–100°C) temperatures although most of them are mesophiles that can function in a temperature between 20°C and 40°C (Garcia *et al*., 2000). Similarly, some grow at low temperatures (psychrophiles; Nozhevnikova *et al*., 2003), increased salt concentration (halophiles; Riffat and Krongthamchat, 2006), acidic environment (acidophiles; Zhou and Nanqi, 2007), high pH (alkaliphilic; Thakker and Ranade, 2002) and increased radiation (radophiles).

Thermophilic methanogens are found in solfataras and hot springs, where temperatures span from 40–100°C. Extreme thermophilic methanogens are found in marine hydrothermal vents having temperatures upto 350°C. Thermophiles like *Methanothermobacter*, *Methanothermus*, *Methanothermococcus*, *Methanocaldococcus*, *Methanotorris* and *Methanopyrus* grow at 80°C and above (Serranosilva *et al*., 2014). Halophiles *Methanohalobium*, *Methanohalophilus* and *Methanolobus* grow at 4.3mol/L NaCl (Nazaries *et al*., 2013). The pH range for most methanogens is 6.0-8.0, although moderately acidophiles could grow in the pH range 5.6–6.2 and the alkaliphiles function at a pH range of 8.6–9.2 (Garcia *et al*., 2000). As per the NCBI, prominent methanogens of phylum Euryarchaeota are *Methanoculleus marisnigr*, *Methanosphaerula palustris*, *Methanococcoides burtonii*, *Methanosaeta thermophile*, *Methanocorpusculum labreanum*, *Methanoplanus petrolearius*, *Methanothermobacter marburgensis*, *Methanospirillum hungatei*, *Methanosarcina acetivorans*, *Methanosarcina mazei*, *Methanohalobium evestigatum*, *Methanosarcina barkeri*, *Methanocella paludicola*, *Methanococcus maripaludis*, *Methanobrevibacter ruminantium*, *Methanocaldococcus fervens*, *Methanocaldococcus vulcanius*, *Methanocaldococcus jannaschi*, *Methanopyrus kandleri*, *Methanothermobacter thermautotrophicus*, *Methanococcus vannielii*, *Methanoregula boonei*, *Methanocaldococcus infernus*, *Methanothermus fervidus*, *Methanohalophilus mahii* etc. Based on their substrate preferences, methanogens are classified into five categorised (Thauer *et al*., 2008; Serranosilva *et al*., 2014) as:

i) Hydrogenetrophs: $4H_2 + CO_2 \rightarrow CH_4 + 2H_2O$ (9)

ii) Formatotrophs: $4HCOOH \rightarrow CH_4 + 3CO_2 + 2H_2O$ (10)

iii) Acetotrophs: $CH_3COOH \rightarrow CO_2 + CH_4$ (11)

iv) Methylotrophs: $CH_3OH + H_2 \rightarrow CH_4 + H_2O$ (12)

$4CH_3NH_2 + 2H_2O \rightarrow 3CH_4 + CO_2 + 4NH_3$ (13)

v) Alkalotrophs:

$4CH_3CHOHCH_3 + CO_2 \rightarrow CH_4 + 4CH_3COCH_3 + 24H_2O$ (14)

Acetatogenic methanogens contribute 70–90% to CH_4 production while others utilising substrates like H_2/CO_2, methanol, methyl amine, formate etc. contribute 10–30%, although the figure may vary. CH_4 formation from compounds like methanol and methylamine is not significant in the most environments. A mathematical model depicting methanogenesis as a function of factors can be represented as (Segers and Leffelaar, 1996):

$$R_{MP} = I \times C \times F \quad (15)$$

where, R_{MP}: methane production rate; I: aeration inhibition function which is 1 in complete anaerobic conditions and 0 under aerobic conditions; C: anaerobic C-mineralisation rate; and F: fraction of the anaerobically degraded C that is transformed into CH_4.

Although methanogens primarily use NH_4 as the nitrogen source, some in the four orders, *i.e.*, *Methanobacteriales*, *Methanococcales*, *Methanomicrobiales* and *Methanosarcinals* reportedly bear the N-fixation (nif) gene (Leigh, 2005).

Biochemistry and bioenergetics of methanogenesis

In general, microbes synthesise ATP by oxidising organic or inorganic electron donors. The electron released from the oxidation is received by a terminal electron acceptor invoking proton motive force leading to ATP synthesis by ATP synthase (respiratory phosphorylation). In aerobes, O_2 acts as the terminal electron acceptor (aerobic respiration) while in anaerobic microbes the terminal electron acceptor is nitrate, sulphate, iron (III), manganese (IV), carbon dioxide, possible formic acid or some other organic molecule (anaerobic respiration). Normally, energy gain in aerobes are significantly high than the energygain by anaerobes(Muller, 2001).

The biology of CH_4 production initiates only when chemical agents with high redox potential, *viz.*, O_2, NO_3^-, SO_4^{2-}, Fe^{3+}, Mn^{4+}, and possibly formic acid, are reduced completely (Garcia *et al.*, 2000; Dalal *et al.*, 2008; Serranosilva *et al.*,

2014). These reduction processes generate higher energy (ATP) than methanogenesis. Once the above terminal electron acceptors are reduced, CO and CH_3COOH then act as electron acceptors to facilitate methanation. Hence, the ecological regions with larger CH_4 production (natural wetlands and flooded rice fields) have significantly high methanogen population (Smith *et al.*, 2003; Conrad, 2009). Others such as methanol, methyl amine are also converted to CH_4 by methanogens, although CH_4 generation from these substances are minimal.

Methane formation from CO_2

The CH_4 formation from CO_2 involves 13 reactions (reactions 2–14, Table 4). During this reduction process, methanofuran (MFR), tetrahydrosarcinapterin (H_4MPT) and coenzyme M (HS-COM) are C_1-unit carriers (carbon carriers) whereas ferrodoxin (Fd), coenzyme F_{420} ($E^{0'}$=–360 mV), coenzyme B (HS-CoB, $E^{o'}$=–140 mV), methanophenazine (MP, $E^{o'}$=–165 mV) are the electron carriers (Thauer *et al.*, 2008).

Reactions 2–6, 8, 10 are catalysed by cytoplasmic enzymes, and 7, 9, 11–14 are catalysed by membrane-bound enzyme complexes. Fig. 6 presents the information regarding the reduction of CO_2 to CH_4 and the cytoplasmic membrane transportation of Na^+ ions and protons moieties across it during CH_4 formation process. The six membrane-bound enzymes involved in the conversion are, CH_3 -H_4MPT Coenzyme M-methyltransferase (MtrA–H), energy converting (NiFe) Hydrogenase (EchA–F), Methanophenazine reducing [NiFe] hydrogenase (VhoACG), Methanophenazine-dependent heterodisulphide reductase (HdrDE), an A_1A_0-ATP synthase/A_1A_0-ATPase (AhaA–K), and Na^+/H^+ antiporter (Thauer *et al.*, 2008).

Involvement of membrane bound enzymes

MtrA-H catalyses the reaction 7 (Table 5) with translocation of two Na^+. The EcHE subunit of nickel-iron-sulphur protein [EchA-F] docks the active site [NiFe] centre and is expectedly involved in translocating two protons (Thauer *et al.*, 2008). VhoACG is a haemeprotein containing nickel-iron-sulphur in which VhoC (Cytochrome-*ab* type) contain the haeme group and the VhoA contain the active [NiFe] centre. VhoACG complex in cytosolic layer is oriented by its active site facing towards the periplasmic membrane. The HdrD subunit of HdrDE (an iron-sulphur containing haemeprotein) comprises a rare FeS cluster active site. The HdrE (Cytochrome-*ab* type) subunit of HdrDE contains the haeme group. VhoACG and HdrDE complexes catalyse the reduction of heterodisulphide of COM-S-S-COB with H_2 coupled with electrochemical proton potential resulting in translocation of about two protons per electron, hence four protons per H_2 molecule. The proton translocating ATP synthase, AhaA-K,

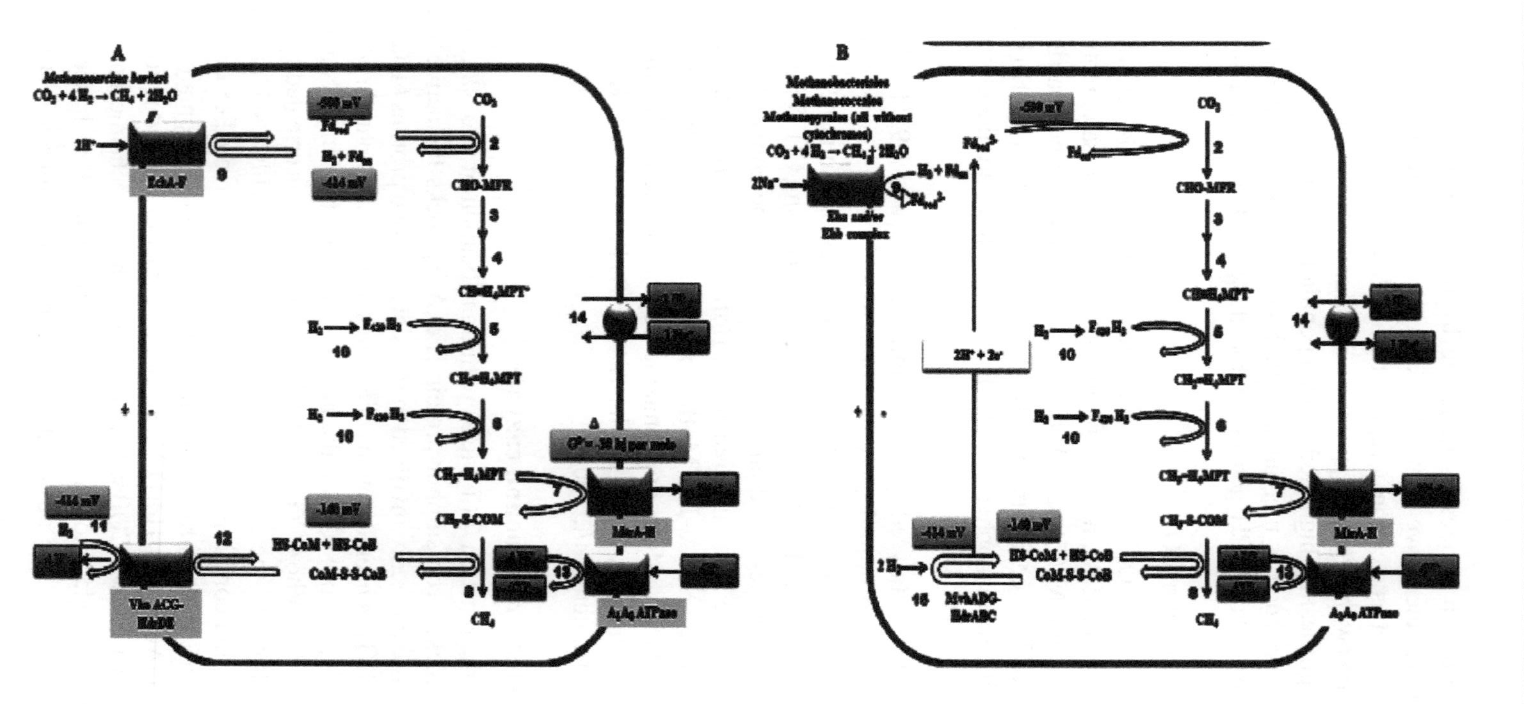

Fig. 6: Energy conservation in hydrogenotrophic methanogen. a: with cytochrome (*e.g.*, *Methanosarcina barkeri*); b: without cytochrome (*e.g.*, Methanococcales) (from Thauer *et al.*, 2008)

Table 4: Chemical reactions involved in the formation of CH_4 from CO_2 or methanol by H_2 (standard free energy change, *i.e.*, ÄG°' was calculated from equilibrium constants or from standard free energies of formation at 25°C with H_2, CO_2 and CH_4 in the gaseous state at 105 Pa, H_2O in liquid state, pH at 7.0 and all other compounds at 1 molar activity (Adopted from Thauer *et al.*, 2008)

Reaction number	Equation	ÄG°' (kJ per mole)
1.	$4\ H_2 + CO_2 \rightarrow CH_4 + 2H_2O$	–131
2.	CO_2 + MFR + Fd_{red}^{2-} + 2 H^+ ↔ CHO-MFR + Fd_{ox} + H_2O	0
3.	CHO-MFR + H_4MPT ↔ CHO H_4MPT + MFR	–5
4.	CHO H_4MPT + H^+ ↔ CHa"H_4MPT^+ + H_2O	–5
5.	CH=H_4MPT^+ + $F_{420}H_2$ ↔ CH_2=H_4MPT + F_{420} + H^+	6
6.	CH_2=H_4MPT + F_{420} H_2 ↔ CH_3 H_4MPT + F_{420}	–6
7.	CH_3 H_4MPT + HS CoM ↔ CH_3 S-CoM + H_4MPT	–30 (coupled with 2 Na^+ translocations)
8.	CH_3 S-CoM + HS CoB ↔ CH_4 + CoM S S-CoB	–30
9.	H_2 + Fd_{ox} ↔ Fd_{red}^{2-} + 2 H^+	+16 (coupled to $2H^+$, or possible 2 Na+, translocations)
10.	H_2 + F_{420} ↔ $F_{420}H_2$ (× 2)	–11
11.	H_2 + MP ↔ MPH_2	–50 (coupled with $2H^+$ translocations)
12.	MPH_2 + CoM S S-CoB ↔ MP + HS CoM + HS CoB	–5 (coupled with $2H^+$ translocations)
13.	ADP + Pi ↔ ATP + H_2O	–32 (coulped with $4H^+$, or posible 4 Na^+ translocations)
14.	2 H^+ (outside) + 1 Na^+ (inside) ↔ 2 H^+ (inside) + 1 Na^+ (outside)	0
15.	2 H_2 + CoM S S-CoB + Fd_{ox} ↔ HS CoM + HS CoB + Fd_{red}^{2-} + 2 H^+	-39
16.	CH_3OH + HS CoM ↔ CH_3 S-CoM + H_2O	-17.5.

efficiently translocate four protons per ATP. The proton to Na^+ stoichiometry for the Na^+/H^+ antiporter has been estimated to be two (Thauer *et al.*, 2008).

Involvement of cytoplasmic enzymes

Reactions 2–6, 8 and 10 are catalysed by the cytoplasmic enzymes (Table 4). The enzymes catalysing the reactions 3–6 contain only one type of subunit and do not possess a prosthetic group. Reaction 2 is catalysed by tungsten-iron-molybdenum protein (Fmd or Fwd; formylmethanofurandehydrogenase) containing five diverse subunits. Methyl CoM reductase (Mcr or Mrt) bears porphinoid F_{430} that catalyses reaction 8. A flavoprotein (Frh; F_{420}-reducing hydrogenease) comprise of nickel-iron-sulphur catalyses reaction 10.

Methanogenesis with cytochromes

Methanogenesis utilising CO_2 and H_2 involves specific enzymes as detailed in Fig. 6a (Thauer *et al.*, 2008). During the pathway, H_2 combines with ferodoxin (Fd_{ox}) to reduce it to Fd_{red} and generate two H^+ that reduces CO_2 to CHO-MFR catalysed by Formylmethanofuran dehydrogenase (Fmd). H_2 reduction is catalysed by EchA-F inwardly translocating two H^+. Fd_{red} is re-oxidised by donating the electrons to CO_2-Methanofuran complex and reducing the later to CHO-MFR. Formylmethanofuran-H_4MPT formyltransferase (Ftr) catalyses transfer of formyl group of formylmethanofuran to H_4MPT (tetrahydromethanopterin) and form formyl-H_4MPT (Breitung and Thauer, 1990; Donnelly and Wolfe, 1986). Formyl-H_4MPT converts to N_5N_{10}-Methenyl-H_4MPT by methenyl-H_4MPT cyclohydrolase (Mch). An example of methanogen producing CH_4 with cytochromes is *Methanosarcina barkeri*.

The N_5N_{10}-Methenyl-H_4MPT converts to methylene-H_4MPT (reaction 5) by F_{420}-dependent or independent pathway. F_{420}-independent pathway is catalysed by H_2-forming methylene-H_4MPT dehydrogenase (Hmd), and the F_{420}-dependent pathway is catalysed by dependent methylene-H_4MPT dehydrogenase (Mtd; Thauer, 1998). The methylene-H_4MPT is then converted to CH_3-H_4MPT (reaction 6) by F_{420}-dependent methelene-H_4MPT reductase (Mer). The CH_3 group from CH_3-H_4MPT transfers to Coenzyme-M, a third C_1 carrier, catalysed by methyl-H_4MPT-coenzyme M methyl transferase (Mtr A-H; DiMarco *et al.*, 1990; Gottschalk and Thauer, 2001). Mtr A-H of 670kDa is an integral membrane protein complex that creates Na_+-membrane potential to translocate two Na^+. Finally, Methyl CoM reductase (Mcr/Mrt) catalyses the reduction of CH_3 group of Methyl-Coenzyme M to CH_4 with Coenzyme B as reductant (Thauer, 1998). During the reaction, heterodisulphide (CoM-S-S-CoB) forms as Coenzyme M and Coenzyme B oxidise (Grabarse *et al.*, 2001). Heterodisulphide acts as a basic material for energy conservation reaction performed by VhoACG and HdrDE complex, so a significant transition molecule in methanogenic energy metabolism. VhoACG and HdrDE complexes reduce heterodisulphide to Coenzyme M (CoM-SH) and Coenzyme B (CoB-SH) while translocating four H^+ (Hedderich *et al.*, 1994).

During CH_4 formation, four H_2 molecules enter the pathway along with one CO_2 molecule to generate one CH_4 molecule and two H_2O molecules. During this reduction of CO_2, AhaA-K (A_0A_1 ATP synthase/A_0A_1 ATPase) catalyses the translocation of four H^+ per ATP molecule. The ATP gain is supposed to be 1.5, *i.e.*, about 1.5 moles ATP from one mole CH_4 produced. This is due to the low (1–10 Pa) H_2 fractional pressure (pH_2) which prevails in majority of the natural methanogenic environment, which results in the standard free energy change (ÄG′) in the range of -17—40kJ/mole which is enough to produce <1 ATP mole (Sirohi *et al.* 2010); formation of ATP from ADP requires at least -50kJ/mole, and the standard free energy change accompanied with the reduction of CO_2 by H_2 to methane is -131kJ/mole is adequate to produce three ATP moles (Thauer *et al.*, 2008). When methanogens grow autotrophically, biomass production estimatedly may not exceed 6.5g from ATP mole (Thauer *et al.* 2008).

Methanogenesis without cytochromes

Methanogenesis without cytochromes happens in the principle as in methanogens with cytochromes except the reduction of heterodisulphide CoM-S-S-CoB by H_2 (reactions 11–12; Fig. 6b). Cytoplasmic MvhADG–HdrABC complex reduces the heterodisulphide CoM-S-S-CoB by H_2 in methanogens with no cytochrome (*e.g.*, *Methanobacteriales*, *Methanococcales*, *Methanopyrales*, *Methanomicrobiales*). Further, EchA-F enzyme catalysing reaction 9 as Eha and Ehb complex in methanogens without cytochromes differs by number of subunits in methanogens with cytochromes. Ech complex has six subunits whereas Eha and Ehb complex contains atleast 16 subunits (Thauer *et al.*, 2008). Methanogens without cytochrome generate about 0.5 ATP, considerably lesser compared to the 1.5 in methanogens with cytochrome.

Methane formation from CH_3OH

Methyl-coenzyme M synthesis is the first step in methanol methanation (Fig. 7a). CH_3-coenzyme M is formed from CH_3OH and Coenzyme M by cytoplasmic enzyme MtaABC (reaction 16, Table 5). The methyl-CoM reductase (Mcr/Mrt) reduces CH_3-coenzyme M to CH_4 (reaction 8). Coenzyme B which acts as a reductant here is the characteristic of archeal methanogen. Coenzyme M and Coenzyme B are oxidised to respective heterodisulphides, CoM-S-S-CoB (Grabarse *et al.*, 2001). The heterodisulphide (CoM-S-S-CoB) is reduced to Coenzyme M (CoM-SH) and Coenzyme B (CoB-SH) by cytoplasmic MvhADG–HdrABC enzyme complex with addition of two H_2 molecules. One out of two H_2 molecules is used for the reduction of CoM-S-S-CoB, while other one combines with Fd_{ox} and reduces it to Fd_{red} with generation of two H^+ (reaction 15, Table 5). The reduction of CoM-S-S-CoB starts another cycle through reactions 7 (CoM-SH) and 8 (CoB-SH) (Table 5). The Fd_{red} is again oxidised with two H^+ forming of H_2 (reaction 9, Table 5). The hydrogenase (Ehb) catlyses the reaction and transports Na^+ as the motive force that can lead to ATP synthesis (reaction 13, Table 5). The CH_4 conversion equation is (Thauer *et al.*, 2008):

$$CH_3OH + H_2 \rightarrow CH_4 + H_2O, \Delta G^{0'} = -112.5 \text{ kJ/mole} \quad (16)$$

Methylamine (CH_3-NH_2) and possibly methylthiol (CH_4-S) enters methanogenesis at the same step as methanol (Fig. 7) and reduce to CH_4 by H_2.

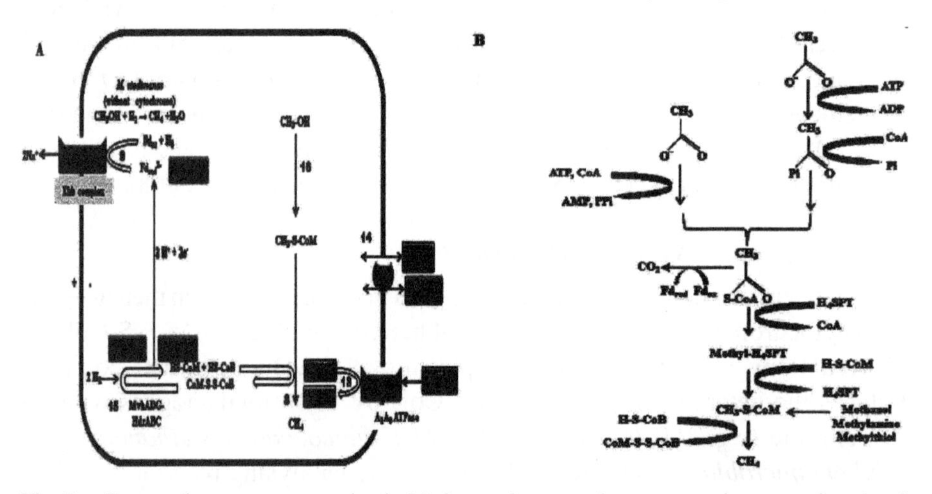

Fig. 7. a: Proposed energy conservation in *Methanosphaera stadtmanae* growing on methanol and H_2; b: Methanogenesis in *Methanosarcina* and *Methanosaeta* growing on CH_3COOH and methyl amine (Modified from Thauer *et al.*, 2008 and Walte *et al.*, 2013)

Methane formation from CH_3COOH

Majority CH_4 formed during methanogenesis is obtained from the acetate (Fig. 7b). Till now only *Methanosarcina* and *Methanosaeta* are known to utilise this substrate for methanation (Walte and Deppenmier, 2014). In *Methanosarcina*, this aceticlastic pathway begins with the activation of COOH group of acetate through ATP-dependent phosphorylation by acetate kinase to convert acetate to acetyl-phosphate. The acetyl-phosphate then converts to acetyl-CoA by phosphotransacetylase. Acetate is activated in obligate aceticlastic *Methanosaeta* with the catalysis by acetyl-CoA synthetase generating acetyl-CoA, AMP (adenosine monophosphate) and PP_i (pyrophosphate) from HS-CoA, acetate and ATP. Acetic acid is converted to CH_4 with a standard free energy change of -36kJ/mole ($\Delta G^{o'}$ = -36 kJ/mole) (Walte and Deppenmier, 2014).

The basic diversity between *Methanosarcina* and *Methanoaceta* is that the latter is associated with lower concentration of acetic acid for suitable growth. Hence, *Methanosarcina* grows faster at higher acetate concentration whereas *Methanosaeta* prevails even at an acetate concentration below 1mM. The acetyl-CoA is cleaved to its methyl and carbonyl moiety by CO dehydrogenase/ acetyl-CoA synthase. *Methanosarcina* has an enzyme CdhABCDE made up of a five-subunit complex, in which CdhC cleaves the C-S bond and transports CH_3CO (acetyl) group to cluster A consisting of a binuclear Ni–Ni site and a [$4Fe_4S$] cluster. After breaking the C-C bond, subcomplex CdhDE receives CH_3 group and transport to the methanogenic cofactor H_4SPT. Subcomplex CdhAE catalyses CO oxidation to CO_2. Ferredoxin (Fd_{ox}) accepts the electrons to reduce to Fd_{red}. The removed CO_2 is assumed to optimise the thermodynamic efficiency of the growth of the methanogen.

In the succeeding reaction, the methyl group of methyl-H_4SPT is transferred to HS-CoM. The above reaction is performed by a membrane associated methyltransferase (MtrA-H) that simultaneously transfers Na^+ across the cytoplasmic membrane resulting in the formation of an electrochemical Na^+ gradient (reaction 7). Subsequently, Methyl CoM reductase catalyses the reduction of CH_3 group of CH_3-S-CoM to CH_4. The HS-CoB combines with the S-CoM moiety of CH_3-S-CoM to form the heterodisulphide CoM-S-S-CoB. The electrons derived from Coenzyme B result in the formation of disulphide bond between coenzyme B and coenzyme M forming the heterodisulphide CoM-S-S-CoB. The heterodisulphide (CoM-S-S-CoB) is reduced to Coenzyme M (CoM-SH) and Coenzyme B (CoB-SH) by H_2 molecule performed by cytoplasmic MvhADG–HdrABC enzyme complex. CoM-S-S-CoB acts as terminal electron acceptor in this anaerobic respiratory chain (Walte and Deppenmier, 2014).

Hence, the substrates like CO_2, CH_3OH, and probably CoM-S-S-CoB for CH_3COOH etc. act as terminal electron acceptors during methanogenesis and the substrates are reduced to CH_4 wherein H_2 molecule(s) act as the electron donors. As a result of this, proton motive force arises due to the electron transfer (electron transfer happens due to H_2 oxidation) leading to translocation of H^+ and/or Na^+ (in some cases) to the periplasmic space through cytoplasmic membrane. Furthermore, some enzymatic reactions not associated with oxidation-reduction (*i.e.*, electron transfer) but only in the product transfer (MtrA-H enzyme) may translocate Na^+ (known as ion gradient translocation). The accumulated H^+ and Na^+ in periplasmic space creates proton and sodium motive force across the cytoplasmic membrane which enables them (H^+ and Na^+) to reach back to the cytoplasm through the transporter A_0A_1 ATP synthase (ATP synthase) that lead to ATP synthesis from ADP and Pi.

Structure and function of Methyl CoM reductase (Mcr)

Methyl CoM reductase is a hexamer composed of two a-, b-g- subunits along with two F_{430} cofactors (Fig. 8a; Ermler *et al.*, 1997; Goncearenco and Berezovesky, 2012). Each of the cofactors contains a Ni atom. Both the α- and β- subunits contain two (C- and N-terminal) structural domains (Fig. 8b).

The N-terminal domains of α- and β- subunits as well as γ-subunit originate from ferredoxin like folds, while the C-terminals domains of α- and β-subunits originate from α folds. In Fig. 8a, the Mcr is separated into two halves each containing one each of α-, β- and γ-subunits, associated with a F_{430} cofactor containing Ni (Goncearenco and Berezovesky, 2012). The N- and C-terminal domains of α-subunits are in marine and dark violet colour respectively, while the N- and C-terminal domains of β-subunit are in light green and dark green

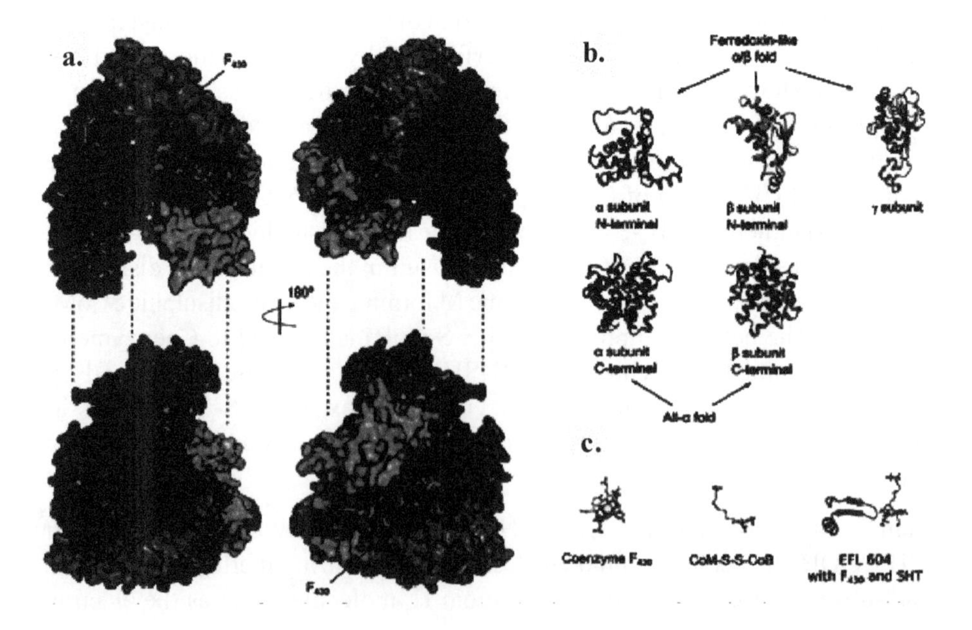

Fig. 8. a: The structure of Methyl CoM reductase (Mcr). The structure of Mcr has been separated into two halves (on the left), each part containing one α-, β- and γ-subunit each. Each α-subunit contains two structural domains, N-terminal (marine blue) and C-terminal (dark violet). Each β-subunit also contains two structural domains, N-terminal (light green) and C-terminal (dark green). The γ-subunit is in orange colour. The two halves in the left are rotated at 180°C around Y axis to give the Mcr appearance on the right. Two molecules of cofactors F_{430} are shown along with heterodisulphide (CoM-S-S-CoB), positions indicated by arrows; **b:** The γ-subunit, N- and C-terminal domains of α- subunit and β-subunit; **c:** The structure of cofactor F_{430} with Ni atom, heterodisulphide (CoM-S-S-CoB) and elementary functional loop (from γ-subunit) with cofactor F_{430} corresponding to the profile 604. (Reproduced from Goncearenco and Berezovsky 2012) (See colour version after references)

respectively and γ-subunit in orange. Mcr constitutes two active sites, each having a Ni containing F_{430} cofactor (Goncearenco and Berezovesky, 2012).

The elementary functional loops (EFLs) associated with F_{430} cofactors possess distinctive amino acid sequences. The EFL having active Tyr367 interacts with CoM-S-S-CoB ligands associated with the Ni atom in F_{430}. Another EFL represented by profile 604 with sequence -RGxDxG [TS] LSGRQxx ExRExDxExxxK- associated with F_{430} takes part in the methyl group transfer (Goncearenco and Berezovesky, 2012). Prior to this methyl group transfer by Mcr to form CH_4, the methyl group transfer from CH_3-H_4MPT to S-CoM is catalysed by Methyltransferase (Mtr), an integral membrane bound complex catalysing the methyl group transfer by Na^+ gradient. The Co^+ in Methyltransferase which acts as a super-reduced nucleophile receiving the methyl group of CH_3-H_4MPT produces CH_3-Co^{3+} in the first of the two partial reactions. The second partial reaction involves the transfer of CH_3 group of CH_3-Co^{3+} to HS-CoM to form CH_3-S-CoM thereby reproducing the activated Co^+ of corrinoid. It is projected that Mtr-linked permease translocates Na^+ and the energy needed to translocate is obtained from a conformational change in Mtr during the methylation–demethylation cycle of Co^+/CH_3-Co^{3+} (Harms and Thauer, 1996).

The next step is associated with the conversion of CH_3-S-CoM to CH_4 by Mcr for which three mechanisms are proposed (Fig. 9), each of which possess different chemistry to initiate catalysis (Wongnate *et al.*, 2016). In mechanism I, nucleophilic attack of Ni(I)-Mcr on the methyl group of CH_3-SCoM forms a CH_3-Ni(III) intermediate. The CH_3-Ni(III) intermediate accepts electron from S-CoM anion to reduce to CH_3-Ni(II) intermediate. In the next step, H moiety from HS-CoB is subsequently abstracted to CH_3 forming and releasing the CH_4. The CoM thiol radical (CoM-S$^{\bullet}$) combines with S-CoB to form CoB-S-S-CoM transferring the excess electron to Ni(II) to form Ni(I) (Ermler *et al.*, 1997, Thauer *et al.*, 2010).

Mechanism II begins with attack and homolytic breakage of Ni(I) on sulphur atom of CH_3-S-CoM of CH_3-S bond, producing a methyl radical ($^{\bullet}CH_3$) and a Ni(II)-S-CoM complex. Abstraction of H atom by methyl radical forms CH_4 and subsequently release the thiol radical Coenzyme B ($^{\bullet}$S-CoB). The thiol radical-CoB reacts with Ni(II)-S-CoM to generate CoB-S-S-CoM whereas the excess electron transfers to Ni(II) forming Ni(I). Mechanism (III) involves the nucleophilic attack of Ni(I)-Mcr on sulphur of methyl-S-CoM to generate highly reactive methyl anion and Ni(III)-S-CoM. The deprotonation of CoB-SH on methyl anion results in CH_4 and S-CoB anion formation. Ni(III)-S-CoM then reacts with S-CoB anion to form heterodisulphide CoB-S-S-CoM, the excess electrons reducing Ni(III) to Ni(I).

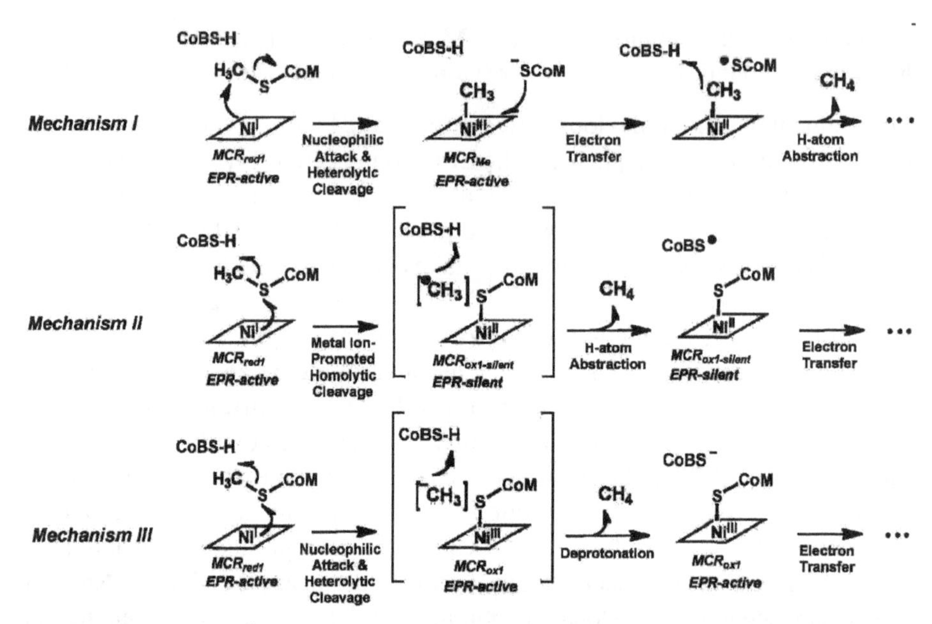

Fig. 9: Three proposed catalytic mechanisms (methyl-N(III) intermediate, Ni(II)-thiolate and Ni(III)-S-CoM pathways) of Methyl CoM reductase (Reproduced from Wongnate *et al.*, 2016)

Evolutionary aspects of methanogens and corresponding enzymes

As discussed, super kingdom Archaea is classified into phyla Crenarchaeoata, Euryarchaeota, Korarchaeota, Nanoarchaeota. Methanogens mainly belong to Euryarchaeaota with orders Methanobacteriales, Methanopyrales, Methanocaccales, Methanomicrobiales and Methanosarcinales etc. These orders contain wide variety of taxa varying significantly in morphology and physiology (Bapteste *et al.*, 2005). They all possess common anaerobic metabolism and produce CH_4 which requires at least 25 genes in addition to for more than 20 proteins involved in the synthesis of coenzymes. The genes encoding an enzyme are clustered together in a genome, however the genes encoding monomers or homopolymers are scattered throughout. Some Methanosarcinales are believed to have more than 250 such genes involve in various aspects of methanogenesis.

Studies on the 3.8 billion years old methane-rich sediments of the Isua district of Greenland suggest the presence of methanogens since then. The genes and their scattered genomoic arrangements are believed to be not acquired through gene transfer with time though some portions of the pathway may have been. For example, the homologous enzymes catalysing the first three steps of CO_2 reduction to CH_4 are used for formaldehyde oxidation in methanogeneic proteobacteria and Panktomycetes. It is very unlike that these enzymes were present in the common archaeal and bacterial ancestor and lost in all but few lineages of prokaryotes, instead inter-domain transfer is more likely.

Fig. 10 provides a glaring possibility of evolutionary trend in hydrogenotrophic methanogens with directionality of the redox-coupled ion translocation. It exhibits an evolutionary pattern representing a series of gene duplication/rearrangement that led to modern hydrogenotrophic methanogens possessing group 3 and 4 NiFe-hydrogenases. The directionality of redox-coupled ion translocation in Eha/Ehb, Mbh/Ech, and Nuo is indicated by positive or negative Na^+/H^+ or H^+ and the change in membrane potential associated with a given reaction is indicated by positive or negative psi (ψ).

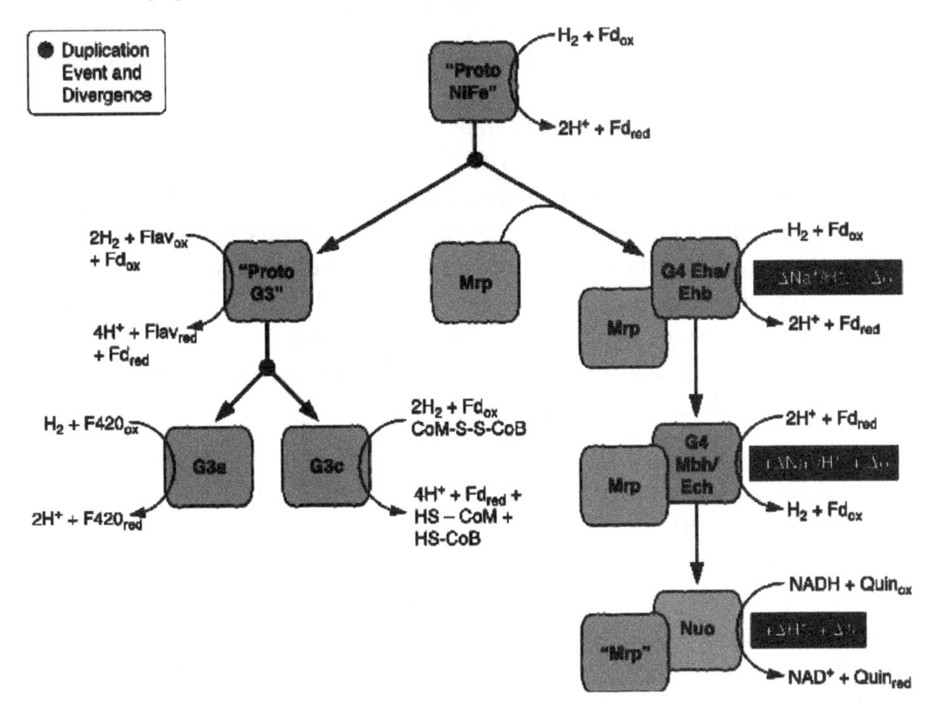

Fig. 10: Phylogenetic reconstruction explaining the series of duplication and gene recruitment events as the modern hydrogenotrophic methanogen [NiFe]-hydrogenase and simple (Mbh/Ech) and complex respiratory systems (Nuo) evolved. (G3: group 3; G3a: group 3a; G3c: group 3c; G4: group 4; Nuo: NADH dehydrogenase; Mrp: Na^+/H^+ antiporter domain; Flav: Flavoprotein; Com-S-S-CoB: Coenzyme M (CoM)-Coenzyme B (CoB) heterodisulphide; HS-CoM: reduced CoM; HS-CoB: reduced CoB; Quin: Quinone)

It shows that the prototype (primitive ancestral) [NiFe]-hydrogenase diverged with time to produce group 3 and group 4 enzymes. Group 3 enzymes underwent further divergence by gene duplication and gene recruitment process resulting in group 3a (8-hydroxy-5-deazafalvin (F_{420}) reducing NiFe-hydrogenases) and group 3c (heterodisulphide reducing NiFe-hydrogenases) subclasses (Boyd *et al.*, 2014). Group 4 {[NiFe]-hydrogenases (Eha/Ehb)} is proposed to be further transformed to Mbh-NiFe-hydrogenase or Ech-NiFe-hydrogenase followed by the transformation to Nuo (NADH dehydrogenase). The large and small subunits of group 4 NiFe-hydrogenase display significant similarity with subunits NuoD and NuoB subunits of NADH dehydrogense respectively. Apart from NuoBD, the small and large subunits of [NiFe]-hydrogenase share homology with FpoBD and Mrp-MbxJL. FpoBD catalyses the reduction of methanophenazine with $F_{420}H_2$. Mrp-MbxJL is reported to be involved in respiring elemental sulphur. Thus, evolutionary trend suggest that, there has been a gradual modification of obligate anaerobic methanogenesis to the facultative.

Recent advances in genomics and proteomics helped determine the evolutionary changes in ancient and latent enzymatic activities and in recombination of functional domains into proteins with new functions. It is reported that closed loops (polypeptide chain returns) with a size of 25–30 residues is a common basic structural element of globular proteins. Elementary functional loops (EFLs) in residues are important in binding, activation and catalysis. In methanogenesis, carbon is transferred between three carbon carriers, methanofuran (MF), tetrahydromethanopterin (H_4MPT) and coenzyme M (CoA-SH), and the major methanogenic enzymes are oxidoreductases and transferases. Although a variety of enzymes is involved in methanogenesis, they all evolved from one ancestral set of enzymes (Bapteste *et al.*, 2005).

Formayl-methanofuran dehydrogenase (Fwd) from *Methanococcus jannaschii* is composed of two subunits B and D (Goncearenco and Berezovsky, 2012). The subunit B (FwdB) is composed of three domains, among which two possess Rossmann-like $\alpha/\beta/\alpha$ folds. The third domain binds to a [4Fe-4S] cluster. The subunit D (FwdD) is a β-barrel fold. The subunits B and D together hold two molecules of molybdenum cofactor dinucleotide (MGD). The EFL of Rossmann-like domains bind to MGD cofactor containing –Rx [TS]AxxADx(6)PG[TS]D. The EFL from the third domain binds to [4Fe-4s] cluster rich in cysteine signature –CxxCxxCxxxCP-. Besides, two psi-loops constituting the core of double-psi β-barrel fold (D subunit) are also involved in MGD binding.

The heterosulphide reductase (Hdr) is a protein complex composed of three subunits HdrA, HdrB and HdrC (Goncearenco and Berezovsky, 2012). In *Methanothermobacter*, HdrABC forms a complex with another enzyme [NiFe]-hydrogenase (Mvh). HdrABC utilises the electrons obtained from [NiFe]-hydrogenase to reduce ferrodoxin and heterodisuphide. Subunit A constitutes several ferreoxidin reductase type FAD-binding motifs. Besides, HdrA contains four motifs for binding [4Fe-4S] clusters with the common signature -CxxCxxCxxxC-. HdrC contains two [4Fe-4S] cluster binding EFLs with same signature as in subunit A. HdrB is the catalytic domain that also contains a [4Fe-4S] cluster having different cysteine rich signature.

Limited reports on the evolutionary divergence of various methanogenic enzymes from their ancestral type exist, though there is substantial information on the evolution of hydrogeneases that perform H_2 metabolism during methanogenesis from various substrates. Two families of hydrogenase enzyme, one composed of iron ([FeFe]-hydrogenases) and other nickel and iron ([NiFe]-hydrogenases) metalloclusters function in the metabolism of H_2 (Boyd *et al.*, 2014). These enzymes can oxidise hydrogen as well as produce it. Although these enzymes possess structural and functional similarity, they are unrelated owing to the result of evolution. [FeFe]-hydrogenases are found in few strict anaerobes and some unicellular eukaryotes but not in archaea. However, [NiFe]-hydrogenases are broadly distributed in bacterial as well as archeal domains but not in eukaryotes. This suggest that [FeFe]-hydrogenases originated after the divergence of bacteria and archaea from the universal common ancestor, whereas [NiFe]-hydrogenases likely originated prior to the divergence of archaea and bacteria.

The [NiFe]-hydrogeneases are significantly found in hydrogenotropic methanogens which mainly contain three phylogenetically different [NiFe]-hydrogenases. Among them, two contain flavin cofactors, one of which is flavin-containing [NiFe]-hydrogenase named 8-hydroxy-5-deazafalvin (F_{420}) reducing NiFe-hydrogenase, catalyses the oxidation of H_2 coupled with the reduction of F_{420}. The other NiFe-hydrogenase (Mvh) functions as a complex with a flavin-containing heterodisulphide reductase, couples oxidation of H_2 to heterodisuphide reduction bond between coenzyme M and coenzyme B. The energy released from this reaction drives the reduction of ferredoxin. The third types are membrane-bound [NiFe]-hydrogenases (Eha/Ehb) that coupled the oxidation of H_2 with reduction of ferredoxin which is coupled with the translocation of Na^+ into the cell. The reduced fereedoxin is then associated with the initial reduction of CO_2 to CH_4.

Occurrence of these phylogenetically related NiFe-hydrogenases suggest that they originated though a series of gene duplications with time. The flavin-containing NiFe-hydrogenases have been considered to have originated from divergence of the gene encoding an ancestral enzyme that also contains a flavin cofactor. Because all the three NiFe-hydrogeneases types oxidise H_2, hypothesis is made that the ancestral enzyme was also involved in H_2 oxidation.

Several [FeFe]- and [NiFe]-hydrogenases are the part of flavin-containing complexes, mainly carrying out fermentation, acetogenesis, methangenesis. NiFe-hydrogenase/heterodisulphide reductase complex of hydrogentrophic methnaogenesis generating reduced Fd from H_2 is a critical energy conserving process. Thus, oxidation-reduction mechanism plays a vital role in maintaining the reduced and oxidised forms of the cofactor in strict anaerobes, and thought to be performed since early time. Anaerobes normally lacking well-defined respiratory system synthesise ATP by metabolising glucose coupled with Fd or NAD^+ reduction during early Earth when O_2 was absent. Fd possess low reduction potential than H^+/H_2 couple regenerate by reducing H^+ to H_2 while NADH can't be regenerated by reduction of protons to H_2.

Several orders of *Thermococcales* overcome such situation by using Fd as the main redox carrier during sugar metabolism. Membrane-bound hydrogenases like Eha/Ehb form multi-subunit complexes with Fe-translocating proteins Mrp. Together; the complex couples the oxidation-reduction of Fd to the oxidation and reduction of H_2. The outcome of this reaction determines the direction of ion translocation. Ions enter the cytosol when Fd reduction is coupled to H_2 oxidation, while ions leave cytosol to periplasmic space when Fd oxidises and protons reduce. Some organisms possess additional Mrp-Mbh complexes that contain carbon monoxide dehydrogenase (CODH), formate dehydrogenase (FDH), or NADPH modules that couple oxidation of these substrates to H_2 production, perform outward ion translocation for ATP synthesis. These complexes are considered respiratory complexes in early existing life.

Conclusion

Biomethanation comprises of four sequential stages, hydrolysis, acidogenesis, acetogenesis and methanogenesis. Hydrolysis is carried out by both aerobic and anaerobic microorganisms those convert the polymers to monomers and/or simpler units that act as substrates for anaerobic microorganisms to produce fermentation products by acidogenesis. The products of acidogenesis are then used as substrates in acetogenesis by strict methanogenic archaea anaerobes which convert them to CH_3COOH, along with CO_2, H_2 and CH_3OH etc. During methanogenesis, H_2 acts as electron donor while substrates CO_2, CH_3OH and probably CoM-S-S-CoB for CH_3COOH acts as terminal electron acceptors and gets reduced to CH_4, an anaerobic respiration process. This conversion is associated with the electrogenic proton motive force along with sodium motive forces to drive ATP synthesis by ATPase through respiratory phosphorylation. Conversion of CH_4 from CO_2 by methanogens with cytochromes is same as those without cytochromes except the reduction of heterodisulphide CoM-S-S-CoB is performed by membrane bound VhoACG-HdrDE complex in the former while by cytoplasmic MvhADG–HdrABC complex in the latter. Methyl CoM reductase catalysing the final step of methanogenesis for all substrates is the characteristic of all methanogens. Methyl CoM reductase (Mcr) is a hexamer composed of two α-, β- and γ-subunits along with two F_{430} cofactors (each containing Ni). The final step of methanogenesis by Mcr is proposed to be proceeding by three mechanisms each differing by the type of intermediates. Mechanism I results in methyl-N(III), mechanism II in Ni(II)-thiolate, while III result in Ni(III)-S-CoM intermediates. Evolutionary trends suggest that the hydrogenotrophic methanogens are the progenitors of all methanogens, and acetogenotrophic methanogens evolved from the former.

Acknowledgements

The authors are thankful to the National Research Foundation (NRF), Korea for the support from the Ministry of Education, Science and Technology (MEST), Korea. Authors gratefully acknowledge the Korea Institute of Geoscience and Mineral Resources (KIGAM).

References

Adekunle KF, Okolie JA, 2015. A review of biochemical process of anaerobic digestion. *Advances in Bioscience and Biotechnology*, **6(3)**: 205–12.

Bapteste E, Brochier C, Boucher Y, 2005. Higher-level classification of the Archaea: evolution of methanogenesis and methanogens. *Archaea*, **1(5)**: 353–63.

Beguin P, Aubert JP, 1994. The biological degradation of cellulose. *FEMS Microbiology Reviews*, **13(1)**: 25–58.

Boyd ES, Schut GJ, Adams MW, Peters JW, 2014. Hydrogen metabolism and the evolution of biological respiration. *Microbe*, **9(9)**: 361–7.

Breitung J, Thauer RK, 1990. Formylmethanofuran: Tetrahydromethanopterin formyltransferase from *Methanosarcina barkeri* Identification of N^5-formyltetrahydromethanopterin as the product. *FEBS Letters*, **275(1-2)**: 226–30.

Conrad R, 2009. The global methane cycle: recent advances in understanding the microbial processes involved. *Environmental Microbiology Reports*, **1(5)**: 285–92.

Dalal RC, Allen DE, Livesley SJ, Richards G, 2008. Magnitude and biophysical regulators of methane emission and consumption in the Australian agricultural, forest and submerged landscapes: a review. *Plant and Soil*, **309(1-2)**: 43–76.

DiMarco AA, Bobik TA, Wolfe RS, 1990. Unusual coenzymes of methanogenesis. *Annual Review of Biochemistry*, **59(1)**: 355–94.

Donnelly MI, Wolfe RS, 1986. The role of formylmethanofuran: tetrahydromethanopterin formyltransferase in methanogenesis from carbon dioxide. *Journal of Biological Chemistry*, **261(35)**: 16653–9.

Doukov TI, Iverson TM, Seravalli J, Ragsdale SW, Drennan, CL, 2002. A Ni-Fe-Cu center in a bifunctional carbon monoxide dehydrogenase/acetyl CoA synthase. *Science*, **298(5593)**: 567–72.

Ermler U, Grabarse W, Shima S, Goubeaud M, Thauer RK, 1997. Crystal structure of methyl-coenzyme M reductase: the key enzyme of biological methane formation. *Science*, **278(5342)**: 1457–62.

Fuchs G, 1986. CO_2 fixation in acetogenic bacteria: variations on a theme. *FEMS Microbiology Letters*, **39(3)**: 181–213.

Garcia JL, Patel BK, Ollivier B, 2000. Taxonomic, phylogenetic, and ecological diversity of methanogenic archaea. *Anaerobe*, **6(4)**: 205–226.

Goncearenco A, Berezovsky IN, 2012. Exploring the evolution of protein function in archaea. *BMC Evolutionary Biology*, **12(1)**: 1–13.

Gottschalk G, Thauer RK, 2001. The Na^+-translocating methyltransferase complex from methanogenic archaea. *Biochimica et Biophysica Acta (BBA)-Bioenergetics*, **1505(1)**: 28-36.

Grabarse W, Mahlert F, Duin EC, Goubeaud M, Shima S, Thauer RK, Lamzin V, Ermler U, 2001. On the mechanism of biological methane formation: structural evidence for conformational changes in methyl-coenzyme M reductase upon substrate binding. *Journal of Molecular Biology*, **309(1)**: 315–30.

Harms U, Thauer RK, 1996. The corrinoid-containing 23-kDa subunit mtrA of the energy-conserving N^5-methyltetrahydromethanopterin: coenzyme M methyltransferase complex from *Methanobacterium thermoautotrophicum*: EPR spectroscopic evidence for a histidine residue as a cobalt ligand of the cobamide. *European Journal of Biochemistry*, **241(1)**: 149–54.

Hedderich R, Koch J, Linder D, Thauer RK, 1994. The heterodisulfide reductase from *Methanobacterium thermoautotrophicum* contains sequence motifs characteristic of pyridine-nucleotide-dependent thioredoxin reductases. *European Journal of Biochemistry*, **225(1)**: 253–61.

Hook SE, Wright ADG, McBride BW, 2010. Methanogens: methane producers of rumen and mitigation strategies. *Archaea*, **2010**: 1–11.

Kangle KM, Kore SV, Kore VS, Kulkarni GS, 2012. Recent trends in anaerobic codigestion: a review. *Universal Journal of Environmental Research and Technology*, **2(4)**: 210–9.

Kröger A, Geisler V, Lemma E, Theis F, Lenger R, 1992. Bacterial fumarate respiration. *Archives of Microbiology*, **158(5)**: 311–4.

Leigh JA, 2005. Genomics of diazotrophic archaea. *In*: R Palacios and WE Newton (Eds) *Genomes and Genomics of Nitrogen-fixing Organisms*. Springer, Dordrecht, Netherlands: 7–12 pp.

Liu X, Zhu Y, Yang ST, 2006. Butyric acid and hydrogen production by *Clostridium tyrobutyricum* ATCC 25755 and mutants. *Enzyme and Microbial Technology*, **38(3-4)**: 521–8.

Ljungdahl LG, 1994. The acetyl-CoA pathway and the chemiosmotic generation of ATP during acetogenesis. *In*: HL Drake (Ed) *Acetogenesis*. Springer, Boston, MA: 63–87 pp.

Müller V, 2003. Energy conservation in acetogenic bacteria. *Applied and Environmental Microbiology*, **69(11)**: 6345–53.

Müller V, Gottschalk G, 1994. The sodium ion cycle in acetogenic and methanogenic bacteria: generation and utilization of a primary electrochemical sodium ion gradient. *In*: HL Drake (Ed) *Acetogenesis*. Springer, Boston, MA: 127–156.

Muller V, 2001. Bacterial fermentation. *Encylopedia of Life Sciences*. Nature Publishing Group: 1–7.

Nazaries L, Murrell JC, Millard P, Baggs L, Singh BK, 2013. Methane, microbes and models: fundamental understanding of the soil methane cycle for future predictions. *Environmental Microbiol*ogy, **15(9)**: 2395–417.

Nozhevnikova AN, Zepp K, Vazquez F, Zehnder Alexander JB, Holliger C, 2003. Evidence for the existence of psychrophilic methanogenic communities in anoxic sediments of deep Lakes. *Applied and Environmental Microbiology*, **69(3)**: 1832–5.

Ragsdale SW, Kumar M, 1996. Ni-containing carbon monoxide dehydrogenase/acetyl-CoA synthase. *Chemical Reviews*, **96(7)**: 2515–40.

Riffat R, Krongthamchat K, 2006. Specific methanogenic activity of halophilic and mixed cultures in saline wastewater. *International Journal of Environmental Science and Technology*, **2(4)**: 291–9.

Segers R, Leffelaar PA, 1996. On explaining methane fluxes from weather, soil and vegetation data *via* the underlying processes. *In*: R Laiho, J Laine and H Vasander (Eds) *Northern Peatlands in Global Climate Change*. The Academy of Finland, Helsinki: 226–241

Serrano-Silva N, Sarria-Guzmán Y, Dendooven L, Luna-Guido M, 2014. Methanogenesis and methanotrophy in soil: a review. *Pedosphere*, **24(3)**: 291–307.

Sirohi SK, Pandey N, Singh B, Puniya AK, 2010. Rumen methanogens: a review. *Indian Journal of Microbiology*, **50(3)**: 253–62.

Smith KA, Ball T, Conen F, Dobbie KE, Massheder J, Rey A, 2003. Exchange of greenhouse gases between soil and atmosphere: interactions of soil physical factors and biological processes. *European Journal of Soil Science*, **54(4)**: 779–91.

Suwannakham S, 2005. *Metabolic engineering for enhanced propionic acid fermentation by Propionibacterium acidipropionici* (Doctoral dissertation). The Ohio State University, Ohio, The United States of America: 258 pp.

Thakker CD, Ranade DR, 2002. An alkalophilic *Methanosarcina* isolated from Lonar crater. *Current Science*, **82(4)**: 455–8.

Thauer RK, 1998. Biochemistry of methanogenesis: a tribute to Marjory Stephenson. *Microbiology*, **144(9)**: 2377–406.

Thauer RK, Kaster AK, Goenrich M, Schick M, Hiromoto T, Shima S, 2010. Hydrogenases from methanogenic archaea, nickel, a novel cofactor, and H_2 storage. *Annual Review of Biochemistry*, **79**: 507–36.

Thauer RK, Kaster AK, Seedorf H, Buckel W, Hedderich R, 2008. Methanogenic archaea: ecologically energy differences in energy conservation. *Nature Reviews Microbiology*, **6(8)**: 579–91.

Walte C, Deppenmier U, 2014. Bioenergetics and anaerobic respiratory chains of acetolastic methanogens. *Biochemica et Biophysica Acta (BBA)-Bioenergetics*, *1837*(7): 1130–47.

Wongnate T, Sliwa D, Ginovska B, Smith D, Wolf MW, Lehnert N, Raugei S, Ragsdale SW, 2016. The radical mechanism of biological methane synthesis by methyl-coenzyme M reductase. *Science*, **352(6288)**: 953–8.

Zhou X, Nanqi R, 2007. Acid resistance of methanogenic bacteria in a two-stage anaerobic process treating high concentration methanol wastewater. *Frontiers of Environmental Science and Engineering in China*, **1(1)**: 53–6.

5

Application of Methanation as an Alternate Energy Technology

Rachita Rana[a] and Sonil Nanda[b]

[a]Department of Chemical and Biological Engineering
University of Saskatchewan, Saskatoon, Canada
[b]Lassonde School of Engineering, York University, Toronto, Canada
[b]Corresponding Author: soniln@yorku.ca

ABSTRACT

Increased environmental pollution and high demand over supply for conventional fossil fuel sources have led to the exploration of other renewable energy sources. Out of the potential alternatives, methane has come up with a promising unconventional fuel source. Several technologies have been developed to establish methanation, including biomethanation, chemical methanation and gas-to-liquid technology. Several factors like process parameters, catalysts and reactors have been studied for process optimisation and making methane production commercially viable. Although methane as a greenhouse gas is a contributor to global warming, its combustion produces relatively 'environment friendly' gas CO_2, and H_2O. The released CO_2 is fixed into biomass by plant through photosynthesis process. Only rarely does methane combustion produces CO when combusted in an oxygen-limited condition. Thus, methane production and combustion are carbon neutral process. This chapter focuses on the application of methane as a source of alternate energy. Various technologies used, recent advances and challenges faced in methanation process are discussed.

Keywords: Methanation, methanogens, catalysts, methane, greenhouse gas, fuel

Abbreviations used

BTU : British thermal unit
GWP : Global warming potential
IPCC : Intergovernmental panel on climate change
LPG : Liquefied petroleum gas
CNG : Compressed natural gas

Introduction

Depleting conventional energy resources are a matter of concern for almost every developed and developing nation. Situation is alarming for global energy crises and on the roll already. It is high time for the policy makers to focus on the unforeseen outcomes of this. Although energy is a basic need for all, 1.2 billion people live devoid of electricity supply and more than 2.7 billion people are still deprived of clean energy supplies for domestic and cooking applications even till this day (IEA, 2017). Poor energy efficiency and governmental policies are a few predominant issues in developing nations whereby most of the energy remains out of reach of the rural poor. In global context, the wastage of the energy in the developed and developing nations add to the problem. The total world energy demand would shoot up from 575 quadrillion BTU (quads) as in 2015 to 736 quads in 2040, and this 28% increase is certainly an alarming situation (Conti *et al.*, 2016).

The world's 80% energy supply depends on coal, oil and gas, based on the non-renewable fossil fuels. Worldwide, one-third of the energy is supplied by coal which typically contributes to 40% of the electricity production. It also significantly meets the growing energy demands of the industries (such as that of iron and steel industries). However, these fossils-driven energy sources are getting to the verge of depletion and are unsustainable in the long run. The global energy demands are increasing exponentially to add to owes. Thus, the gorge between the supply and demand is deepening day by day. This essentially proves that the change in the energy consumption pattern as per the drastic policy changes cannot be easy and smooth.

Oil comprises of a wide range of energy sources including crude oil, natural gas, refinery feed, condensates, other hydrocarbons (including synthetic crude oil, oils extracted from bituminous sands, shale oil etc.), and petroleum products such as refinery gas, light and heavy oils, ethane, liquefied petroleum gas, jet fuels, diesel oil, naphtha, etc. These products find wide applications in industrial, domestic and commercial/transport sector. The environmental and geopolitical concerns of the oil resources have been the topics of debate since decades. In addition to the concern of sustainable oil resources, use of oil has also put the environmental concerns in focus. Countries need to lay emphasis on improving the energy efficiency, eradicate wastage of energy resources and address the environmental concerns due to conventional energy sources. This has called for a need to switch to alternate sustainable and environment-friendly products.

Notwithstanding their preferential use, the constant depletion of the conventional resources and the growing environmental concerns of their use have raised questions. Studies have suggested that the global energy demand will rise by almost half over the next 20 years. Conventional oil reserves are not sufficient to sustainably meet this demand. In the late 1960s, the US government and the energy sector realised the future shortage of the existing natural gas as the source of energy due to the escalating demands. There were regulations imposed on the usage and prices of the natural gas along with the search for developments towards an efficient and economical methanation process. There is an immediate urge to find rather sustainable alternatives to the growing energy needs and switching to unconventional energy sources. A conscious shift in energy consumption calls for a focus-shift to unconventional sources that currently form a meagre 10% of the global energy supply. These unconventional energy sources include hydroelectric, geothermal, biomass-based, solar and wind.

Out of the various options explored for alternative energy resources, the biomass-based energy sourcing is an excellent one and has proven its proficiency over time. A prominent benefit of using biomass as a feedstock to produce biodiesel or any other form of biofuel is its qualification as a clean fuel (Nanda *et al.*, 2014). Hydrogen-rich synthetic gas (Syngas) is comparatively less harmful to the environment unlike otherwise caused through greenhouse gas emissions (Nanda *et al.*, 2017). Biofuels are produced from biomass available in almost every country and it can reduce the dependence of non-oil producing nations on their oil-producing counterparts.

Another significant benefit of biofuels is that they are predominantly produced from agricultural and biomass waste, thereby bioconverting the wastes from these sectors to wealth. Significant focus has been drawn to the efficient production of biofuel to replace the conventional fossil-oil in the energy market (Nanda *et al.*, 2015). Methods have been developed, validated and put to commercial production, and are being further fine-tuned. Blends of jet fuels and biofuels have been introduced, and commercialisation and economical production of biofuel are further rapidly gaining pace. Energy efficiency is the core of a safe, reliable, sustainable and affordable energy supply for future.

Global warming and the GHGs

Global warming is an alarming global situation in the current scenario. The foreseen climatic changes and its effects will vary from region to region across the globe. The predicted impacts comprise the elevated global temperatures, sea level rise, and increase in the desert area and alteration in the precipitation pattern. Global climate changes due to anthropogenic activities in the ecosystem have time and again been reported. Human encroached land (especially in the Arctic) has been observed to get warmer than the oceans. Melting of the glaciers, permafrost and sea ice has reached alarming limits. Other consequences of global warming include frequent extreme weather changes and extinction of endangered species due to alterations in temperature patterns. Greenhouse gases emissions caused by humans seem to be beyond the carrying capacity of the natural geochemical cycles alone, and thus global warming issue needs a mindful human intervention.

Greenhouse gases are capable of absorbing and emitting the radiations that lie in the thermal infrared range to the atmosphere. The various ways in which the humans emit greenhouse gases into nature through fossil fuel combustion are industries, automotives and other power needs. The gases that majorly contribute to global warming are, carbon dioxide (CO_2), methane (CH_4), chlorofluorocarbons (CFCs), hydrochlorofluorocarbons (HCFCs), ozone (O_3) and nitrous oxide (N_2O). As plants would consume CO_2 during photosynthesis, depleting forest cover has further triggered an increase in carbon dioxide in the environment. Methane is naturally released from agricultural fields and landfills, and nitrous oxide is released by the gases used in refrigerators, fertilisers and industrial syntheses.

Defined by the nature of the gas, its quantity and indirect impacts it is likely to cause, each greenhouse gas has a specific contribution to the greenhouse effect. For instance, methane has 72 times stronger radiative property compared to CO_2 of same mass for approximately 20 year timeline (Nanda *et al.*, 2016). None-the-less, the quantity of atmospheric methane is way less than CO_2 and has comparatively shorter atmospheric life. Thus, an overall stronger greenhouse contribution comes from carbon-di-oxide, and therefore is coined as the GHG unit for measuring the impact of other GHGs. Efficiency of a gas molecule as a greenhouse gas contributor and its lifetime in the atmosphere, dictate its global warming potential (GWP). Carbon dioxide is assumed to be 1.0 GWP for all time periods. GWP of a gas is measured by its comparison to the equivalent mass of CO_2 released for the same period of time. It is understood that if a gas particle has high radiative force along with a short lifespan, it shall have greater GWP on the scale of 20 years, compared to a gas with smaller particles for a scale of 100 years. It implies that a molecule in atmosphere with longer life than CO_2 has an increased GWP on timescale basis. CH_4 is considered to have an atmospheric lifespan of about 12–15 years.

Intergovernmental Panel on Climate Change (IPCC) reports suggest that though the impact of methane is initially 100 times more than that of CO_2, its shorter atmospheric lifetime (it rapidly degrades in the atmosphere to CO_2 and H_2O) makes GWP of methane relatively declining compared to CO_2. The indirect radiative impact of methane cannot be neglected as it contributes to ozone formation.

The less GWP of methane over years and its easy production in infrastructure-wise deprived locations such as rural areas make it a fit candidate as an alternate fuel. Several methods to produce methane from domestic wastes have been standardised. Grass, a suitable material for methane-rich biogas production, can produce upto 1–0.5ft^3 biogas per pound (grass silage is an even better). Thus, 20 pounds of grass can generate cooking fuel for an hour of domestic use. Other potential methane feedstock is the food waste with better methane (biogas) yields. Animal manure is another alternative that is well-suited for methane production on farms; though the yield is less still it is relatively an easier option. Adult cattle can potentially produce close to 140 pounds of manure daily, which can produce 85ft^3 of biogas (three hours of cooking fuel).

Chemical technologies for methane generation

Though biomethanation route is uncomplicated and can serve for short-term interests, chemical route is the preferred one when it comes to its mass production. There are several uses of methane and producing large quantities of methane to resolve the energy crisis seems logical. This has paved the way for the establishment of chemical methanation process. Thus, chemical production of methane is a more industrially-acceptable process unlike biomethanation.

The energy sector realised the imminent shortage of existing natural gas and regulations were imposed on the usage and prices of the natural gas. Around this time, the search to develop an efficient and economical methanation process to convert coal to synthetic natural gas emerged (Kopyscinski *et al.*, 2010). Chemical methanation converts carbon dioxide and carbon monoxide to methane in presence of hydrogen. The credit of this discovery goes to Sabatier and Senderens in 1902. The major use of the methanation of CO_x is its use as an alternative fuel. Methanation has been considered a major technology to produce synthetic natural gas since the 1970s. Recent technological advances have discovered methanation as a route to store solar or wind energies, and promote the potential carbon capture mechanism. Methanation also finds application in ammonia synthesis. Catalyst fouling during its hydrogen producing steps (wherein carbon oxides otherwise accumulate) can be avoided by adopting the methanation route. Haldor Topsoe's process is a famous methanation reaction. Here, hydrogen and carbon monoxide (3:1) is converted to methane, usually in presence of nickel-based catalysts. Chemical methanation is a traditional approach for fuel production, there have been several advancements made in this field though. Several advanced technologies employed to produce methane are, gasification, hydromethanantion, supercritical water gasification, and absorption enhanced reforming.

All these technologies do not directly focus to produce methane but methanation reaction is rather an integral part of the processes. The biomass used is gasified at high temperature and pressure to produce hydrogen-rich syngas, and methane is often a component of the product gases. Similar to gasification, when water in its supercritical conditions is used to react with the feedstock it is termed as supercritical water gasification. The product efficiency here is improved compared to simple gasification process yielding hydrogen-rich syngas. Hydromethanation involves the injection of pressurised steam into the fluidised mixture of catalyst and carbon-rich feedstock. The catalysed multiple reactions take place between the carbon and steam to generate methane and carbon-di-oxide. Absorption-enhanced reforming is an enhancement of gasification process. It adds a bed of burnt lime in the indirect gasification technology.

In chemical methanation, the prominent role is played by the catalyst used and therefore several advances towards the use of efficient catalysts to enhance methane yields have been made. Kruissink *et al.* (1981) used the coprecipitated nickel-alumina catalyst for high-temperature methanation. Ruthenium catalysts have been extensively used to produce methane from CO_2. Lunde and Kester (1974) studied chemical methanation of CO_2 using ruthenium at different feed ratios of H_2-to-CO_2 and derived an empirical expression from CH_4 yields from CO_2 reacted with H_2 in the presence of 0.5 wt% ruthenium on alumina support.

Selective methanation of carbon monoxide was studied by Rehmat and Randhava (1970). Feed gases CO, CO_2 (~3000 ppm) and H_2 were tested with different catalysts at temperatures ranging from 125–300°C. Ruthenium most efficiently catalysed CO methanation reducing the CO concentration to as low as 50–100ppm without impacting the yield of the product gases, the next efficient being Raney nickel-type catalyst. However, catalysts with precipitated nickel particles were not well suited for the process (Rehmat and Randhava, 1970). Along with conventional catalysts, nano-scale catalysts have proven effective in methanation. Nickel, molybdenum, ruthenium, tungsten and other noble-metals are doped inside or on the surface of nanotubes, and the catalysts are used in methanation to enhance methane yields. Chen *et al.* (2002) used end-opened MoS_2 nanotubes for methanation of CO in the presence of H_2. Such catalysts were demonstrably effective due to reduced temperature requirements for methanation.

Supercritical water gasification has also been useful in methanation technology contributing significantly. Here, methanation occurs along with water-gas shift reaction and steam reforming to produce H_2 during supercritical water gasification (Reddy *et al.*, 2014). Several techniques are being tested towards catalytic supercritical water gasification to yield methane-rich product gas. During supercritical water gasification reaction, water acts as the reactant as well as the reaction medium enabling the breakdown and dissolution of organic species to produce a homogeneous supercritical phase. The product gases from supercritical water gasification contain hydrogen, methane, and traces of CO, CO_2 and C_{2+}. The major advantage of the process is the drying of the biomass that ensures process efficiency. Catalytic processes typically use the above-mentioned catalysts, nickel and ruthenium being the most common. Ruthenium has better resistance towards sintering and gives high activity for both steam reforming and methanation that significantly convert biomass molecules to methane gas (Elliot, 2008; Vogel, 2007; Waldner *et al.*, 2007).

Sulphur adsorption on surface of the catalysts affects the methanation efficiency of the catalyst thus affecting product formation. Dreher *et al.* (2013) studied the Ru/Co catalytic supercritical water gasification to address sulphur poisoning issues during biomass reforming methanation. Methane was produced by sequential supercritical gasification of organic compounds in aqueous form and selective CO_2 methanation. Gaseous products from the supercritical reactor, *viz.*, CH_4, H_2 and CO_2 were used as feed stream for a catalyst-loaded fixed bed system to further promote selective hydrogenation of carbon dioxide to enhance methane yield (Frusteri *et al.*, 2017).

It is essential to produce a high-performance catalyst that could promote reverse water-gas shift reaction, and Fischer-Tropsch process (FTP) could contribute to usable hydrocarbons production. Meiri *et al.* (2015) found improved product yields by adding potassium as a promoter. Potassium improved the catalytic performance in terms of better reaction rates, improved product (higher hydrocarbons) selectivity, and enhanced process stability. It also improved the exposure of carbide iron at the catalyst surface which increased CO_2 adsorption. There are several processes that are accompanied with methanation as the side reaction and it is essential to identify and best use these process to extract value-added fuel gases and other such products.

Biological technologies for methane generation

Typically, biogas is produced by subjecting the organic matter to microbial fermentation under anaerobic conditions. The process is precisely referred as 'anaerobic digestion of biodegradable materials that takes place in the presence of microorganisms'. Biogas production from organic wastes is assumed to contribute to about 25% of the overall renewable energy budget (Holm-Nielsen *et al.*, 2009). Various dung, sewage sludge, several biodegradable wastes from farms and households are good feedstock for biomethane production. These feedstock include dung, sewage sludge, municipal waste, plant shreds, agri-wastes, kitchen waste etc. Many such organic wastes and dung are abundantly available in rural areas, usually dumped as waste. With increased urbanisation and consumerism, the dumping of organic wastes has increased in urban set-ups as well.

Biogas production from organic wastes involves a digester, anaerobic microbes and biodegradable feedstock. The route of anaerobic digestion is considered as the most significant technology to convert organic matter to methane (biomethane) as a renewable energy source. Anaerobic digestion that has benefits like the abilities to handle high organic loading, low cell yield, low nutrient demand is a well-established and extensively studied technology which is low on operational and maintenance expenses for the reactor. Vrieze *et al.* (2012) reported great potential of *Methanosarcina* sp. in biomethanation. Compared to other known methanogens, these are comparatively robust species and adjust to various impairments better.

A biogas digester is fed with organic waste (ideally of small size) slurry, stirred and provided with optimal heat condition to initiate feedstock digestion by microbes. The biogas thus produced is very handy to store and finds its application as a cooking fuel. Its refined form is demonstrably useful as automotive fuel (either alone or blended with LPG to enhance its calorific value). Along with blending of biogas with LPG, research to upgrade it to natural gas standards is increasingly gaining pace. Upon upgrading and compression, biogas becomes biomethane, a preferred fuel that can be useful as transportation fuel in cars, ships, trains etc. like compressed natural gas (CNG).

Biomethane is certainly an environment-friendly alternative as compared to coal, kerosene and other traditionally used fuels. For remote locations in the countryside where gas supply is hard to reach, biomethane production could resolve energy-related limitations and challenges. Along with productive waste disposal that would ensure a cleaner environment, establishing biomethane plants could provide energy as well as employment to contiguous population. Biomethane is also an economic fuel resource for the villagers as the installation and running cost of digesters is meagre. Fig. 1 gives a typical process flow of anaerobic digestion route.

Use of crop wastes as feedstock for biomethanation is a widely accepted and explored route. In particular, rice straw is reportedly an excellent alternate feedstock for biomethanation. Somayaji and Khanna (1994) studied anaerobic digestion of wheat/rice straw mixed with cattle manure for biomethanation. High cellulose content in plant residues can have combustion energy as high as 50kJ/g when converted to biomethane. Gunnarson *et al.* (1985) studied the potential of Jerusalem artichoke for biogas production. The aerial parts of Swedish *Helianthus tuberosus* L. were investigated for possible biogasifiability. Biogas yields from three varieties were determined at different harvesting times. It showed that, these fresh plant parts yielded 480-680 L biogas per kg biomass when subjected to anaerobic digestion.

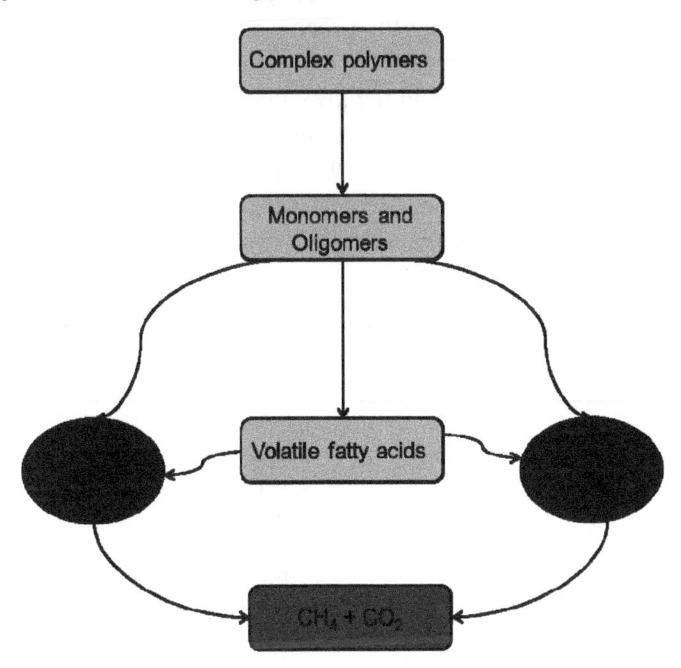

Fig. 1: A typical anaerobic digestion process

Dar and Tandon (1987) blended pretreated wheat straw, lantana residue, apple and peach leaf litter with cattle manure to produce biogas. At ambient temperature (28–31°C), pretreated plant residues showed improved biodegradability of their cellulosic and organic contents. This also increased the biogas and methane production by two-fold. A maximum (63.6%) methane yield was obtained for lantana slurry followed by apple litter (59.6%), wheat straw (58%), and peach litter (57.7%). It was interesting to observe that, methane yield from all these was better than cattle manure (56.1%). Further, the anaerobic digestion of these pretreated biomass was 31–42% more efficient than cattle slurry. There are several other plant and animal residues that find application in biomethane production. Several rural community-scale biogas plants provide daily fuel supplies for the local population that have been running smoothly for appreciable duration.

Challenges and future prospects

Considering the significance of methane in the energy domain, it is extremely important to explore and develop techniques for effective methane production. The four major challenges that such industries face are related to the catalyst, reactor, environment and economics. The catalyst in methanation often undergoes sintering, sulphur poisoning, and fouling, ultimately impacting its performance. Also, the cost of the catalyst considering the industrial-scale operation is a concern. Hence, there is a wide scope to develop effective catalysts to address all the above issues and making the process more effective through reduced temperatures requirements. One of the other prominent limitations is the reactor corrosion at high temperatures in the presence of water. The commonly used materials for cost-effective reactor fabrication are not corrosion resistant, and the alternative non-corrosive quartz material is an expensive proposition. This calls for research to find a viable alternative material that outlasts methanation reaction conditions. Table 1 summarises the benefits and drawbacks of biological and chemical methanation routes.

Table 1: Comparison between chemical and biological methanation processes

Chemical methanation	Biomethanation
• Faster process	• Time consuming
• Energy intensive process	• Less energy input is required
• High purity gas products are obtained	• Products need further refining and purification
• Catalyst is usually required	• Microbes and enzymes catalyse the process
• Catalyst coking, sintering and poising can occur which adds to the cost	• No catalysts are required, though ideal microbial conditions are necessary
• Reactor corrosion could occur due to the erosive activity of the catalysts	• Usually, no alarming corrosion of the bioreactor or digester
• Reaction may reverse due to thermodynamics, reaction kinetics and adverse catalytic effects	• Natural/synthetic inhibitor may inhibit/ mutate the microbes thereby lowering product yield
• Catalyst removal and regeneration is required which adds to the process cost	• Reuse of microbes is usually not required, but can easily be accomplished
• Costly due to high maintenance, energy needs	• Relatively economical process
• Chemical catalyst is product/process selective	• undesirable product may form by contaminated microbes
• Chemical methanation is limited to urban set-up	• Biomethanation is suitable across various locations, thereby boosting the economy
• No malodour formation is observed	• Nauseating odour may form depending on the input material
• Residual waste needs pre-discharge treatment being environmentally hazardous	• Residual waste is usually biodegradable and ecofriendly; can be used as a soil conditioner

For any industrial operation, the reusability and regenerability of the materials used are important from economics viewpoint. It is yet to be attained for methanation reaction. Biomethanation waste usually has fertiliser use, which is not the same case with chemical methanation. The treatment and disposal of the refuse of industrial chemical methanation into the environment, therefore, is another challenge to address. Overall, the process economics for industrial scale chemical methanation seems infeasible thereby making room for a great deal of research in this field addressing the pressing issues. Further, the greenhouse gas issue while using methane as a fuel source also cannot be neglected.

Conclusion

The use of methane as an alternative energy source is significant in response to the growing energy demand and depleting oil reserves. Advances in research on fuel cells operating on methane and other such technologies have been recorded. Further, the increasing environmental carbon dioxide concentration is believed to be a major contributor to greenhouse gas emissions leading to global warming. Converting this carbon dioxide to methane thereby reducing its harmful effects and to lessen the burden on the fossil fuels is very alluring for the energy sector. The current scenario focuses on the use of waste for valuable product formation.

Both chemical and biological routes for methanation have been explored successfully, although the electrochemical process is demonstrably more efficient than thermochemical methanation out of the chemical routes in converting carbon dioxide to methane. With the use of apt metal electrodes and organic solvents as electrolytes, electrochemical process has higher conversion rates. From available literature it is understood that catalysts play a significant role in carbon dioxide methanation. Allegedly, multi-metallic catalysts are more efficient in this. Although a simpler and easily-scalable technique, biomethanation faces certain challenges like feedstock supply and its commercialisation. Though researches on methanation process are significant, there is a huge scope to further optimise the process and standardise the operation.

References

Chen J, Li S, Xu Q, Tanaka K, 2002. Synthesis of open-ended MoS_2 nanotubes and the application as the catalyst of methanation. *Chemical Communications*, **16**: 1722–3.

Conti J, Holtberg P, Diefenderfer J, LaRose A, Turnure JT, Westfall L, 2016. *International energy outlook 2016 with projections to 2040* (No. DOE/EIA-0484 (2016)). USDOE Energy Information Administration (EIA), Washington, DC (United States). Office of Energy Analysis.

Dar GH, Tandon SM, 1987. Biogas production from pretreated wheat straw, lantana residue, apple and peach leaf with cattle dung. *Biological Wastes*, **21(2)**: 75–83.

De Vrieze J, Hennebel T, Boon N, Verstraete W, 2012. *Methanosarcina*: The rediscovered methanogen for heavy duty biomethanation. *Bioresource Technology*, **112**: 1–9.

Dreher M, Johnsosn B, Peterson AA, Nachtegaal M, Wambach J, Vogel F, 2013. Catalysis in supercritical water: pathways of methanation reaction and sulfur poisoning over a Ru/C catalyst during the reforming of biomolecules. *Journal of Catalysis*, **301**: 38–45.

Elliott DC, 2008. Catalytic hydrothermal gasification of biomass. *Biofuels, Bioproducts and Biorefining*, **2(3)**: 254–265.

Frusteri F, Frusteri L, Costa F, Mezzapica A, Cannilla C, Bonura G, 2017. Methane production by sequential supercritical gasification of aqueous organic compounds and selective CO_2 methanation. *Applied Catalysis A: General*, **545**: 24–32.

Gunnarson S, Malmberg A, Mathisen B, Theander O, Thyselius L, Wünsche U, 1985. Jerusalem artichoke (*Helianthus tuberosus* L.) for biogas production. *Biomass*, **7(2)**: 85–97.

Holm-Nielsen JB, Al Seadi T, Oleskowicz-Popiel P, 2009. The future of anaerobic digestion and biogas utilization. *Bioresource Technology*, 100(22): 5478–84.

Kopyscinski J, Schildhauer TJ, Biollaz SM, 2010. Production of synthetic natural gas (SNG) from coal and dry biomass-A technology review from 1950 to 2009. *Fuel*, **89(8)**: 1763–83.

Kruissink EC, van Reijen LL, Ross JR, 1981. Coprecipitated nickel–alumina catalysts for methanation at high temperature. Part 1.—Chemical composition and structure of the precipitates. *Journal of Chemical Society, Faraday Transactions 1: Physical Chemistry in Condensed Phases*, **77(3)**: 649–63.

Lunde PJ, Kester FL, 1974. Carbon dioxide methanation on a ruthenium catalyst. *Industrial & Engineering Chemistry Process Design and Development*, **13(1)**: 27–33.

Meiri N, Dinburg Y, Amoyal M, Koukouliev V, Nehemya RV, Landau MV, Herskowitz M, 2015. Novel process and catalytic materials for converting CO_2 and H_2 containing mixtures to liquid fuels and chemicals. *Faraday Discussions*, **183**: 197–215.

Nanda S, Azargohar R, Dalai AK, Kozinski JA, 2015. An assessment on the sustainability of lignocellulosic biomass for biorefining. *Renewable and Sustainable Energy Reviews*, **50**: 925–41.

Nanda S, Mohammad J, Reddy SN, Kozinski JA, Dalai AK, 2014. Pathways of lignocellulosic biomass conversion to renewable fuels. *Biomass Conversion and Biorefinery*, **4(2)**: 157–91.

Nanda S, Rana R, Zheng Y, Kozinski JA, Dalai AK, 2017. Insights on pathways for hydrogen generation from ethanol. *Sustainable Energy and Fuels*, **1(6)**: 1232–45.

Nanda S, Reddy SN, Mitra SK, Kozinski JA, 2016. The progressive routes for carbon capture and sequestration. *Energy Science and Engineering*, **4(2)**: 99–122.

Reddy SN, Nanda S, Dalai AK, Kozinski JA, 2014. Supercritical water gasification of biomass for hydrogen production. *International Journal of Hydrogen Energy*, **39(13)**: 6912–26.

Rehmat A, Randhava SS, 1970. Selective methanation of carbon monoxide. *Industrial & Engineering Chemistry Product Research and Development*, **9(4)**: 512–5.

Somayaji D, Khanna S, 1994. Biomethanation of rice and wheat straw. *World Journal of Microbiology and Biotechnology*, **10(5)**: 521–3.

Vogel F, Waldner MH, Rouff AA, Rabe S, 2007. Synthetic natural gas from biomass by catalytic conversion in supercritical water. *Green Chemistry*, **9(6)**: 616–9.

Waldner MH, Krumeich F, Vogel F, 2007. Synthetic natural gas by hydrothermal gasification of biomass: Selection procedure towards a stable catalyst and its sodium sulfate tolerance. *The Journal of Supercritical Fluids*, **43(1)**: 91–105.

6

Biogas Generation from Kitchen Wastes and its Purification Technology

Amar Kumar Das

Department of Mechanical Engineering, Gandhi Institute For Technology (GIFT)
Gramadiha, Gangapada, Bhubaneswar – 752 054, India
Corresponding Author: amar.das@gift.edu.in

ABSTRACT

Recent challenges for fossil fuels, associated environmental issues with disposal of organic garbage, and the rising costs for energy has encouraged researchers to search for alternate energy sources. Massive uncontrolled biogas from landfills, sludge and organic material degradation under anoxic conditions is an environmental management issue. Methane emission to atmosphere has a tremendous green house effect, but this can also be a valuable renewable energy resource. Biogas from kitchen wastes through anaerobic digestion by methanogens is such an alternate technology. Kitchen wastes are considered one of the best raw materials for biogasification. Biogas is mainly composed of 50-74% methane (CH_4), 25-28% carbon-di-oxide (CO_2), other volatile compounds including hydrogen sulphide (H_2S), water vapour and other trace gas compounds. The impurities in it, like Hydrogen sulphide (H_2S) which is corrosive, downgrade its quality. Purified biogas of higher calorific value attracts attention as a substitute fuel for additional applications. Its purification using various suitable scalable and economically-viable scrubbing technologies to enhance its quality and performance as an alternate fuel is an important step in this direction. Thus, a technology to generate and purify biogas from kitchen waste has significant prospects in the energy sector, and also as a promising enterprise to employ the youth.

Keywords: Biogas, kitchen waste, biogas purification, alternate fuel

Abbreviations used

CH_4 : Methane
CO_2 : Carbon-di-oxide
H_2S : Hydrogen sulphide
GHG : Greenhouse gas
TS : Total solids
MSW : Municipal solid waste
CSTR : continuously stirred tank reactor
NH_3 : Ammonia
H_2O : Water
ZnO : Zinc oxide
FeS : Iron sulphide
Fe(OH) : Ferrous hydroxide

Introduction

Energy is a tool for economic growth, social progress, human wellbeing and a better comfortable life (Nijaguna, 2002). Depleting fossil fuels and increasing demand on the existing energy sources has evoked interest in the search for alternate energy source. The global energy needs for agriculture, industry, transport, commercial and domestic uses have increased exponentially. Excess dependency and consumption of energy from fossil fuels poses a threat to the environment. Fossil-fuel based energy systems have several environmental issues including global warming and greenhouse gases (GHG) effects. Further, the dwindling fossil fuel reserves to meet the global energy demand beyond the next few decades is under pressure due to the increasing energy demands due to the rapidly growing population and modernisation. The depleting fossil fuels and increasing demand for energy sources has lead to the search for alternate sources of energy.

Fuel alternatives as energy inputs point towards global economic development, economic prosperity and quality of life (Grande and Rodrigues, 2007). In developing and large nations like India, the need for fossil fuels is very high and to meet such demands the government spends huge Forex which adversely affects the national economic growth. Due to its availability in abundance, usefulness as a renewable resource, lesser pollution threat and ecofriendly nature, biofuel has been a promising alternate fuel source to meet global demands (Grande and Rodrigues, 2007). Biomass fuel sources are easily available in rural and urban areas all over the world and can be suitably exploited for the purpose. The chapter discusses the various aspects of biogas as an alternate fuel for household and commercial applications, benefits of biogas concerning its effect on the environment, economic viability, and commissioning of plant compared to conventional fuel sources. In light of economic benefits and technical feasibility, biogas is more promising as an alternate transportation fuel (Murphy *et al.*, 2004).

Food wastes

There are various reasons responsible for the generation of food wastes, the remaining uneaten and waste bulk of food received from a food point, starting from its stage of production, processing, retailing to consumption (Anon, 2018). Wastage of foods is encountered at all stages of the food supply chain, from farm to the fork, in the farm-to-kitchen chain in low income countries and in the kitchen-to-garbage chain in the developed nations. Global surveys reveal that about 30-50% of total food produced is wasted.

Being organic and biodegradable, these food wastes evolves a mixture of carbon dioxide (CO_2), water vapour, methane (CH_4) and some other impurities including hydrogen sulphide (H_2S) upon degradation in controlled conditions. These also degrade naturally by microbes following various other pathways.

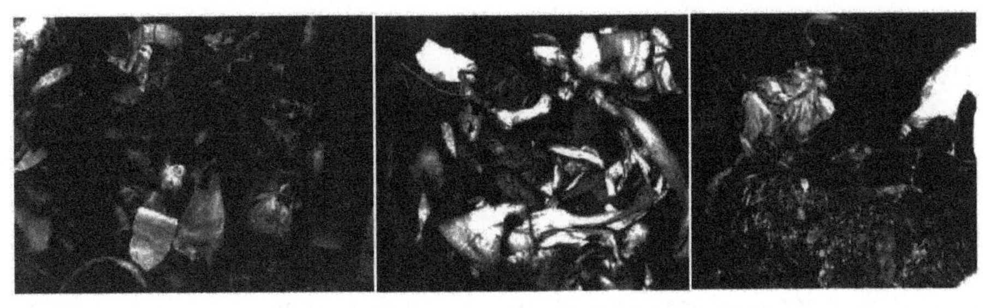

Fig. 1: Various forms of biodegradable wastes from Indian household kitchen and restaurant

Methods of biogasification

Biogas is produced when organic waste is anaerobically digested in altered operational conditions. Biogas generation follows natural biogeochemical carbon cycle which is an acceptable technology for both rural and urban set-up (Cavenati *et al.*, 2005; Igoni *et al.*, 2008; Abhishek *et al.*, 2015; Singh *et al.*, 2001; Joy *et al.*, 2014). Taleghani and Kia (2005) have outlined several financial and community benefits of biogas production. Such benefits include, effective municipal solid waste (MSW) management, reduced threat of soil and water pollution, curbing the rampant usage of chemical fertilisers and herbicides/pesticides, encouraging organic firming and generating alternate sources of income, reducing pressure on land acquisition for landfill. The organic wastes that contain complex carbon compounds breakdown through depolymerisation into simpler substances, followed by methanogenesis. The microbial action in this can be divided into aerobic degradation and anaerobic digestion.

Aerobic degradation

The digestion process usually occurs in the presence of atmospheric oxygen that produces a mixture of gases, the primary constituent being the greenhouse gas carbon dioxide (CO_2), responsible for global warming.

Anaerobic digestion

The gas mixture of anaerobic digestion occurring in the absence of oxygen contains CH_4 as the most usable gas in power generation, cooking and other applications. The calorific value of this gas is around 5200–5800KJ/m^3 when burnt at normal atmospheric condition. The gas is treated as an environment-friendly energy source that could potentially replace fossil fuel.

Mechanism of anaerobic digestion

Anaerobic digestion occurs inside the digester with certain bacterial fermentations. The entire process of digestion is accomplished in a sequence of stages, hydrolysis, acidogenesis, acetogenesis and methanogenesis.

Hydrolysis

Hydrolysis, the process of decomposition of complex organic chains of carbohydrates, proteins, and lipids into simpler substrates like sugars and amino acids, is initiated and/or expedited in the presence of microbial enzymes. One such hydrolysis process involving carbohydrates is (Ostrem and Themelis, 2004):

$$(C_6H_{10}O_4)_n + 2H_2O \rightarrow C_6H_{12}O_6 + H_2$$

Different microbial enzymes active in such processes include various extracellular enzymes, cellulase, amylase, protease and lipase.

Acidogenesis

This process involves acid-producing bacteria that accelerate fermentation toward the formation of various organic acids. The acidic environment inside the anaerobic reaction chamber helps faster and selective growth of the specialised bacteria to carryout the later steps in biogasification. Such microorganisms grow faster under anaerobic conditions and reduce the compounds with low molecular weight alcohols, organic acids, amino acids, CO_2, H_2S and traces of CH_4. The process is partially endergonic (*i.e.*, it needs energy inputs).

Acetogenesis

Once sufficient amount of organic acids accumulate, there is a risk of very low pH conditions inside the digestor. This situation may become worse which may lead to a system (digestion) failure if the VFA (volatile fatty acids) accumulation is high. It is often observed that, such a situation is encountered if the input material has high fatty acids content. All these acids are then converted to acetic acid in the presence of partial dissolved oxygen through bacterial action. Acetic acid being the preferred raw material for methanation, this step ensures that all organic acids are available for methanogenesis.

$$2CH_3CH_2OH + CO_2 \rightarrow 2CH_3COOH + CH_4$$

Methanogenesis

Methanogenic bacteria decompose low molecular weight compounds anaerobically and utilise hydrogen, CO_2 and acetic acid, under otherwise normal ambient conditions to yield CH_4. The various mechanisms involved are:

Acetic acid conversion to methane: $CH_3COOH \rightarrow CH_4 + CO_2$

Carbon dioxide reduction by hydrogen to methane: $CO_2 + 4H_2 \rightarrow CH_4 + H_2O$

Methanal conversion to methane: $CH_3OH + H_2 \rightarrow CH_4 + H_2O$

The methane production stoichiometry is deduced by the reaction (Parkin and Owen, 1986):

$$C_nH_aO_bN_c + \left[n - \frac{a}{4} - \frac{b}{2} + 3\left(\frac{c}{4}\right)\right] H_2O = \left[\frac{n}{2} - \frac{a}{8} + \frac{b}{4} + 3\left(\frac{c}{8}\right)\right] CO_2 + \left[\frac{n}{2} + \frac{a}{8} - \frac{b}{4} - 3\left(\frac{e}{8}\right)\right] CH_4 + cNH_3$$

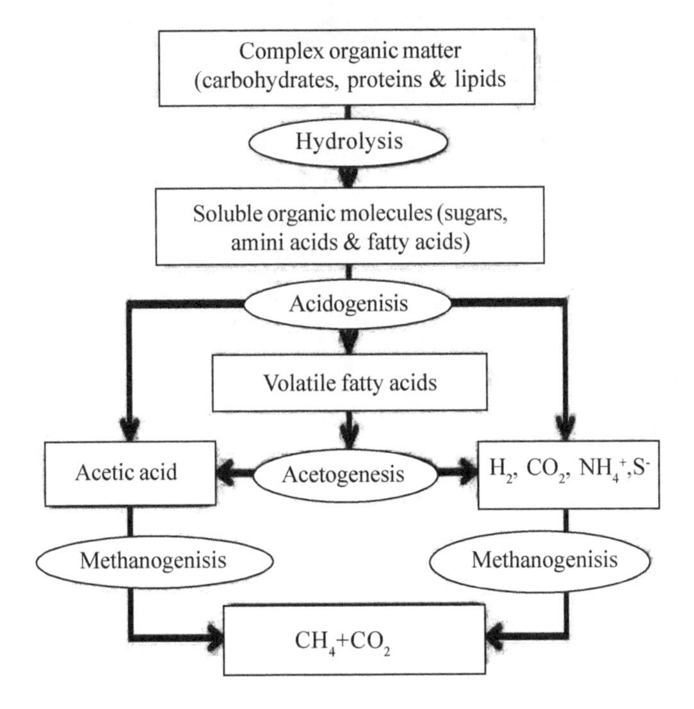

Fig. 2: Different stages of anaerobic digestion

Types of biogas plants

According to the mode of material-feeding to the digester, it is classified as single- or double-chamber type. Biogasification is a continuous process in a series that includes charging feed to the digester, producing a volume of gas and the fermented material overflowing from it. There are two types of successful biogas plant designs most commonly in use, *viz*., fixed-dome and floating-drum, for biogas production from simple biogasifiable substrates.

Fixed-dome biogas plant

This plant consists of a digester with a fixed, non-movable gas holder on top of the digester. Slurry is fed to the digester and reacts inside the drum. The generated gas pressure displaces the slurry into the compensating tank. The volume of gas in the upper level of the drum is measured by the difference in height of slurry level in the digester and the compensating tank.

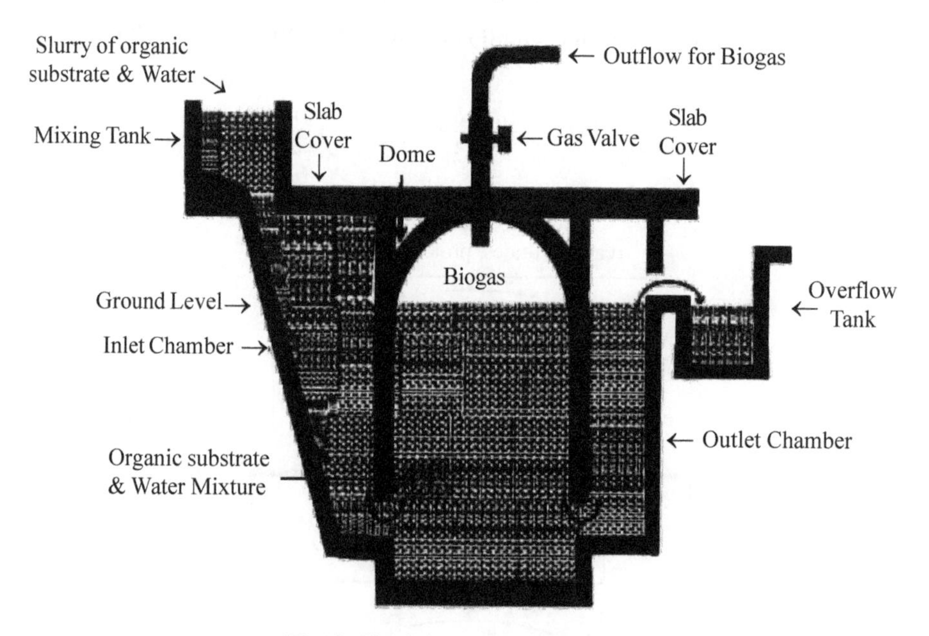

Fig. 3: Fixed-dome type biogas plant

Floating-drum biogas plant

Floating-dome biogas plant is cylindrical or dome-shaped with a moving floating gas-holder, or drum. The gas-holder floats either directly in the feed slurry or in a separate water jacket. The biogas-collection drum has an internal or external guide frame called as spider that provides stability and keeps the drum upright for ease in dome movement and gas collection.

The biogas production is ensured by the upward movement of dome inside the digester and after collection or consumption of gas, the gas-holder sinks down. It is very much experienced that the operation of Floating-dome plants is very easy and safe. The plant supplies a non-toxic, odourless and inflammable gas at an unvarying pressure.

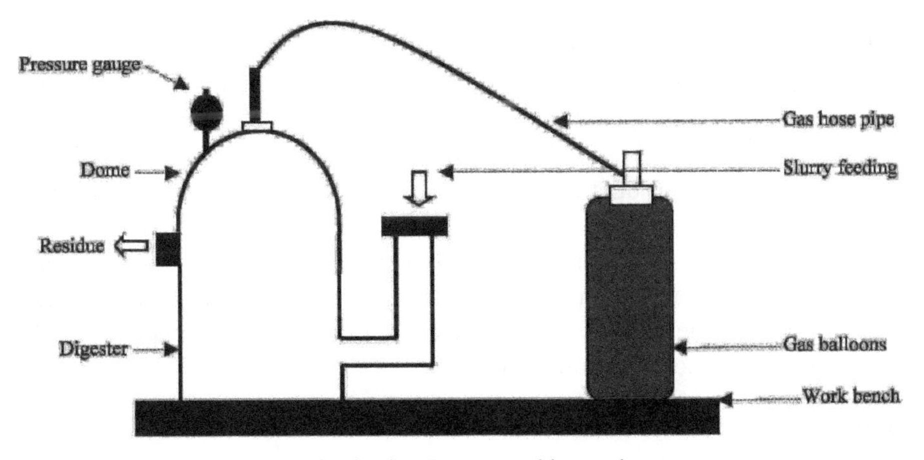

Fig. 4: Floating-Drum type biogas plant

Composition of biogas

The composition of the gas is associated with the contents of the slurry fed to the digester. The gas produced is very light compared to air and has an ignition temperature ranging from 650–750°C. It is odourless and colourless and burns with a blue flame similar to LPG gas. Its calorific value is about 20MJ/m^3 and usually burns with 60% efficiency in a conventional biogas stove. It can be treated as a replacement to firewood, cattle dung, petrol, LPG, diesel and electricity, depending on the requirements, availability and economic limitations.

Table 1: The various chemical composition and physical properties of Biogas

Composition		Properties	
Component	Concentration	Feature	Specification
Methane (CH_4)	55–85%	Energy Content	6–6.5kWh/m^3
Carbon dioxide (CO_2)	35–40%	Fuel Equivalent	0.6–0.651 oil/m^3 biogas
Water (H_2O)	2–7%	Explosion Limits	6–12% biogas in air
Hydrogen sulphide (H_2S)	2% (20-20000 ppm)	Ignition Temperature	650–750°C
Ammonia (NH_3)	0–0.05%	Critical Pressure	75–89 bar
Nitrogen (N)	0–2%	Critical temperature	82.5°C
Oxygen (O_2)	0–2%	Normal Density	1.2kg/m^3
Hydrogen (H)	0–1%	Odour	Rotten egg

Factors affecting biogas yield

The important factors affecting the fermentation process of organic wastes under anaerobic conditions and thereby the biogas yield are, ambient temperature and pressure, concentration of solids and loading rate, pH of the slurry, supplementary nutrients, and the reaction period.

Ambient temperature and pressure

Temperature is vital in effective biogas production. Methane is produced progressively in a temperature range of 29–41°C and the absolute pressure range of 1.1–1.2bars. Thermophilic bacteria are sensitive to ambient conditions. The volume of gas production increases with the rise of ambient temperature up to a certain level. However, CH_4 content in the gas decreases as the temperature rises further inside the digester. Mesophilic conditions are favourable for large-scale production of biogas in a temperature range of 32–35°C. However, the stability and efficiency of the plant purely depends on the parametric and process optimisations.

Solid concentration and loading rate

Organic matters, such as, cattle dung, kitchen wastes, agricultural wastes, etc. are primarily fed to the biogas plant digester as slurry after mixing with water. The amount of water needed to prepare the slurry varies in proportion depending on the waste type, it usually being 1:1. The amount of slurry fed into the digester daily is its loading rate, and depends on the plant size, capacity and retention time, the average loading rate being about 0.2kg/m^3. The required loading rate needs to be maintained for maximum production of the gas.

pH value of the slurry

The pH is the antilogarithm of the hydrogen ion (H^+) concentration in the solution. The main objective of measuring the pH value of slurry is to maintain the acidity level of the digester medium. For optimal digester productivity, the pH is maintained between 6.5 and 7.5.

Supplementary nutrients

Supplementary nutrients and inocula are required to be added to a few organic matters to enhance the rate of methanogenesis. As cattle dung contains all potential nutrients in required proportions to produce CH_4, there is no need to add further nutrients to it. However, kitchen waste lacks some nutrients which are needed to be added with different proportions. This may include inorganic chemicals to microbial cells consortium.

Reaction period

In a most favourable condition depending on the digester tank capacity, 80–90% of total biogas volume is produced in 3–4 weeks. It was experimentally observed that the biogas production per unit volume was high with the diameter to depth ratio between 0.66 and 1.00.

From Fig. 5, it is seen that the retention time is also affecting the biogas production and the slurry prepared with different ratio of bio-wastes. The gas production is found higher when the slurry is prepared by mixing kitchen waste and poultry waste.

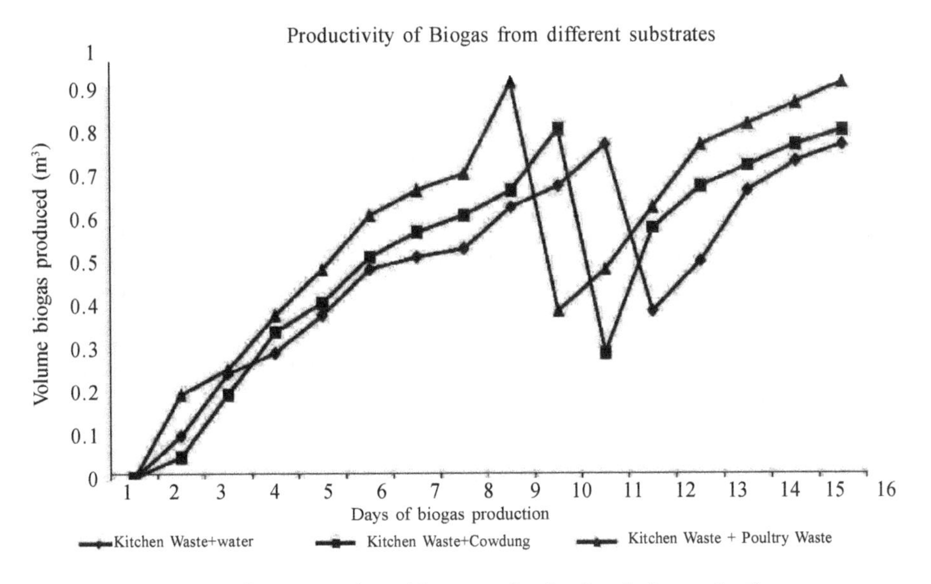

Fig. 5: The mean volume biogas production in relation to the time

From Fig. 6, it is observed that %CH_4 was higher when slurry was prepared by mixing kitchen waste and poultry waste with water, with 9-day retention time. The maximum CH_4 content of the total volume of biogas produced was 85%. When the slurry was prepared by mixing kitchen waste and the water, lowest %CH_4 was obtained. The maximum %CH_4 in the biogas produced was 60. When the slurry was a mix of kitchen waste and cattle dung, there was medium %CH_4 production. The maximum %CH_4 in the produced biogas was 75. The total gas collected was around 0.95m^3 with 85% methane content at pH 7.3. Gas production minimised when the slurry was prepared by mixing kitchen waste and water only.

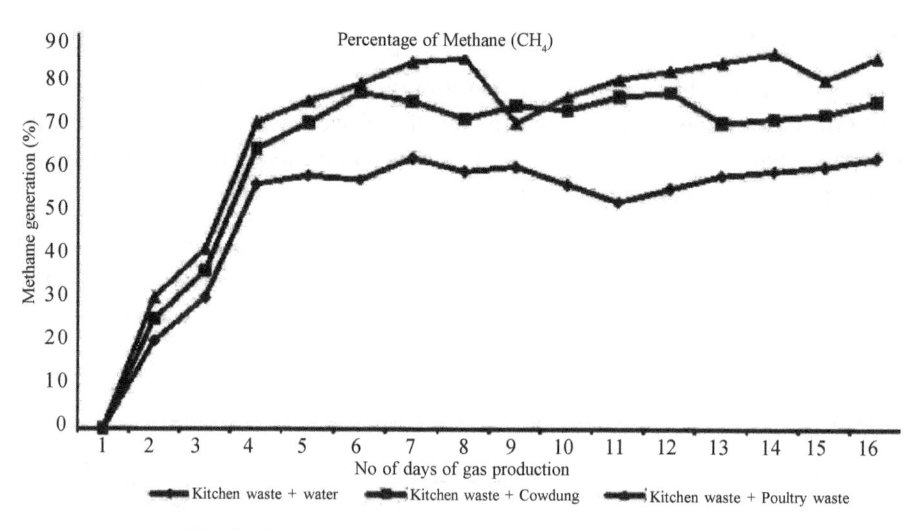

Fig. 6: The mean % CH_4 produced during a study period

Fig. 7 explains that the best pH value of slurry was obtained with slurry prepared by mixing kitchen waste and poultry waste with water. Maximal pH of the slurry was 7.3. When the slurry was prepared by mixing kitchen waste with water, the highest pH was 7.9. The pH of the slurry was medium when prepared by mixing water with cattle dung. Maximal pH of the slurry was 7.6. A constant pH of the digester slurry is important to maximise biogas quantity.

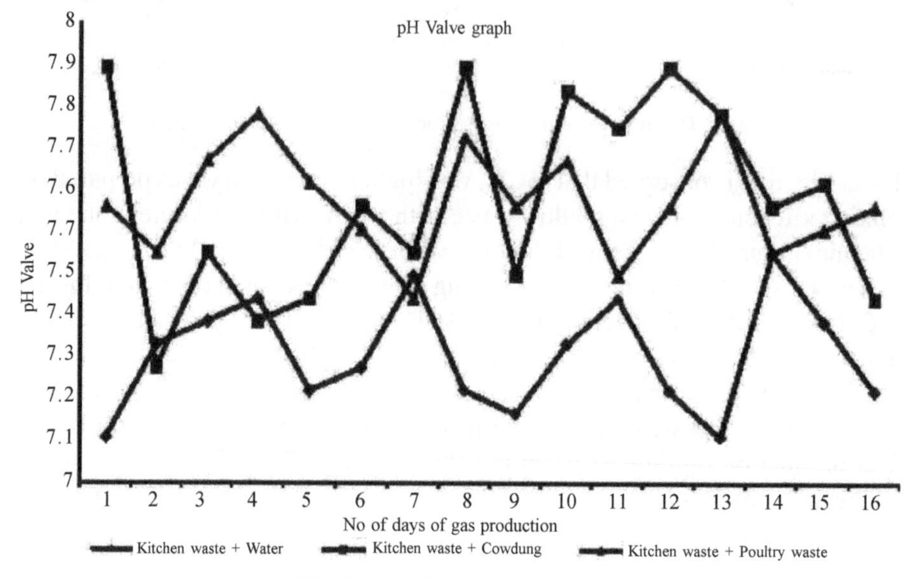

Fig. 7: pH value in length of time

The total solids (TS) content of MSW affected biogas production by anaerobic digestion. The TS critically influence parameters like pH, temperature and efficacy of microbial consortia in the decomposition process. Considering the TS content in MSW, higher TS contents in MSW produced higher volume of biogas in a continuously stirred reactor/digester (CSTR) under anaerobic condition (Igoni *et al.*, 2008). It may be due to faster methanogenesis rate. However, the volume of the biogas maximised when a TS saturation level was achieved.

Limited fossil fuel reserve and related issues associated with combustion process encourage research in new and renewable energy resources. Amongst various alternative fuel sources, biogas is a promising alternative fuel, more so due to the added importance in collection, control and conversion of organic waste materials into value-added products (Abhishek *et al.*, 2015). Also, biogas is generated with simple technology and low skill-set requirement.

Microbial stimulants positively affect the biogas production and yield. Addition of microbial stimulants to cattle dung and the combined slurry of cattle dung and kitchen waste maximises biogas production. Addition of stimulants like 'aquasan' and 'teresan' in the slurry resulted in the rapid increase of 55% and 15% respectively in biogas production by volume (Singh *et al.*, 2001).

Techniques to enhance biogas yield (upto 90% of total digester slurry weight) in large-scale include lab-scale mesophilic reactors, up-flow biofilm reactor and fibre liquefaction reactor involving *Fibrobacter succinogenes* integrated with continuous watering system (Joy *et al.*, 2014). Though highly promising, process standardisation for consistent favourable conditions is suggestible.

Wet-oxidation with decreased O_2 pressure on digester slurry remarkably enhanced (85–90%) the CH_4 yield (Joy *et al.*, 2014). Such findings reinforce the confidence in the economic benefits of biogas technology. Wastewaters and organic wastes can effectively be biogasified with optimised parameters that affect gas production rate, like pH, temperature, quality and quantity of substrate, mixing, organic loading, and volatile fatty acids (Jantsch and Mattiasson, 2004). However, inadequate research and lack of updated technology affects its stability and sustainability.

Researchers produced high CH_4 value biogas using kitchen waste (Das *et al.*, 2017), attributable to its high calorific and nutritive values for microbial growth compared to conventional substrates. Such research approach validates biogas as alternate fuel while successfully addressing the MSW disposal issue. In the process, the uncontrolled CH_4 emission from the indisposed garbage that causes greenhouse gas effect and global warming is satisfactorily controlled.

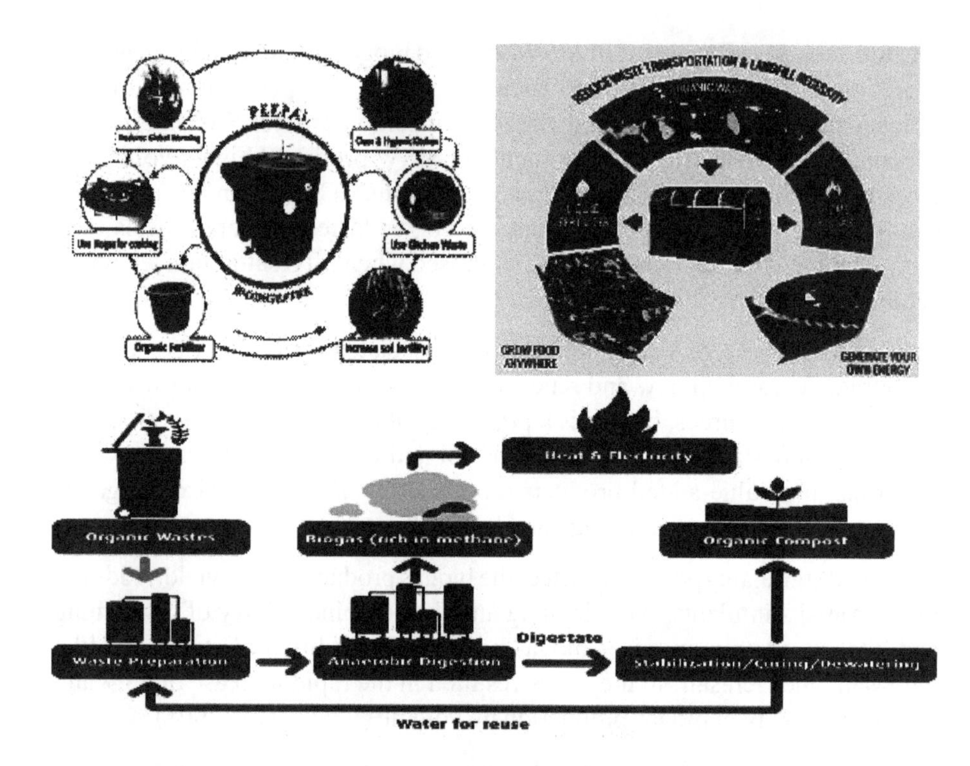

Fig. 8: Overview of usage of biodegradable wastes

Biogas purification technology

Biogas from food wastes for use as alternate fuels contains undesirable impurities. Presence of impurities can also lead to undesired emissions during utilisation. Thus, biogas cleaning technologies represents the most crucial step before utilising biogas as fuel source. Standard techniques are to eliminate these impurities to enhance the quality of biogas are developed. The quality requirements (concentrations of constituent gases) of biogas as a vehicular fuel are (Bauer *et al.*, 2013): CH_4>80–96%, CO_2<2–3%, O_2<0.2–0.5%, H_2S<5mg/m^3, NH_3<3–20mg/m^3, and Siloxanes<5–10mg/m^3. Impurities in raw biogas, like CO_2, H_2S, NH_3, H_2O, siloxanes, etc., need removal before its ready to utilise as fuel in high-grade applications.

Impurity removal from biogas follows a cleaning and evaluation process. Purifying biogas *via* absorption and separation of contaminants significantly improves the purity of CH_4 and, as a consequence, due to the absence of acidic compounds the upgraded gas is much less corrosive to pipelines. Approaches to clean/remove main biogas impurities are classified as (Chen *et al.*, 2015), (temperature-pressure dependent) physical removal wherein absorption occurs at high pressure and low temperature, and (acid-base neutralisation dependent;

caustic solvents) chemical removal wherein impurities are absorbed. Some techniques are detailed below.

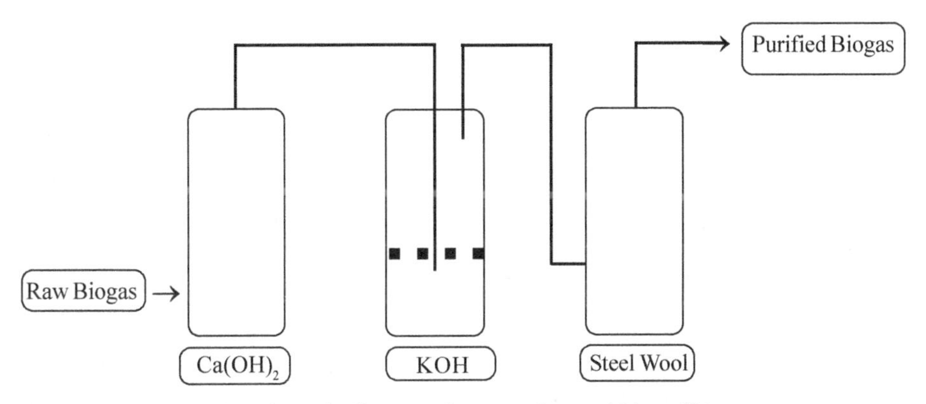

Fig. 9: Schematic diagram of a set-up for scrubbing of biogas

Solid absorption technique

A conventional way to separate gas mixture is to use a solid material for selective absorption. Absorption is more attractive from an energy point of view than gas compression by high pressure or low temperature. The absorbent denotes the solid that takes up guest molecules, while the latter are referred to as the absorbate, the guest molecules (gas) adhering to the surface or pores. The two processes of adsorption (guest molecules absorb on the surface only) and absorption (guest molecules absorb and transfer inside the material) are occurring simultaneously since most applied solid materials are porous allowing gases to easily diffuse. A variety of materials have been measured as useful absorbents. Generally, porous materials are selected either organic or inorganic for absorbents. Organic absorbents includes different plant residues, weeds, crop residues like maize stalks, rice straw, cotton stalks, wheat straw, microbial cultures, etc. which are available naturally can enhance biogas production (Gunaseelan, 1987). Powdered leaves of plants and legumes (like Gulmohar, *Leucacena leucocephala*, *Acacia auriculiformis*, *Dalbergia sisoo* and *Eucalyptus tereticonius*) are found to stimulate biogas production between 18% and 40% (SPOBD, China, 1979; Chowdhury *et al.*, 1994). Shimizu (1992) claimed that higher concentration of bacteria could be retained in the digester by adding metal cations since cations increase the density of the bacteria that are capable of self-aggregation. Wong and Cheung (1995) found that the plant with a higher content of heavy metals (Cr, Cu, Ni, Zn) could enhance CH_4 yield. Solid absorbents are quite promising for the purpose, may be due to its lower heat capacity and higher chemical and thermal stability (Sreekrishnan *et al.*, 2004).

Water-scrubbing

The availability of water at low cost and in sufficient quantity often makes the water scrubber configuration an easy and economical proposition. The amount of water required is related to pressure and temperature operations, and can be estimated as (Muñoz, 2015):

$\frac{Q_{biogas}}{H.P}$ where, Q_{biogas} = raw molar biogas flow rate, kmol/hr; H = Henry's Law constant, m/atm, and P = total atmospheric pressure, atm.

Organic solvent absorption

Organic absorption is very similar to water-scrubbing, and uses organic solvents like dimethyl ethers, polyethylene glycol or solvent mixtures etc. Due to the low freezing point of the organic mixture, the system can be operated at temperatures less than 20°C without any additional heat management. However, pipelines and equipment need to be constructed of stainless steel due to the corrosive nature of the organic mixture.

Amine absorption: This technology uses aqueous solutions of various alkyl amines to absorb the impurities. Using amines is a common approach to absorb sour (acid) gas, such as CO_2, H_2S, etc. Sour gas is selectively absorbed by the aqueous basic solvent in normal atmospheric conditions. The solvent is recycled by simply heating the solution at above 100°C (consuming energy). Low sorption capability and huge energy demand for organic regeneration indicates that the process needs further improvement.

Chemical sorption to remove $\boldsymbol{H_2S}$: H_2S gas requires extensive care for removal due to its specific characteristics of it being colourless, poisonous, pungent smelling and flammable. Combining with water contained in biogas the gas produces sulphuric acid, which is corrosive in nature and causes cracking of metallic surface. This method applies to fixed-bed reactors wherein the raw gas source passes through the sorbent (gas solid contactors).

Chemical absorption process using Fe_2O_3, $Fe(OH)_3$ and ZnO-based filters has become popular since the material is simple, highly efficient, and has fast kinetics. The absorbing material consists of oxide/hydroxide iron compounds that react with the gas converting into simpler form as:

$$2Fe(OH)_3 + 3H_2S \rightarrow Fe_2S_3 + 6H_2O$$

$$Fe(OH)_2 + H_2S \rightarrow FeS + 2H_2O$$

Hydrogen sulphide may be controlled in the biogas mixture using iron oxide in the form of oxidised steel wool or chips of iron as catalyst. The reaction is further explained as:

$Fe_2 O_3 + 3H_2S \rightarrow Fe_2S_3 + 3H_2O$

$Fe_2 S_3 + 3O_2 \rightarrow 2 Fe_2O_3 + 6S$

Operational conditions exhibit gas residence time (ranging from 1–15min) as an important factor with H_2S concentrations of 100ppm and further continues regeneration with a low air flow through the gas stream. However, these processes are still limited for the up-scaling of the desulphurisation. Removal of H_2S from biogas can also be done by passing through NaOH solution.

$2NaOH + H_2S \rightarrow NaHS + 2H_2S$

Hydrogen sulphide (H_2S) reacts with sodium carbonate in the solution forming sodium hydrosulphide as described by the following reaction:

$H_2S + Na_2CO_3 \rightarrow NaHS + NaHCO_3$

$Fe_2 O_3 + 3NaHS + 3NaHCO_3 \rightarrow Fe_2 S_3 + 3H_2O + 3Na_2 CO_3 + 3H_2O$

Chemical sorption to remove CO_2: In order to eliminate CO_2 from raw biogas, water column or caustic scrubbing may be used. Pressurised water circulation absorbs the gas, becoming more effective in biogas purification. Solution of NaOH is prepared with water in a proportion of 30:70 by weight, and the biogas is allowed to pass through the solution. The solution absorbs the leftover CO_2 in the biogas, and form sodium carbonate and water (Grande CA, Rodrigues AE, 2007; Cavenati *et al.*, 2005) as:

$2NaOH + CO_2 \rightarrow Na_2 CO_3 + H_2O$

Moisture removal

It is a usual practice to remove moisture from raw biogas by passing it through a bed of silica gel as it has a good moisture absorbing capacity. Normally, silica gel is available in blue and white colours. The silica gel eliminates moisture contained in raw biogas while passing through. Saw dust is also used as another absorbent to eliminate moisture from biogas.

Conclusion

Production of clean biogas may be achieved by using different purification techniques like chemical scrubbing and the purified gas may be used for different applications like fuel for automobile engines, domestic cooking and lighting purposes, etc. Measures have been taken to produce purified biogas in a large-scale from the available biomass to meet the challenges of fossil fuel deficit and ensure energy security. The economic standard of both rural and urban areas shall be elevated by resorting to such energy drives from waste. Such approach of waste to energy would not only provide an alternative source to meet the energy demand but also protect our environment from various kinds of pollutions.

References

Abhishek J, Murali G, Ravishankar M, Sibichakravarthy M, Sundhirasekar A, 2015. Performance analysis of anaerobic digestion to extract biogas from kitchen waste. *International Journal of Scientific & Engineering Research*, **6(3)**: 703-708.

Anon, 2018. https://en.wikipedia.org/wiki/Food_waste, accessed on 3rd April 2018.

Bauer F, Persson T, Hulterberg C, Tamm D, 2013. Biogas upgrading–technology overview, comparison and perspectives for the future. *Biofuels, Bioproducts and Biorefining*, **7(5)**: 499–511.

Cavenati S, Grande CA, Rodrigues AE, 2005. Upgrade of methane from landfill gas by pressure swing adsorption. *Energy & Fuels*, **19(6)**: 2545–55.

Chen YP, Bashir S, Liu J, 2015. Carbon capture and storage. *In*: JL Lui, S Bashir (Eds) *Advanced Nanomaterials and Their Applications in Renewable Energy*. Elsevier, Amsterdam, Netherlands: 329–66.

Chowdhury SR, Gupta SK, Banerjee SK, 1994. Evaluation of the potentiality of tree leaves for biogas production. *Indian Forester*, **120(8)**: 720–8.

Das AK, Nandi S, Behera AK, 2017. Experimental study of different parameters affecting biogas production from kitchen wastes in floating drum digester and its optimization. *International Journal of Latest Technology in Engineering, Management & Applied Science*, **6(7S)**: 98–103.

Grande CA, Rodrigues AE, 2007. Biogas to fuel by vacuum pressure swing adsorption I. Behavior of equilibrium and kinetic-based adsorbents. *Industrial & Engineering Chemistry Research*, **46(13)**: 4595–4605.

Gunaseelan VN, 1987. *Parthenium* as an additive with cattle manure in biogas production. *Biological Wastes*, **21(3)**: 195–202.

Igoni AH, Abowei MFN, Ayotamuno MJ, Eze CL, 2008. Effect of total solids concentration of municipal solid waste on the biogas produced in an anaerobic continuous digester. *Agricultural Engineering International: The CIGR EJournal*, Manuscript EE 07 010, **X**: 1-11.

Jantsch TG, Mattiasson B, 2004. An automated spectrophotometric system for monitoring buffer capacity in anaerobic digestion processes. *Water Research*, **38(17)**: 3645–50.

Joy G, Das J, Das P, 2014. Generation of bio gas from food waste using an anaerobic reactor under laboratory conditions. *International Journal of Scientific Research and Education*, **2(7)**: 1453–65.

Muñoz R, Meier L, Diaz I, Jeison D, 2015. A review on the state-of-the-art of physical/chemical and biological technologies for biogas upgrading. *Reviews in Environmental Science and Bio/Technology*, **14(4)**: 727–59.

Murphy JD, McKeogh E, Kiely G, 2004. Technical/economic/environmental analysis of biogas utilisation. *Applied Energy*, **77(4)**: 407–427.

Nijaguna BT, 2002. *Biogas Technology*, New Age International (P) Limited, Kolkata, ISBN 81-224-1380-3: 289 pp.

Ostrem K, Themelis NJ, 2004. Greening waste: Anaerobic digestion for treating the organic fraction of municipal solid wastes. *Earth Engineering Center, Columbia University*, 54 pp.

Parkin GF, Owen WF, 1986. Fundamentals of anaerobic digestion of wastewater sludges. *Journal of Environmental Engineering*, **112(5)**: 867–920.

Shimizu C, 1992. Holding anaerobic bacteria in digestion tank. *JP Patent*, 4341398.

Singh S, Kumar S, Jain MC, Kumar D, 2001. Increased biogas production using microbial stimulants. *Bioresource Technology*, **78(3)**: 313–6.

SPOBD, 1979, Biogas Technology and Utilization, *Chengdu Seminar*, Sichuan Provincial Office of Biogas Development, Sichuan, People's Republic of China.

Sreekrishnan TR, Kohli S, Rana V, 2004. Enhancement of biogas production from solid substrates using different techniques – a review. *Bioresource Technology*, **95(1)**: 1–10.

Taleghani G, Kia AS, 2005. Technical–economical analysis of the Saveh biogas power plant. *Renewable Energy*, **30(3)**: 441–6.

Wong MH, Cheung YH, 1995. Gas production and digestion efficiency of sewage sludge containing elevated toxic metals. *Bioresource Technology*, **54(3)**: 261–8.

7

Biomethanation of Rice Straw - Feasibility Assessment

Sneha Tapadia-Maheshwari, Pranav Kshirsagar and Prashant K. Dhakephalkar[1]

Bioenergy group, MACS-Agharkar Research Institute[#]*, GG Agarkar Road Pune–411 004, Maharashtra, India*
[#]*Affiliated to Savitribai Phule Pune University, Ganeshkhind, Pune–411 007, India*
[1]*Corresponding Author: pkdhakephalkar@aripune.org, pkdhakephalkar@gmail.com*

ABSTRACT

Rice straw is one of the most abundant lignocellulosic agricultural wastes in India and can be used as a major resource for biogas production. The structure of rice straw is complex and is made up mainly of cellulose, hemicelluloses and lignin. The lignin physically shields the cellulose and hemicelluloses, the main components which can be converted to energy, and hinders the access of microbial enzymes to these carbohydrate moieties. To overcome this making the straw a potential resource of biogas production, effective pretreatment is necessary that can efficiently hydrolyse the straw and make it accessible to cellulolytic enzymes and methanogens. Various pretreatment methods which are employed currently in the market are discussed here. For a process to be economically feasible it is necessary that the pretreatment be efficient as well as cost effective. To gain some insights into this aspect, the techno-commercial feasibility of Biomethanation of rice straw has been discussed.

Keywords**:** Rice straw, biogas, pretreatment, feasibility

Abbreviations used

GHG	: green house gas
HMF	: hydroxyl methyl furfural
sCOD	: soluble chemical oxygen demand
VS	: Volatile Solids
Mt	: million tone
Gg	: gigagram
GJ	: gigajoules
kWh	: kiloWatt hour
MPa	: megapascal
CNG	: Compressed natural gas
MMT	: Million Metric Tonne
MTOE	: Million tonne of oil equivalent
EJ	: Exajoules
EIA	: Environmental Impact Assessment

Introduction

The ever-increasing population has led to quantum increase in energy demand leading to energy crisis, posing food security and climate change issues. Energy has always been indispensable to anthropogenic activities at domestic, agriculture, industry and transport fronts. Economic growth and accompanying structural changes influence world energy consumption. With development and improvement in living standards, the energy demand has grown rapidly. Increased demand for appliances and transportation, and growing capacity to produce goods and services for both local and global markets lead to higher energy demand. At present, world energy consumption is annually growing by 2% (Mustafa *et al*., 2016). Total world consumption of energy is expected to expand from 549 quadrillion British thermal units (Btu) in 2012 to 629 quadrillion Btu in 2020 and to 815 quadrillion Btu in 2040, a 48% increase between 2012 and 2040 (EIA, 2016).

In 2014, the world energy consumption was 12928.4MTOE (541.3EJ), of which oil, natural gas coal, nuclear energy, hydroelectricity and other renewable energy consumption were 32.6, 23.7, 30.0, 4.4, 6.8 and 2.4%, respectively (Statistical Review of World Energy, June 2015). Fossil fuel consumption accounted for 86.3% whereas renewable energy consumption accounted for 9.2% of global energy consumption. Coal, oil and natural gas are the three major fossil fuels for our energy needs. But these fuels are limited in supply, and their burning causes green house gas emissions (EIA, 2016). As anthropogenic carbon-di-oxide (CO_2) emissions result primarily from the combustion of fossil fuels, energy consumption is at the center of the climate change debate. Energy-related CO_2

emissions would increase from 32.3 billion MMT in 2012 to 35.6 billion MMT by 2020 and to 43.2 billion MMT by 2040 globally (International Energy Outlook, 2016). Concerns over depleting fossil fuels and increasing energy demands have led to the need to look for renewable sources of energy, produced from various sources including solar, wind, biomass, hydropower, and geothermal. The advantage of deriving energy from these sources is that there is an unlimited supply of sun, wind, and water and also they are minimally polluting. The European Commission set the goal that 20% of energy consumed should come from renewable energy sources by 2020 (Renewable energy – Targets by 2020, European Commission: http://ec.europa.eu/energy/en/topics/renewable-energy).

First generation and second generation biofuels

Based on the feedstock source, biofuels are categorised into first, second and third generation. First generation biofuels are produced from food crops like grains, sugar-beet and oil seeds, the second generation biofuels make use of non-food crops like lignocellulosic biomass from cereal, bagasse, forest residue and purposefully grown energy crops, whereas third generation biofuels exploit oleaginous microbes (algae and fungi). First generation biofuels are more energy efficient though their use has several drawbacks. They impact the food prices competing with the existing food, land and water resources (Bioenergy IEA, 2008). The main advantage of second generation biofuels is that it limits 'food versus fuel' competition associated with the first generation biofuels (Nigam *et al*., 2016). Research on the use of residues and wastes for biofuel generation, therefore, has attained much importance due to its ease of availability globally. Lignocellulosic biomass is considered to be the most abundant renewable biomass available for exploitation.

Need for crop residue based renewable energy

Agriculture forms the backbone of Indian economy. Around 60% of the country's land is utilised for cultivation of a wide range of crops (Hiloidhari *et al*., 2014). According to The Ministry of New and Renewable Energy, Govt. of India, about 500Mt of crop residues are generated every year (Gupta *et al*., 2012). Among the various crops, cereals (rice, wheat, maize, millets) account for maximum residue production (352MMT), followed by fibres (66MMT), oilseeds

(29MMT), pulses (13MMT) and sugarcane (12MMT). Cereal crops contribute majorly (70%) to the total crop residues and rice production generates maximum (34% of total crop) residues (Fig. 1).

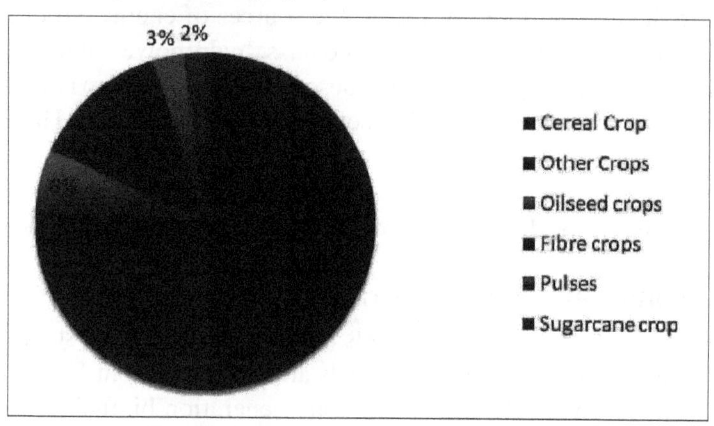

Fig. 1: Contribution of different crops in residue generation in India

These crop residues are used for various purposes such as animal feed, fuel, thatching, packaging and composting. The remaining residues are left unused or burned. Around 91-141MMT residues is surplus (Gupta *et al.*, 2012). The bioenergy potential of the surplus residues is estimated to be 4.15EJ, equivalent to 17% of India's total primary energy consumption (Hiloidhari *et al.*, 2014). Around 82MMT surplus residues are generated from cereal crops, with rice straw as the largest (44MMT) contributor (Gupta *et al.*, 2012; Fig. 2).

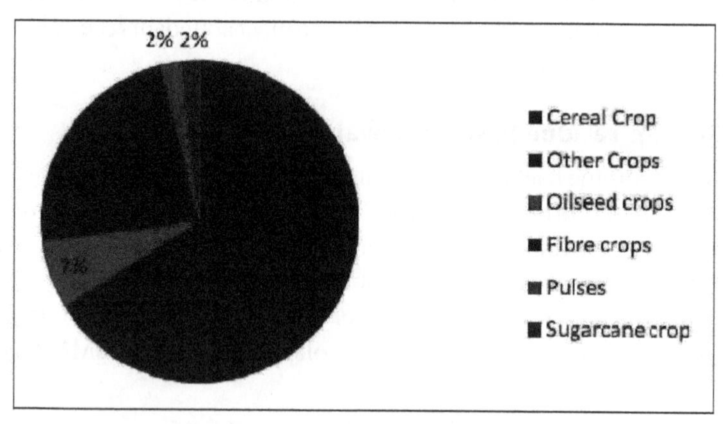

Fig. 2: The share of unutilised residues generated by different crops in India

Rice straw generation in India

Rice is an important staple food for nearly half of the world population. Asia contributes 90.1% to global rice production (Trivedi *et al.*, 2017). China (30%)

and India (21%) are the leading countries contributing to about half of world's total rice production (Gadde *et al.*, 2009). In India, around 43.95 million hectares of land in utilised for rice cultivation and 106.54MMT of Rice is produced every year. Rice straw is produced at a ratio of around 1:1.5 (rice:rice straw) which is approximately 160MMT per year (Trivedi *et al.*, 2017) and can serve as an important resource for biorefining (Abraham *et al.*, 2016). According to Gadde *et al.* (2009) 23% of the rice straw is surplus and amounts to about 22289Gg/year. Haryana and Punjab are the leading rice straw producing states accounting to about 48% of total straw production in India.

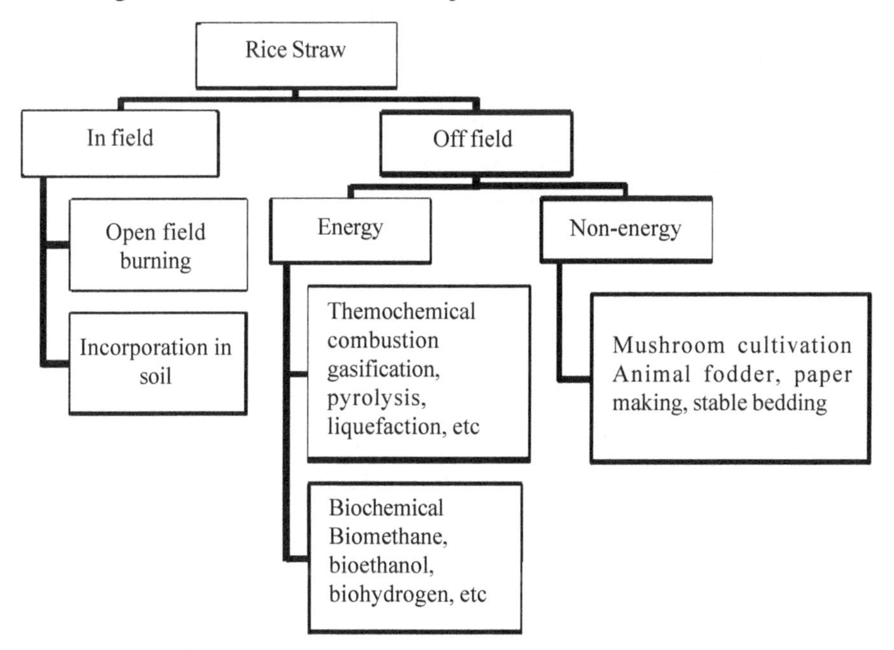

Fig. 3: An overview of potential applications of rice straw

Rice straw management

Rice straw is utilised for various purposes in India as mentioned in Fig. 3. The applications can be categorised broadly into in-field and out-of-field uses.

In-field use

Incorporation in soil: Incorporation of rice straw in soil helps return the plant nutrients and has short term positive effects on crop yield and improvement of physical, chemical and biological properties of soil (Dobermann *et al.*, 2002). On the other hand, incorporation of excess of straw in soil can lead to temporary sequestration of soil and applied nitrogen and increase in CH_4 emissions and in turn may contribute to green house gas emissions (Sander *et al.*, 2014).

Open field burning: Open field burning is an anthropogenic activity which is done to free the soil for next crop, remove weeds and residues. In India, out of the 22,289Gg surplus rice straw generated, 13,915Gg is subjected to open field burning (Gadde *et al.*, 2009). This process leads to loss of useful nutrients present in the crop residues and also release toxic pollutants such as CO_2, CO, NO_x, SO_2, polycyclic hydrocarbons, furans, dioxins, fine particulate matter, etc. (Romasanta *et al.*, 2017). Trivedi *et al.* (2017) reported that burning one ton rice straw releases 3kg particulate matter, 60kg CO, 1460kg CO_2, 199kg ash and 2kg SO_2. Biogenic CO_2 emissions from burning of rice straw are considered to be zero because the bulk of CO_2 released after combustion is utilised for photosynthesis (Soam *et al.*, 2017). But other toxic pollutants released from burning of residues lead to various health hazards. This process also leads to the loss of essential soil nutrients (upto 80% N, 25% P, 21% K and 4-60% S) and beneficial soil insects and microbes (Satlewal *et al.*, 2017). Unburnt rice straw can be used for biogas production yielding energy equivalent to 8.0GJ/tonne rice straw, and in turn reduce net global warming potential by 2750kg CO_2e emissions/tonne (Trivedi *et al.*, 2017).

Out-of-field options

Mushroom cultivation: The ability of mushrooms to colonise lignocellulosic wastes (like wheat and rice straw) and produce edible fruiting bodies, makes it an efficient way of handling such residues in an eco-friendly manner. Zhang *et al.* (2002) produced mushroom at 5–10% yield (50–100kg mushroom per tonne dried substrate) from rice straw.

Animal fodder: In Asia, rice straw is one of the principal fibrous residues fed to over 90% of the ruminants (Devendra *et al.*, 2002). Rice straw is difficult to digest because of its lignin and silica contents. Although numerous physical, chemical and biological pretreatments are available to improve the digestibility of rice straw, the practical use of such pretreatments is usually restricted due to the high costs and negative environmental consequences (Kumar *et al.*, 2014). Hence, there is an urgent need to discover value-added conversion technologies to harness the energy potential of rice straw.

Energy production: Rice straw is converted to energy by either thermochemical or biochemical processes. Thermochemical processes include direct combustion, gasification, pyrolysis, liquefaction, etc. to generate heat and energy. Biochemical processes include conversion of straw into methane, hydrogen or ethanol (Lim *et al.*, 2102). Biomethanation generated 777kWh electricity, and bioethanol fermentation produced 544.25kWh electricity per tonne rice straw (Trivedi *et al.*, 2017). Hence, biomethanation through anaerobic digestion is more efficient in extracting energy from straw.

Calorific value of rice straw: It is defined as the quantity of heat generated per unit sample on its combustion at constant pressure and at standard conditions (Satlewal *et al*., 2017). This is an important index to indicate the useful energy content of straw as a fuel. Patel *et al.* (year unknown) described calorific values of 20 different rice straw varieties grown in Chhattisgarh ranging from 3510–6810kcal/kg (14.7–28.5MJ/kg) with mean value of 5531±278kcal/kg (23.1MJ/kg). A heating value (HHV) of 16.2MJ/kg of rice straw from North India region is reported (Raj *et al.*, 2015). Jenkins *et al.* (1988) reported HHV of rice straw as 15.09MJ/kg. Theoretically, the energy potential of rice straw in India (with LHV of 14MJ/kg) would be 644.2PJ per annum if all the surplus rice straw including the 50% of the rice straw currently used for numerous other applications is diverted for energy production (Gadde *et al*., 2009).

Structure of lignocellulose (rice straw)

Lignocellulosic biomass is a complex structure having 25-55% cellulose, 8-25% hemicelluloses and 10-35% lignin along with small amounts of pectin, protein, extractives and ash (Bajpai, 2016; Saha, 2003). An artistic interpretation of the structural details of lignocellulose is provided in Fig. 4.

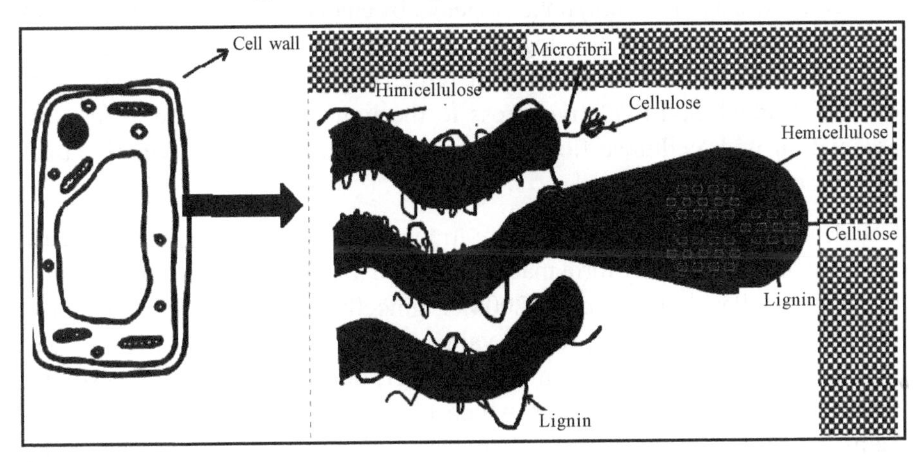

Fig. 4: Structural details of lignocellulosic biomass

Cellulose: The most abundant component of lignocellulose, it is a linear polymer of D-glucose subunits, linked by β-1,4 glycosidic bonds. Cellobiose, a disaccharide, is the smallest repeating unit of cellulose which can be converted to glucose. The cellulose chains are grouped together to form microfibrils, which in turn are assembled to form cellulose fibres that are stabilised by hydrogen and van der Waals bonds. The inter-chain hydrogen bonds determine the ordered (crystalline) or disordered (amorphous) structure of the cellulose (Agbor *et al.*, 2011). Cellulose is more susceptible to enzymatic degradation in its amorphous form (Brodeur *et al.*, 2011). In order to breakdown cellulose, three groups of enzymes are necessary, endoglucanases that randomly cleave the internal bonds of the cellulose chains, cellobiohydrolases that attack at the chain ends releasing cellobiose units, and β-Glucosidases that act on cellobiose to release glucose monomers (Kumar *et al.*, 2008).

Hemicelluloses: These second most abundant component of lignocellulose are branched, hetero-genous polymers of pentoses and hexoses (glucose, arabinose, galactose, mannose and xylose and acetylated sugars). These are linked together by β-(1,4)-glycosidic bonds, and occasionally by β-(1,3)-glycosidic bonds. Xylan, the most abundant sugar in hemicelluloses, is broken down by different enzymes such as endo-1,4-β-xylanase, β-xylosidase, β-glucuronidase, β-L-arabinofuranosidase and acetyl-xylan (Kumar *et al.*, 2008). Hemicelluloses 'coat' the cellulose fibers similar to the lignin. It is stated that, about 50% hemicelluloses needs to be removed to improve cellulose digestibility (Agbor *et al.*, 2011). Hemicelluloses removal increases the mean pore size of the substrate thereby increasing the hydrolysability of cellulose (Alvira *et al.*, 2010).

Lignin: It is an amorphous heteropolymer of phenyl propane units (*p*-coumaryl, coniferyl and sinapyl alcohol) linked together by ether bonds. It provides structural support to the plant by binding together different components of lignocelluloses together. However, in the process it renders impermeability, resistance to microbial attack and oxidative stress to the lignocellulose. Due to its close association with cellulose fibers, it serves as a barrier for enzymatic attack (Kumar *et al.*, 2009).

Cellulose, hemicelluloses and lignin are arranged into structures called microfibrils which in turn collectively form macrofibrils and together give structural stability to the plant cell wall (Rubin *et al.*, 2008). Lignicelluloses with high sugar content can be suitable substrate to produce value-added products like ethanol, biogas, lignin, organic acids and enzymes. However, due to the compact crystalline structure and presence of lignin which physically shields the cellulose and hemicelluloses part of biomass, these materials are recalcitrant to anaerobic digestion.

Thus, a suitable pretreatment method is necessary to reduce the recalcitrance of such a biomass by altering the cellulose-hemicellulose-lignin interactions, reducing crystallinity of cellulose, removing the lignin and thereby improving the digestibility of cellulose. Such alterations would improve the yield by improving enzymes' accessibility in subsequent anaerobic digestion (Satlewal *et al.*, 2017). Attaining high sugar levels through pretreatment is important for commercial competitiveness of lignocellulose biomass (Chandel *et al.* 2012).

Pretreatment of rice straw

Lots of studies have been conducted on various pretreatments measures for lignocellulosic biomass either individually or in combination in last few decades. Although there have been many successful attempts reported, it is evident that the 'one-size-fits-all' hypothesis to pretreat the various lignocelluloses seems oblivious. Lignocellulose pretreatment is broadly classified into mechanical, physical, chemical and biological methods (Fig. 5).

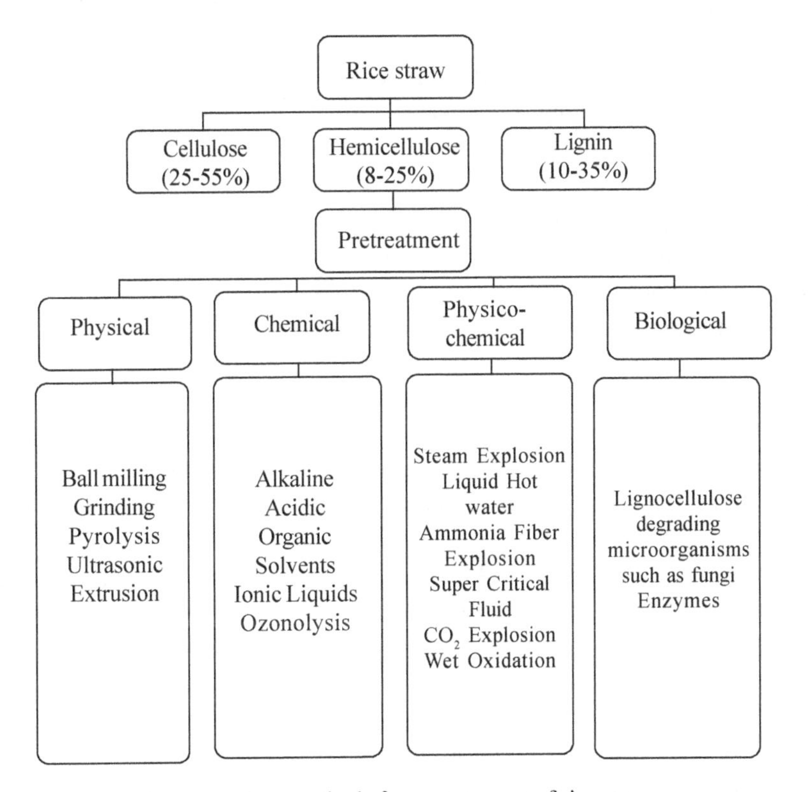

Fig. 5: Various methods for pretreatment of rice straw

Physical pretreatment

Various physical pretreatments such as milling, grinding, pyrolysis, extrusion, microwave, etc. allegedly could facilitate biological degradation thereby enhancing biogas production. Rice straw is reduced to smaller particles using any or a combination of these physical treatment methods. The objective of pretreatment is size reduction, which increases the surface area of the straw and reduces the crytallinity and degree of depolymerisation. These structural changes increase the accessibility of enzymes to the substrate, thus enhancing the biomethanation process. Methane yield increase with decrease in particle size (Krishania *et al.*, 2013).

Sharma *et al.* (1988) studied the effect of particle size (ranging from 0.088–30mm) of various agricultural residues on biogas production and reported maximum biogas production from 0.088 and 0.40 mm particles size. The study showed that the increase in biogasification was negligible with particles size between 0.088 and 0.4 mm which indicates that grinding below 0.4mm neither economical nor desired. An extruder is continuous thermomechanical treatment equipment for size reduction. During extrusion, the biomass undergoes shearing, heating and mixing. These combined forces disrupt the lignocellulosic structure and cause defibrillation of the fibres thereby making the cellulose and hemicelluloses accessible to enzymatic attack. Chen *et al.* (2014) reported 72.2% enhanced methanation of extruded rice straw compared to the untreated one.

Chemical pretreatment

Acid Pretreatment: Concentrated acids like H_2SO_4 and HCl are used to pretreat lignocellulosic biomass due to their high potential to hydrolyse the biomass. Yet, because of various drawbacks such as their toxicity, hazardous nature and the need for corrosion-resistant reactors due to corrosiveness, they are seldom used (Sun and Cheng, 2002). Contrarily, dilute acid hydrolysis is one of the most promising and widely studied pretreatment as it is relatively inexpensive, convenient and effective against a wide variety of substrates (Wei *et al.*, 2012). Acids attack the hemicellulose component releasing sugars such as xylose, glucose, arabinose, etc. and open up the remaining structure for enzymatic attack (Jensen *et al.*, 2010). Also, when the biomass is subjected to high temperature acid pretreatment, lignin reportedly gets chemically modified and recondenses on the biomass as an altered lignin polymer that is less obstinate to digestion. Deacetylation of the xylan backbone also enhances biomass digestibility (Zhu *et al.*, 2008). The sugars released on acid treatment are used to synthesise industrially important products, such as xylitol, ethanol, biogas, citric acid, lactic acid, 2,3-buanediol, etc. One drawback of the use of acids for pretreatment is the formation of inhibitors like furfural, HMF, weak acids, phenolics, etc. (Chandel *et al.*, 2012). However, methanogens reportedly can tolerate furfural and hydroxymethyl furfural to a large extent, within acclimatisation period (Hendriks *et al.*, 2009).

Various acids used for biomass pretreatment are hydrochloric, phosphoric, nitric, and sulphuric acids, etc., amongst which HCl and H_2SO_4 are most widely used. However, acid concentration, biomass quantity, pretreatment time and temperature are factors that affect the concentration of the released sugars. Thus, optimising the process conditions for efficient hydrolysis of biomass is important (Chandel *et al.*, 2012).

Hosseini *et al.* (2012) optimised dilute H_2SO_4 pretreatment using RSM. A soluble COD (sCOD) of 18.2mg/L was attained at optimised pretreatment conditions of 10% H_2SO_4 at 200-°C for 170 min. Chen *et al.* (2011) reported an integrated process of dilute acid and steam explosion to pretreat rice straw. Under optimised conditions of 2% H_2SO_4 at 165°C for 2min followed by steam explosion at 180°C for 20 min, about 75.9% of xylan and 77.1% of glucan were converted to xylose and glucose respectively. Laopaiboon *et al.* (2010) found out that HCl was more effective than H_2SO_4 in breaking the hemicelluloses and releasing the sugars.

Alkali Pretreatment: Alkaline pretreatment is also a major pretreatment method to pretreat rice straw. Various alkalis such as sodium hydroxide, potassium hydroxide, lime, aqueous ammonia are used for the purpose. Primarily, alkalis delignify the rice straw by breaking the ester bonds that crosslink the lignin and hemicelluloses (Hosseini *et al.*, 2012). They also cause swelling of particles due to delignification of intermedullar ester bonds. This causes increase in surface area and gives better accessibility to the enzymes for hydrolysis. Alkali pretreatment is more effective with agricultural residues biomass containing low lignin (Ong *et al.*, 2010).

Rice straw pretreated with 6% NaOH at ambient temperature for 3 weeks showed 44.4% higher biogas yield compared to untreated one. Pretreatment enhanced cellulose, hemicelluloses and lignin contents reduction by 16.4, 36.8, and 28.4% (He *et al.*, 2008). Hosseini *et al.* (2012) extracted 26.232mg/L sCOD under optimised pretreatment conditions of 30g/L KOH, 200°C, 151.23min. Lime pretreatment at optimised conditions of 9.81% $Ca(OH)_2$ at ambient temperature for 5.89 days effectively bettered rice straw digestibility, enhancing the methane yield by 67.55% (225.3 ml/g VS) as compared to the control (Song *et al.*, 2012). Ko *et al.* (2009) attained 71.1% enhanced enzymatic digestibility of lignin selectively from rice straw by treating it with 21% ammonia at 69°C for 10h.

NaOH pretreatment is an effective alkaline pretreatment for enhanced digestibility of rice straw but it has major drawback of high cost. Therefore, there is the need for recovery which is again a costly proposition, which make it unsuitable for large scale applications. Its alkali counterparts such as lime and aqueous ammonia, although are little less effective compared to NaOH, are preferred due to their low costs.

Physicochemical pretreatment

It involves combinatorial pretreatment methods such as steam explosion, hot water, ammonia fibre explosion, supercritical fluid, CO_2 explosion, wet oxidation, organic solvent etc. These methods employ the use of conditions and compounds that alter the physical and chemical structure of lignocellulosic biomass (Agbor *et al.*, 2011).

Steam explosion: It is one of the most common physicochemical pretreatment methods. Here, the biomass is treated with high pressure saturated steam (0.7-4.8MPa) at high temperature (160-240°C) and a short duration pressure. At such high temperature, the hemicellulose is hydrolysed and lignin transformed. The cellulose fraction becomes more accessible for enzymatic attack and this in turn increases digestibility of the biomass. The effectiveness of steam explosion is enhanced by adding H_2SO_4, CO_2 or SO_2 to catalyse. Chen *et al.* (2013) reported 73% of total saccharification yield from rice straw in a combined acid and steam explosion pretreatment under optimised pretreatment conditions (1.3% H_2SO_4, 100-110°C, 1.7rpm, followed by steam explosion at 180°C, 2min). This process has advantage of limited use of chemicals and low input energy requirements. However, it also has various drawbacks like incomplete destruction of lignin-carbohydrate matrix, generation of inhibitors at high temperatures and precipitation of solubilised lignin components thereby compromising on the digestibility of the biomass (Agbor *et al.*, 2011).

Wet oxidation: It is a process wherein air or oxygen along with water or hydrogen peroxide pretreat the biomass at high temperatures (above 121°C, 30min). When temperature is raised above 170°C, water behaves like an acid and catalyses hydrolytic reactions in the biomass. Lignin undergoes oxidation, hemicellulose breaks to monomers while cellulose majorly remains unaffected (Kumar and Sharma, 2017). Banerjee *et al.* (2009) pretreated rice hulls by wet oxidation in statistically optimised process conditions of 85°C, 0.5MPa, and 15 min. and achieved 67% cellulose retention, 70% hemicelluloses solubilisation and 89% lignin removal in the solid fraction. Although this pretreatment is promising, it is less likely to be used at industrial scale because of the high costs of H_2O_2 and combustible nature of O_2 (Kumar and Sharma, 2017).

Hot water treatment: This method is similar to steam explosion but it uses water at elevated temperature instead of steam, without extraneous pressure. Hot water breaks the hemiacetyl bonds and releases acids, which enhances the breakage of ether bonds in the biomass. Although inhibitors are formed due to degradation of monosaccharides at high temperature, it is overcome by maintaining an acidic pH between 4 and 7 during the pretreatment (Agbor *et al.*, 2011). Trivedi *et al.* (2016) found that the methane and biogas yield were twice more after liquid hot water pretreatment of rice straw compared to the control. One major disadvantage of this method is the high energy demand in downstream processing due to large quantity of water required.

Biological pretreatment

Significant drawback of physical, chemical and physicochemical pretreatment is requirement of high energy input, use of expensive chemicals and instruments with significant environmental impact. As against this, biological pretreatment requires less input energy, is cost effective and also environment friendly. These methods employ use of micro-organisms, mainly fungi, actinobacteria, bacteria and enzyme preparations which act mostly on the recalcitrant lignin. However, the rate of biological pretreatment is too slow for industrial purposes. It also requires careful growth conditions and large amount of space required for pretreatment. Also some carbohydrate fraction of the biomass is consumed by the microbes for its growth. Due to these disadvantages, biological pretreatment is usually used in combination with some other pretreatment (Agbor *et al.*, 2011). Fungi such as white-rot and brown rot fungi, capable of producing lignocellulolytic enzymes, are usually used in biological pretreatment of rice straw. White rot fungi are reported to be most useful in the pretreatment of plant biomass. They attack lignin and cellulose by producing various enzymes such as lignin peroxidases, polyphenol oxidases, manganese-dependent peroxidases, laccases, etc. which attack and degrade the lignin. Examples of fungi used for biological pretreatment are *Phanerochaete chrysosporium, Trametes versicolor, Tramates hirsuta, Dichmitus squalens,* etc. Rice straw treated with white-rot fungus *Phanerochaete chrysosporium* and brown-rot fungus *Polyporus ostreiformis*, increased methane production by 46.19 and 31.94%, respectively (Ghosh and Bhattacharya, 1999). A summary of various pretreatment methods along with their mode of action are mentioned in Table 1.

Anaerobic digestion – the process

Biomethanation, also known as anaerobic digestion is a process of conversion of organic matter by a consortium of micro-organisms in the absence of oxygen, into energy (methane). It naturally occurs in anoxic environments such as waterlogged soil, sediments, mammalian gut, etc. It can also be produced from a wide range of sources such as municipal wastes, agricultural wastes, industrial wastes, food wastes, etc. (Krishania *et al.*, 2013). The gas can be used for heat, electricity production, or can be upgraded to vehicular fuel. Furthermore, the process generates nutrient-rich digestate which can be used as organic fertiliser on agricultural land.

This way, nutrients originating from plants are recycled and returned back to the nature. Biogas is a mixture of 55-65% CH_4, 30-45% CO_2, 1-5% H_2, 0.5-2.0% N_2, 0.1-0.5% H_2S, 0-0.3% CO and traces of water vapour (Kaur and Phutela, 2016). The anaerobic digestion is a multi-step complex process which can be divided into four stages: Hydrolysis, Acidogenesis, Acetogenesis, and

Table 1: Different reported pretreatment technologies

Pretreatment	Process condition(s)	Proposed mode of action	Saccharification/ Yield	Reference(s)
Dilute acid	0.35 wt % H_2SO_4/162 C/10min	Hydrolysis of hemicellulose fraction	77% total sugar recovery	Kapoor *et al.*, 2017
Organic acid	0.75 mol/L Acetic acid-propionic acid/ 2h/ 121°C	Can also be utilised by methanogens as substrate for biogas production	Untreated methane yield: 0.25L/g VS Pretreated methane yield: 0.28L/g VS Methane production increased by 35.84% after pretreatment 34.19% lignin was removed	Zhao *et al.*, 2010
Alkali	6% NaOH/3 weeks/ ambient temperature	Swelling of biomass, increase in surface area, lignin removal	Pretreated biogas yield: 520L/g VS Untreated biogas yield: 360L/g VScellulose, hemicellulose, and lignin reduced by 16.4, 36.8 and 28.4%	He *et al.*, 2008
Extrusion + Ca $(OH)_2$	8% $Ca(OH)_2$/room temperature/72 h	Swelling of biomass, decomposition of polysaccharides, increase in enzyme accessibility	Biogas: 564.7ml/g VS Methane: 330.9ml/g VS	Gu *et al.*, 2015
Hydrothermal + NaOH	1) Hydrothermal: 200°C/10min/ 1.55 MPa 2) NaOH: 3% NaOH/37°C/120h	Water under high pressure and temperature can penetrate into the biomass, hydrate cellulose, and removes hemicellulose and part of lignin. NaOH added to neutralise the acidity produced within digester	1) Untreated Biogas yield: 140L/kg VS 59.8 L/kg VS 2) Hydrothermal + NaOH Biogas: 315.9L/kg VS Methane: 132.7L/kg VS	Chandra *et al.*, 2012

Contd.

Pretreatment	Process condition(s)	Proposed mode of action	Saccharification/ Yield	Reference(s)
H_2O_2	3%/ambient temperature (25±2)/ 7 days	Strong oxidisability of H_2O_2	Biogas yield 319.7ml/g VS	Song *et al.*, 2012
Grinding	0.4mm	Increase in surface area due to size reduction	Biogas yield of 487 l/ kg TS (methane content 60%) (VS content of rice straw- 79.4%)	Sharma *et al.*, 1988
Alkali+ Microwave	4% NaOH for 24h/ 30 min microvave	Delignification, Change of cellulose structure to a denser and thermodynamically more stable form	65% reduction in lignin; Biogas yield: 297L/kg dry rice straw (54.7% enhancement over control)	Kaur and Phutela, 2016
Fungal	*Pleurotus ostreatus* (10:1 w/w Rice straw to biomass)/ 60 days	Release of various lignocellulolytic enzymes	41% Klason lignin degradation; 32.7% net sugar yield	Taniguchi *et al.*, 2005

Methanogenesis (Fig. 6.). Each of these steps is carried out by different specialised microbial consortia. These four microbial steps are described below.

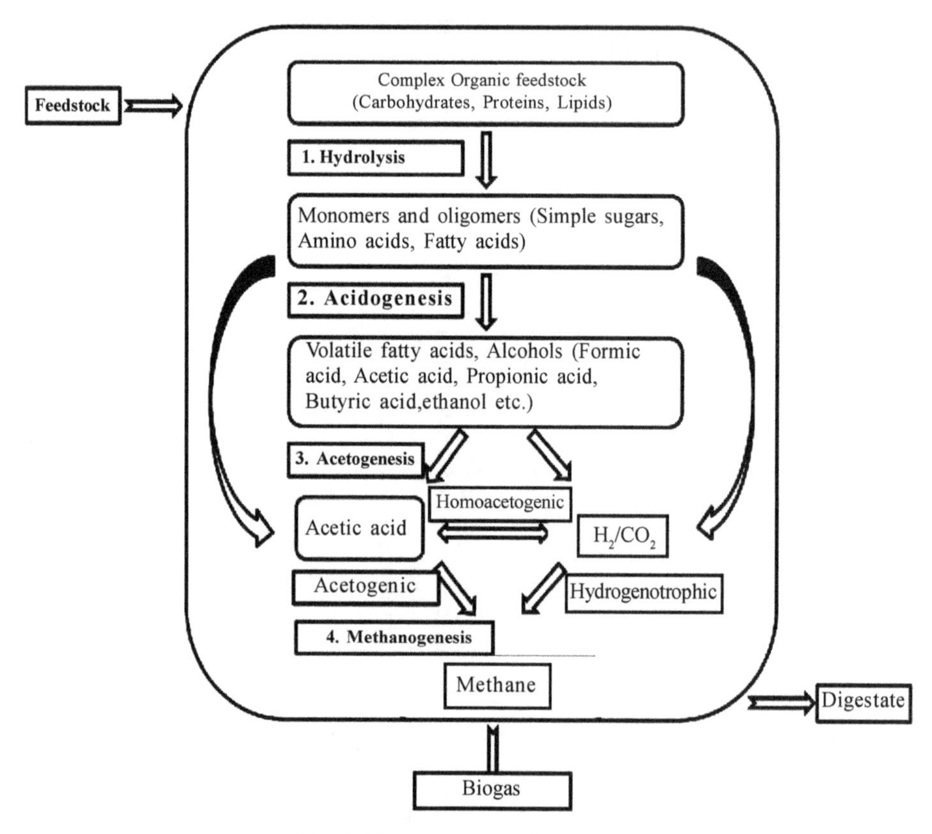

Fig. 6: Phases in anaerobic digestion

Hydrolysis

In this stage, the microbial consortium breaks down the complex organic matters like proteins, carbohydrates and lipids into simpler soluble molecules like amino acids, sugars and fatty acids. The process is carried out by hydrolytic bacteria which excrete extracellular enzymes, such as, cellulase, cellobiase, xylanase, amylase, lipase, protease, etc. Hydrolysis is regarded as the rate-limiting step of the entire process of anaerobic digestion as the degradation of lignocellulose and lignin is usually slow (Chandra *et al.*, 2012). Majority of hydrolytic bacteria such as *Bacteroides*, *Clostridium* sp., *Bifidobacterium* sp., etc, are strict anaerobes.

Acidogenesis

During this process, monomers produced in the hydrolysis stage are further converted to various organic acids, such as, formic acid, acetic acid, propionic acid, butyric acid etc., along with alcohols, hydrogen and carbon-di-oxide by fermentative bacteria.

Acetogenesis

Some simple degradation products of acidogenesis step can directly utilised by methanogens. However, fatty acids longer than two carbon atoms, alcohols longer than one carbon atom and branched chain fatty acids are broken down further to acetic acid, H_2 and CO_2 before methanation. During this phase, homoacetogenic microbes, *viz.*, *Acetobacterium woodii* and *Clostridium aceticum*, constantly reduce H_2/CO_2 to acetic acid (Chandra *et al.*, 2012). Accumulation of hydrogen inhibits acetogenic bacteria. Under high hydrogen partial pressure, acetate formation is reduced and the substrate is converted to propionic acid, butyric acid and ethanol rather than methane, and hence, maintenance of low partial pressure of H_2 is necessary for acetogenic bacteria to remain active (Weiland, 2010).

Methanogenesis

Methanogenesis is the final step in anaerobic digestion where methane is formed from acetate and/or from H_2/CO_2, formate, alcohols, methylated C_1 compounds. The terminal electron acceptor in methanogenesis is carbon-di-oxide. Methane formation can take place in one of the three pathways (Angelidaki *et al.* 2011) as below:

a. Acetoclastic methanogenesis, where acetate is converted to methane and carbon-di-oxide:

$$CH_3\,COO^- + H^+ \rightarrow CH_4 + CO_2$$

b. Hydrogenotrophic methanogenesis, where carbon-di-oxide is reduced to methane:

$$4H_2 + CO_2 \rightarrow CH_4 + 2H_2O$$

c. Methylotrophic methanogenesis, where methylated C_1 compounds (methanol, methylamines, dimethylsulphide, etc.) are converted to methane.

$$4CH_3\,OH \rightarrow 3CH_4 + CO_2 + 2H_2O$$

Acetoclastic methanogens (*e.g.*, *Methanosarcina*) form about 70% methane. Hydrogenotrophic methanogenesis is performed by members from orders Methanobacteriales, Methanomicrobiales, Methanococcales, Methanocellales, Methanopyrales. Methylotrophs are from Methanosarcinales, and *Methanosphaera* sp. from order Methanobacteriales (Angelidaki *et al.*, 2011).

As discussed, the entire process of biomethanation is carried out with the help of microbial consortia, the growth of which depends on the various parameters such as pH, temperature, C/N ratio, organic loading rate, substrate/inoculum ratio, hydraulic retention time, race element supplementation, etc. In order to harness the full anaerobic digestion potential, it is necessary to maintain these parameters in their optimum range (Krishania *et al.*, 2013).

Economic feasibility of the use of rice straw for biogas generation

Field applications of biomethanation of rice straw need techno-economic feasibility assessment. Such assessment involves information on the capital expenditure, operational expenditure (raw material cost, pretreatment cost, process cost, utility cost, and post biomethanation treatment cost), pricing of methane gas (price of the electricity generated using biomethane, or price of the compressed biogas, and price of the digestate sold as organic fertiliser).

These data would partly help to decide the feasibility of a biogas plant project. Being dependent on several factors like pH, temperature, C/N ratio, organic loading rate, substrate/inoculum ratio, hydraulic retention time, race element supplementation, etc., biogasification process yields are highly variable as reported in literature. The two types of biomethanation processes, no thermochemical pretreatment (pulverised rice straw only) and thermochemical pretreatment (using alkali at high temperature), which are widely popular currently (Trivedi *et al.*, 2017) are considered for cost-benefit analyses. For the former one, the cost of rice straw, pulverisation, substrate feeding, etc. have been considered (Table 2).

Table 2: Input cost for non-thermochemically pretreated pulverised rice straw biomethanation

Input	Cost (lakh/100 tonne rice straw)
Paddy straw	1.5
Pretreatment (pulverisation) cost	0.705
Substrate feeding unit	0.173
H_2S scrubbing unit	0.103
Total input cost	**2.48**

Table 3: Input cost for biomethanation of rice straw with thermochemical pretreatment

Input	Cost (lakh/100 tonne rice straw)
Rice straw	1.5
NaOH	0.24
Water (daily 20% top-up)	0.52
Cattle-dung	0.0315
Shredding	0.029
Heating	10.076
Pumping + cooling tower	0.208
Agitation	0.11
Neutralisation	1.337
Solid-liquid separator	0.021
Total input cost	**14.07**

Similarly, various input parameters for thermochemical pretreatment biomethanation, heads such as cost of rice straw, NaOH, heating, neutralisation, etc. have been considered for cost calculations (Table 3).

The biogas yields reported in literature is between 250-450m^3/tonne VS of rice straw, on an average. This value has been considered for cost calculations for both. Considering the biogas equivalent of power and bioCNG and their market rate (Table 4), the output cost of the biogas and the digestate have also been calculated. From Table 4 it is evident that the revenue generated from converting biogas to bioCNG is much more than the same amount of biogas utilised directly for power generation. So, upgradation and conversion of biogas to bioCNG appears to be a sustainable and profitable option.

Table 4: Biogas equivalency with electricity and bioCNG

	Power (electricity)	bioCNG
1 m^3 biogas	2.1 kWh	0.45 kg
Net rate of sale	Rs. 5/kWh	Rs. 50/kWh

Table 5: Output value of electricity and biofertiliser from processing of 100 tonne rice straw

Revenue generation	Biogas produced (m^3)/100 tonne rice straw	
	25000	45000
Output electricity value (lakh)	3.94	7.1
Output digestate value (lakh)	3.5	3.5
Total output cost (lakh)	7.44	10.6
Output-input (net profit) for thermochemical process (lakh)	-6.63	-3.47
Output-input (net profit) for pulverisation process (lakh)	+2.48	+8.12

The net profits of both the processes are presented in Tables 5 and 6. It is safely deduced that the biomethanation process using pulverised rice straw can be a profitable and sustainable process to generate energy. Along with energy, this process also gives nutrient-rich and revenue generating digestate which can be used as an organic fertiliser.

Table 6: Output values of bioCNG and biofertiliser from processing of 100 tonne rice straw

Revenue generation	Biogas produced (m^3)/100 tonne rice straw	
	25000	45000
Output electricity value (lakh)	3.94	7.1
Output digestate value (lakh)	3.5	3.5
Total output cost (lakh)	7.44	10.6
Output-input (net profit) for thermochemical process (lakh)	-6.63	-3.47
Output-input (net profit) for pulverisation process (lakh)	+2.48	+8.12

Conclusion

Several published literature have illustrated the efficiency of biomethanation of rice straw at 400 and 250m^3 biogas/tonne VS of rice straw with or without physicochemical pretreatment. Operational expenditure for biomethanation with or without pretreatment would be about 1.4 and 0.48 million Indian Rupees for 100 tonnes of rice straw. This means that for the process to be economically feasible, biomethanation without any pretreatment seems to be the logical approach even if the process efficiency of such process was significantly lower than that for the chemically pretreated rice straw. If biogas was used for electricity generation, then the net profit would be 2.48 and 8.12 lakh (from 100 tonne rice straw) considering biogas yield of 250m^3/tonne VS rice straw and 450m^3/tonne VS rice straw respectively, after considering the conversion losses as heat. Alternately, if biogas was enriched and converted into bioCNG the cost could be 6.65 and 11.15 lakh (from 100 tonne rice straw) considering a biogas yield of 250m^3/tonne VS rice straw and 450m^3/tonne VS rice straw, respectively. Considering these statistics, converting biogas into bioCNG would be more economical and sustainable.

Acknowledgements

The first author is grateful to the University Grants Commission for the fellowships received.

References

Abraham A, Mathew A, Sindhu R, Pandey A, Binod P, 2016. Potential of rice straw for biorefining: an overview. *Bioresource Technology*, **215**: 29–36.

Agbor V, Cicek N, Sparling R, Berlin A, Levin D, 2011. Biomass pretreatment: fundamentals toward application. *Biotechnology Advances*, **29(6)**: 675–85.

Alvira P, Tomas-Pejo E, Ballesteros M, Negro M, 2010. Pretreatment technologies for an efficient bioethanol production process based on enzymatic hydrolysis: a review. *Bioresource Technology*, **101(13)**: 4851–61.

Angelidaki I, Karakashev D, Batstone D, Plugge M, Stams A, 2011. Biomethanation and its potential. *Methods in Enzymology*, **494**: 327–51.

Bajpai, P., 2016. Pretreatment of lignocellulosic biomass for biofuel production. Singapore: Springer.

Banerjee S, Sen R, Pandey A, Chakrabarti T, Satpute D, Giri S, Mudliar S, 2009. Evaluation of wet air oxidation as a pretreatment strategy for bioethanol production from rice husk and process optimization. *Biomass and Bioenergy*, **33(12)**: 1680–6.

Bioenergy IEA, 2008. *From 1st- to 2nd-Generation Biofuel Technologies. An Overview of Current Industry and RD&D activities.* International Energy Agency, Paris, France.

Brodeur G, Yau E, Badal K, Collier J, Ramachandran B, Ramakrishnan S, 2011. Chemical and physicochemical pretreatment of lignocellulosic biomass: a review. *Enzyme Research*, **2011**: 1–17.

Chandel AK, Antunes FA, De Arruda PV, Milessi TS, Da Silva SS, De Almeida FMDG, 2012. Dilute acid hydrolysis of agro-residues for the depolymerization of hemicellulose: state-of-the-art. *In*: SS Da Silva, AK Chandel (Eds) *D-xylitol*, Springer, Berlin, Heidelberg: 39–61 pp.

Chandra R, Takeuchi H, Hasegawa T, 2012. Hydrothermal pretreatment of rice straw biomass: a potential and promising method for enhanced methane production. *Applied Energy*, **94**: 129–40.

Chandra R, Takeuchi H, Hasegawa T, 2012. Methane production from lignocellulosic agricultural crop wastes: A review in context to second generation of biofuel production. *Renewable and Sustainable Energy Reviews*, **16(3)**: 1462–76.

Chen W, Pen B, Yu T, Hwang W, 2011. Pretreatment efficiency and structural characterization of rice straw by an integrated process of dilute-acid and steam explosion for bioethanol production. *Bioresource Technology*, **102(3)**: 2916–24.

Chen W, Tsai C, Lin C, Tsai Y, Hwang S, 2013. Pilot-scale study on the acid-catalyzed steam explosion of rice straw using a continuous pretreatment system. *Bioresource Technology*, **128**: 297–304.

Chen X, Zhang Y, Gu Y, Liu Z, Shen Z, Chu H, Zhou X, 2014. Enhancing methane production from rice straw by extrusion pretreatment. *Applied Energy*, **122**: 34–41.

Devendra C, Thomas D, 2002. Crop–animal interactions in mixed farming systems in Asia. *Agricultural Systems*, **71(1-2)**: 27–40.

Dobermann A, Fairhurst T, 2002. Rice straw management. *Better Crops International*, **16(1)**: 7–11.

EIA, 2016. Levelized cost and levelized avoided cost of new generation resources in the annual energy outlook 2016. http://www. eia. gov/outlooks/aeo/pdf/electricity_generation. pdf.

Energy Information Administration (US), and Government Publications Office, 2016. *International Energy Outlook 2016: With Projections to 2040*. Washington, The United States of America.

Gadde B, Bonnet S, Menke C, Garivait S, 2009. Air pollutant emissions from rice straw open field burning in India, Thailand and the Philippines. *Environmental Pollution*, **157(5)**: 1554–8.

Ghosh A, Bhattacharyya C, 1999. Biomethanation of white rotted and brown rotted rice straw. *Bioprocess Engineering*, **20(4)**: 297–302.

Gu Y, Zhang Y, Zhou X, 2015. Effect of $Ca(OH)_2$ pretreatment on extruded rice straw anaerobic digestion. *Bioresource Technology*, **196**: 116–22.

Gupta H, Dadlani M, 2012. *Crop Residues Management with Conservation Agriculture: Potential, Constraints and Policy Needs*. Indian Agricultural Research Institute, New Delhi.

He Y, Pang Y, Liu Y, Li X, Wang K, 2008. Physicochemical characterization of rice straw pretreated with sodium hydroxide in the solid state for enhancing biogas production. *Energy & Fuels*, **22(4)**: 2775–81.

Hendriks M, Zeeman G, 2009. Pretreatments to enhance the digestibility of lignocellulosic biomass. *Bioresource Technology*, **100(1)**: 10–8.

Hiloidhari M, Das D, Baruah C, 2014. Bioenergy potential from crop residue biomass in India. *Renewable and Sustainable Energy Reviews*, **32**: 504–12.

Hosseini M, Aziz A, Syafalni M, 2012. Enhancement of rice straw biodegradability by alkaline and acid thermochemical pretreatment process: Optimization by response surface methodology (RSM). *Caspian Journal of Applied Sciences Research*, **1(12)**: 8–24.

Jenkins B, Baxter L, Miles R, Miles R, 1998. Combustion properties of biomass. *Fuel Processing Technology*, **54(1-3)**: 17–46.

Jensen R, Morinelly E, Gossen R, Brodeur-Campbell J, Shonnard R, 2010. Combustion properties of biomass. *Bioresource Technology*, **101(7)**: 2317–25.

Kapoor M, Soam S, Agrawal R, Gupta P, Tuli K, Kumar R, 2017. Pilot scale dilute acid pretreatment of rice straw and fermentable sugar recovery at high solid loadings. *Bioresource Technology*, **224**: 688–93.

Kaur K, Phutela G, 2016. Enhancement of paddy straw digestibility and biogas production by sodium hydroxide-microwave pretreatment. *Renewable Energy*, **92**: 178–84.

Ko K, Bak S, Jung W, Lee H, Choi G, Kim H, Kim H, 2009. Ethanol production from rice straw using optimized aqueous-ammonia soaking pretreatment and simultaneous saccharification and fermentation processes. *Bioresource Technology*, **100(19)**: 4374–80.

Krishania M, Kumar V, Vijay K, Malik A, 2013. Analysis of different techniques used for improvement of biomethanation process: a review. *Fuel*, **106**: 1–9.

Kumar A, Singh V, Kumar N, Kumar R, 2014. Utilization of paddy straw as animal feed. *Forage Res*, **40(3)**: 154–8.

Kumar K, Sharma S, 2017. Recent updates on different methods of pretreatment of lignocellulosic feedstocks: a review. *Bioresources and Bioprocessing*, **4(1)**: 1–19.

Kumar R, Singh S, Singh V, 2008. Bioconversion of lignocellulosic biomass: biochemical and molecular perspectives. *Journal of Industrial Microbiology & Biotechnology*, **35(5)**: 377–91.

Laopaiboon P, Thani A, Leelavatcharamas V, Laopaiboon L, 2010. Acid hydrolysis of sugarcane bagasse for lactic acid production. *Bioresource Technology*, **101(3)**: 1036–43.

Lim S, Manan A, Alwi W, Hashim H, 2012. A review on utilisation of biomass from rice industry as a source of renewable energy. *Renewable and Sustainable Energy Reviews*, **16(5)**: 3084–94.

Mustafa M, Poulsen G, Sheng K, 2016. Fungal pretreatment of rice straw with *Pleurotus ostreatus* and *Trichoderma reesei* to enhance methane production under solid-state anaerobic digestion. *Applied Energy*, **180**: 661–71.

Nigam P, Singh A, 2011. Production of liquid biofuels from renewable resources. *Progress in Energy and Combustion Science*, **37(1)**: 52–68.

Ong A, Chuah C, Chew L, 2010. Comparison of sodium hydroxide and potassium hydroxide followed by heat treatment on rice straw for cellulase production under solid state fermentation. *Journal of Applied Sciences*, **10**: 2608–12.

Patel KS, Sahu PK, Combustion Characteristics of rice straw – calorific value of rice straw. *Journal of Environmental Science*. http://environment.scientific-journal.com/.

Raj T, Kapoor M, Gaur R, Christopher J, Lamba B, Tuli K, Kumar R, 2015. Physical and chemical characterization of various Indian agriculture residues for biofuels production. *Energy & Fuels*, **29(5)**: 3111–8.

Romasanta R, Sander O, Gaihre K, Alberto C, Gummert M, Quilty J, Wassmann R, 2017. How does burning of rice straw affect CH_4 and N_2O emissions? A comparative experiment of different on-field straw management practices. *Agriculture, Ecosystems & Environment*, **239**: 143–53.

Rubin M, 2008. Genomics of cellulosic biofuels. *Nature*, **454(7246)**: 841–5.

Saha B.C., 2003. Hemicellulose bioconversion. Indian Microbiology and Biotechnology, 30: 279–291.

Sander O, Samson M, Buresh J, 2014. Methane and nitrous oxide emissions from flooded rice fields as affected by water and straw management between rice crops. *Geoderma*, **235-236**: 355–62.

Satlewal A, Agrawal R, Bhagia S, Das P, Ragauskas J, 2017. Rice straw as a feedstock for biofuels: Availability, recalcitrance, and chemical properties. *Biofuels, Bioproducts and Biorefining*, **12(1)**: 83–107.

Sharma K, Mishra M, Sharma P, Saini S, 1988. Effect of particle size on biogas generation from biomass residues. *Biomass*, **17(4)**: 251–63.

Soam S, Borjesson P, Sharma K, Gupta P, Tuli K, Kumar R, 2017. Life cycle assessment of rice straw utilization practices in India. *Bioresource Technology*, **228**: 89–98.

Song Z, Yang G, Han X, Feng Y, Ren G, 2012. Optimization of the alkaline pretreatment of rice straw for enhanced methane yield. *BioMed Research International*, 2013: 1–9.

Sun Y, Cheng J, 2002. Hydrolysis of lignocellulosic materials for ethanol production: a review. *Bioresource Technology*, **83(1)**: 1–11.

Taniguchi M, Suzuki H, Watanabe D, Sakai K, Hoshino K, Tanaka T, 2005. Evaluation of pretreatment with *Pleurotus ostreatus* for enzymatic hydrolysis of rice straw. *Journal of Bioscience and Bioengineering*, **100(6)**: 637–43.

Trivedi A, Verma R, S Kaur, Jha B, Vijay V, Chandra R, Prasad R, 2017. Sustainable bio-energy production models for eradicating open field burning of paddy straw in Punjab, India. *Energy*, **127**: 310–7.

Trivedi A, Vijay K, Chandra R, 2016. Liquid hot water pretreatment of paddy straw for enhanced biomethanation. *In*: S Kumar, SK Khanal, YK Yadav (Eds) *Proceedings of the First International Conference on Recent Advances in Bioenergy Research*, Springer, New Delhi, India: 15–22 pp.

Wei W, Wu S, Liu L, 2012. Enzymatic saccharification of dilute acid pretreated eucalyptus chips for fermentable sugar production. *Bioresource Technology*, **110**: 302–7.

Weiland P, 2010. Biogas production: current state and perspectives. *Applied Microbiology and Biotechnology*, **85(4)**: 849–60.

Zhang R, Li X, Fadel G, 2002. Oyster mushroom cultivation with rice and wheat straw. *Bioresource Technology*, **82(3)**: 277–84.

8

Biogas Production from Dairy Effluents

Preseela Satpathy and Trupti Das[1]

Department of Environment and Sustainability, Institute of Minerals and Materials Technology, Bhubaneswar, Odisha, India
[1]Corresponding Author: trupti.sreyas@gmail.com; truptiimmt@gmail.com

ABSTRACT

There is a quantum global growth in both number and size of dairy industries in recent times. These industries discharge thousands of liters of wastewater characterised by a heavy chemical oxygen demand (COD), biological oxygen demand (BOD), nutrients, inorganic salts, besides detergents and sanitisers each day (Kushwaha et al., 2011). High carbohydrates, fats and proteins contents in such waste streams originating from milk pose challenge to conventional municipal sewage treatment systems (Demirel et al., *2005). The easily biodegradable dairy effluent leads to precipitation of casein which is toxic and has a strong odour. Without adequate treatment, these effluents could pose serious threats to the receiving water bodies and disturb the ecosystem (Satpathy et al., 2017). For such heavily loaded organic effluents, biogasification through anaerobic digestion seems a promising strategy. This cost-effective energy generation process is also effective in environmental management compared to its other eco-friendly renewable source counterparts, such as, hydro, solar and wind (Satpathy, 2017). The Upflow Anaerobic Sludge Blanket (UASB) reactors are the most popular anaerobic systems to treat the fluctuating organic loads in dairy effluents. Anaerobic filter (AF), Upflow Anaerobic Filter (UPAF), Fixed-bed anaerobic reactor, Anaerobic Hybrid Reactor (AHR) etc. are few other systems that are demonstrably competent removing COD upto 90% and generating biogas. Successful anaerobic systems reported yielded (with upto about 85% CH_4 content) 0.32-0.34m^3 CH_4/kg COD removed. Biomethanation process and reactors designs are constantly being optimised for effective treatment of the effluents while maximising the biogas production. The energy-rich gas thus obtained could further be utilised for heat and electricity at the domestic level, and also as a vehicular fuel. Residual digestate can also be used as organic manures (Angelidaki* et al., *2007).*

***Keywords*:** Anaerobic Digestion, Dairy effluent, Upflow Anaerobic Sludge Blanket (UASB)

Abbreviations used

SBR : sequencing batch reactor
UASB : upflow anaerobic sludge blanket
COD : chemical oxygen demand
BOD : biochemical oxygen demand
CIP : cleaning in process
MMT : million metric tonne
MNES : Ministry of nonconventional energy sources
HRT : hydraulic retention time
CBM : compressed biomethane
LBM : liquefied biomethane
CNG : compressed natural gas
AF : anaerobic filter
UPAF : upflow anaerobic filter
FBAR : fixed-bed anaerobic reactor
AHR : anaerobic hybrid reactor

Introduction

Dairy industries are regarded as one of the most polluting food industries, characterised with a rich organic load (typical to a food industry) that has an immediate and high oxygen demand (Karthikeyan *et al.*, 2015). Studies have reported these as one of the highest effluent producers (Chavda and Rana, 2014; Shete and Shanker, 2013; Anon, 2017), generating as high as 10L effluent per liter of processed milk (Shete and Shanker, 2013). Milk production increasing from 17MMT to nearly 155.5MMT during 1950-2016 (with a growth rate of 6.27% per annum), India is global leader in milk production (Anon, 2017). Considering two to three times effluent volume produced on an average of the volume of the milk produced as per a study (Bhadouria and Say, 2011), the damaging effects of these effluent cannot be overlooked.

Dairy effluent is widely treated biologically through activated sludge processes by employing sequencing batch reactor (SBR), upflow anaerobic sludge blanket (UASB) reactor, anaerobic filters etc. Biological treatment, involving both aerobic and anaerobic processes, is favoured over physicochemical methods due to the reduced cost and better performance in COD removal from the effluents (Demirel *et al.*, 2005). While aerobic systems successfully reduce soluble organic matter (BOD) along with the nitrogen and phosphorous content, their high energy requirement is a major setback. On the other hand, anaerobic systems are favourable as they require no aeration, produce a low sludge quantity and need relatively less area (Jutta and Gilbert, 2016). Thus, anaerobic technology has been recommended as an effective treatment method to remove a large fraction of the organic content in dairy effluents converting them into the energy-rich biogas.

What is dairy effluent?

Dairy effluent is wastewaters generated from dairy industries. These are one of the major sources of wastewaters globally due to the large volumes laden with heavy organic loading (Tikariha and Sahu, 2014). Farm dairy effluent (cattle-shed refuse) is rich in dairy animals' faeces and urine, subsequently diluted with the wash-down water (Hawke and Summers, 2006). Industrial milk processing units, on the other hand, involve a continuous flow of water along with acid and alkaline sanitisers and detergents (as recommended CIP) for rinsing and washing (Tikariha and Sahu, 2014). The wash-down water also contains large quantities of milk constituents like lactose, fat, casein, inorganic salts etc., generated during production of cheese, yoghurt, butter, ice-cream etc. (Vidal *et al.*, 2000). About 2% of the processed milk lands up as wastewater (Bharati *et al.*, 2013). Spoilt raw or manufactured products, overflow due to inefficient equipments, poor workmanship, etc. further add to the organic loading in the effluent. Due to the presence of such organic components, the effluent is characterised with a heavy COD (Karthikeyan *et al.*, 2015; Bharati *et al.*, 2013), usually 10-100 times higher than domestic wastewaters (Demirel *et al.*, 2005).

Environmental challenges by dairy wastewaters

The high load of pollutants in dairy effluents could disrupt (aquatic and terrestrial) ecosystems. Since these are rich in soluble organics, suspended solids etc. they release malodorous gases, impart colour and turbidity to the receiving waters and promote eutrophication (Tikariha and Sahu, 2014). Dairy waste is usually white in colour and bears a slightly alkaline characteristic. The use of alkalis and detergents render these effluents heavily alkaline (Noorjahan et al., 2004). The carbohydrate fraction rapidly ferments to lactic acid which decreases the pH immediately. The casein precipitates as the pH decreases, which further decomposes to a highly malodorous toxic black sludge (Bharati *et al.*, 2013). Rapid decomposition of the dairy effluents leads to the depletion of the dissolved oxygen levels of the receiving water bodies resulting in an immediate release of strong foul odour. This further promotes breeding grounds for disease-carrying flies and mosquitoes.

Biogas production from dairy effluents

Biogas has earned the reputation of the most versatile and efficient renewable energy source. Formed by anaerobic digestion, it mainly comprises of methane and carbon-di-oxide (Satpathy *et al*., 2013). Its production primarily involves a two-stage (acidogenesis and methanogenesis) biological reaction. At least two different microbial groups, *i.e.*, the acid-forming bacteria and the methane-forming bacteria involve actively. The acid phase includes conversion of complex compounds to simpler organic acids, principally acetic acid. The methanogenesis phase produces the energy-rich methane from acetic acid, carbon-di-oxide and hydrogen. While acetoclastic methanogens convert acetic acid to methane (Equation 1), the hydrogenotrophic methanogens utilise hydrogen and carbon-di-oxide (Equation 2) (Satpathy *et al*., 2016a).

$$CH_3COOH \rightarrow CH_4 + CO_2 \quad (1)$$

$$4H_2 + CO_2 \rightarrow CH_4 + 2H_2O \quad (2)$$

Anaerobic processes are most effective method to treat the high-strength dairy effluents with subsequent conversion to biogas (Shete and Shanker, 2013). Subjected to anaerobic digestion, a COD removal of 75% and methane content of 60% on an average can be expected from dairy effluents (Rao *et al*., 2010). The erstwhile Ministry of Nonconventional Energy Sources (MNES), India reported a biogas generation potential of nearly 220,000m^3/d from effluents from 342 dairy units (Rao *et al*., 2010; Anon, 2004). Krich *et al*. (2005) estimated an annual biomethanation potential of nearly 14.6 billion ft^3 from 1.7 million cattle in the US, roughly corresponding to about 140 Mega Watt electricity.

Codigesting dairy effluents with biomass like agricultural residues, manures, food processing industrial waste etc. could further optimise biogas production yield. Codigesting animal manure and dairy effluents with food processing wastes in community digestion facilities producing electricity in Mega Watt range is widely practiced in many European countries (Krich *et al*., 2005).

Anaerobic reactors for biogasification of dairy effluent

Current studies report that, the UASB reactors have been the most widely used for dairy effluents for full-scale applications (Birwal *et al*., 2017). Gotmare *et al*. (2011) attained COD and BOD removal of upto 87% and 94.5% respectively from such effluent using UASB reactors, with average biogas production of 179.35m^3/d containing upto 70% methane. Using upflow anaerobic fixed-bed reactor, Deshannavar *et al*. (Deshannavar *et al*., 2012) reported 87% COD removal with 77% methane containing 9.8L/d biogas production from dairy effluents. 80.88% COD reduction from 1500-4700mg/L could be achieved by fixed-bed anaerobic reactor (Ramesh *et al*., 2012), with maximal gas conversion ratio of $0.27m^3/kg$ COD removed. Such fixed-bed reactor is widely applied for dairy waste streams due to its capacity to retain microbes thereby reducing hydraulic retention time (HRT). Study on synthetic dairy effluent attained 85% and 90% removal of COD and BOD respectively using fixed-bed reactor. Upflow anaerobic filter unit generated nearly 85% methane, with a yield of 0.32-0.34m^3 CH_4/kg COD removed, producing nearly 770L CH_4/d from the system (Ince, 1998). Biological treatment using UASB in a cheese-producing dairy unit with influent concentrations of 12-60g COD/L showed a COD removal of upto 99% (Gavala et al., 1999). Anaerobic filter reactors in general are validated to be suitable for dairy effluents with low-strength (Demirel *et al.*, 2005). Anaerobic hybrid reactor, a combination of UASB reactor at the bottom and anaerobic filter reactor at the top was effective for low-strength dairy effluent removing upto 85% COD (Sathyamoorthy and Saseetharan, 2012). Efficient COD removal was found associated with the nature and properties of the support material, and a nearly 96% enhanced COD removal by changing the biofilm support media was reported (Demirel *et al.*, 2005). A study with the upflow anaerobic packed bed bioreactor that was packed with seashell displayed COD and lactose removal from high-strength dairy effluent by 94.5% and 99% respectively (Najafpour *et al.*, 2008), attributable to biosorption phenomenon.

Up-gradation of biogas

Biogas thus generated can further be upgraded to biomethane by removing moisture, hydrogen sulphide, carbon-di-oxide etc. (Plombin, 2003). With a methane content of more than 95%, biomethane is equivalent to natural gas. Thus, besides domestic heating and cooling, this gas could be utilised to generate electricity, pumping, or even as vehicular fuel. Many European countries are already using organic waste derived biomethane as fuel to operate public transport. Biomethane can be stored as compressed biomethane (CBM) and can be pumped into natural gas supply pipeline, or can also be subjected to low temperatures to produce liquefied biomethane (LBM) (Krich *et al.*, 2005).

Turning dairy wastes to profits in India

Livestock sector in India has a significant 4.1% contribution to GDP (Islam *et al.*, 2016). While national policies focus on milk production, waste management of the cattleshed and dairy processing units has taken a backstage into the informal economy (Marek, 2014). Study reports a 3,00,000L dairy effluent-based plant to generate nearly 530-1000 m3/d biogas that could potentially provide nearly 668-1,335 units electricity daily (Velmurugan *et al.*, 2017). This green electricity could account upto Rs. 18.3 lakh annually (considering an unit price of Rs. 5) (US EPA, 2011). This electricity could be utilised for on-farm cooling and storing helping the farmers prevent spoilage thereby adding to the farmers' revenues (Anon, 2017). On-farm electricity generation reduces the pressure on fossil fuel, and also the farmer could qualify to receive renewable energy credits or trade as carbon credits in global greenhouse gas market (US EPA, 2011). On a local scale, biogas production could improve the energy availability among the rural poor, especially for the 25% of India which still has no access to electricity (Marek, 2014).

Generating renewable energy from waste like biogas that reduces the carbon footprints could also benefit in obtaining environmental clearance for the industries. The pollution control boards lay stringent standards and measures for dairy sector due to their high consumption of resources and the severe environmental impacts. The Ministry of Environment, Forest and Climate Change (MoEFCC) classifies dairy processor under red category, *i.e.*, under major polluting industry (Javedkar, 2016). Methanation by anaerobic digestion could solve this challenge along with enabling upto 90% removal of the harmful elements before discharge (Anon, 2017). Such strategy not only reduces the environmental impact and improves the overall economy of the milk processing industries, but also generates substantial employment opportunities (US EPA, 2011). The International Labour Organisation (ILO) estimates that nearly 12 million small-scale biogas digesters could be set up all over India generating up to 5,000MW electricity and providing nearly 1,00,000 jobs (Velmurugan *et al.*, 2017).

The Ministry of New and Renewable Energy (MNRE) continues to promote the construction of biogas plants by providing subsidies. Market-based approaches could be complemented with the regulatory mandates by combining dairy industries with biogas systems, and further encouraging the construction of such plants. The Kaira District Cooperative Milk Producers Union Limited, commonly known as Amul Dairy, has recently installed a fully automated bio-CNG plant. This plant utilises 2 million litre dairy effluent daily to generate 2,500m^3 60-65% pure methane. The administration presented a calculation for their mega project where the payback period was only 2 years (saving on their fuel cost) with an investment of 1.75 crore (Rupera, 2016).

Advancements in science have further enabled the technology using mathematical models like the Anaerobic Digestion Model No. 1 (ADM1) that could virtually design, plan and operate biogas systems with various loads and composition of dairy effluents with more than 90% reliability (Satpathy *et al.*, 2016b). The models also help to calculate the electrical and heat energy output that could help in analysing the economic feasibility of a plant before constructing it. Thus, these models help save both time and money. Fig. 1 presents such a model set-up. Such tools are intelligent systems that could be utilised to predict reactor failures or inhibitions and optimise the process to maximise methane generation.

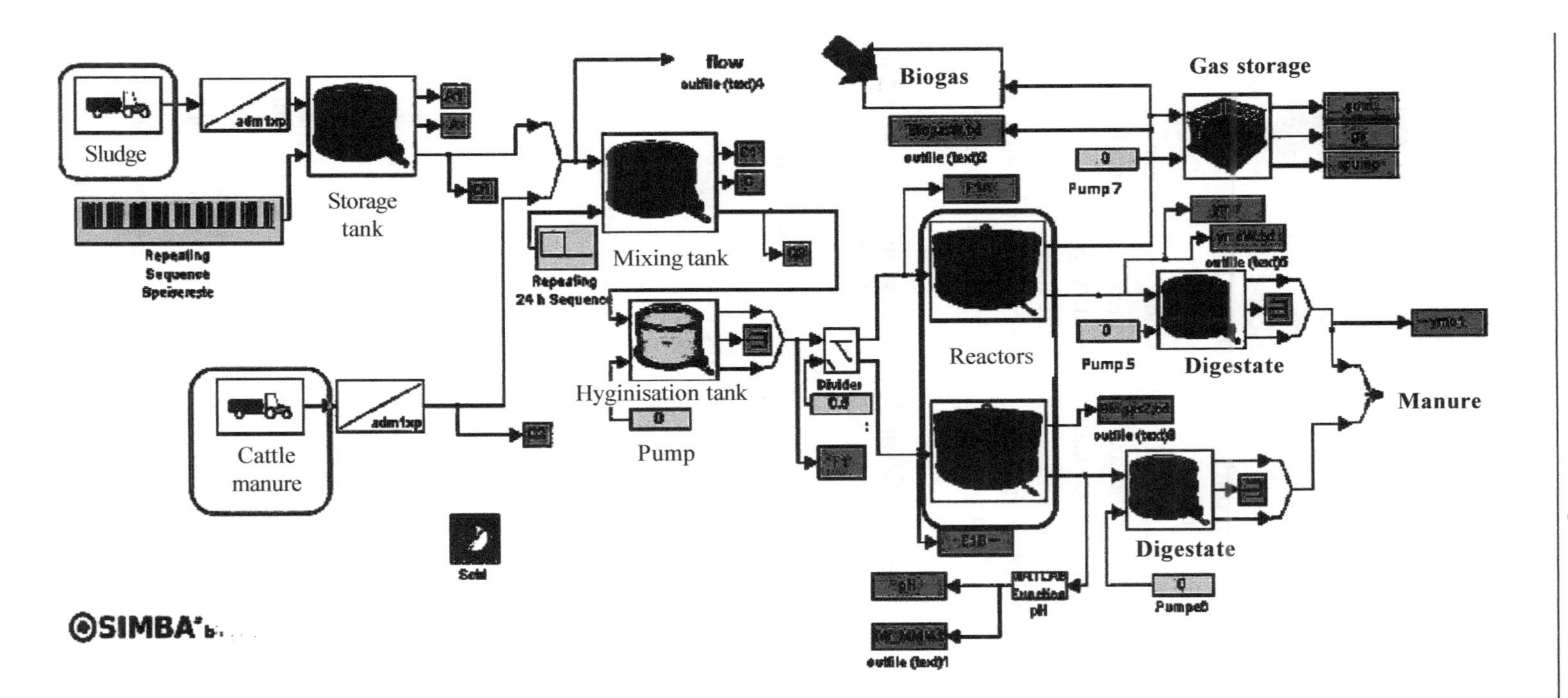

Fig. 1: Virtual set-up of a real biogas plant prepared with SIMBA® simulation software (derived from Satpathy *et al.*, 2016b)

Conclusion

As energy demands continue to grow worldwide, biogas achieves significant popularity for sustainable heat and electricity production (Gu *et al.*, 2016). Energy conservation and waste mitigation attains increased attention with time. Efforts to reduce COD, BOD in dairy effluents to ideally achieve the target of 'zero-waste' discharge in industries continue. Despite the availability of efficient treatment methods, full-scale operations still present challenges due to the variable and complex composition of dairy effluents. The constitution, flow, physicochemical characteristics of the effluent vary depending on the operation methods and systems in use (Vidal *et al.*, 2000). An increased understanding of the processes involved is largely warranted wherein the advanced tools like the mathematical models are gaining increasing popularity. Such cost and time effective virtual set-ups to simulate and determine the ideal design, operating conditions, energy output etc. from biogas digesters fed with varying types of dairy effluents with changing conditions are proving beneficial (Satpathy *et al.*, 2016c).

Anaerobic digestion is a validated efficient treatment strategy to convert the high organic loads in dairy wastewaters (Shete and Shanker, 2013). UASB reactors and their modified variants are most commonly employed due to their capability to treat large volumes of effluents in a short duration of time with remarkable competence (Demirel *et al.*, 2005). Methane in biogas could replace fossil fuels thereby positively impacting not only the industries by reducing the costs imposed by government but also the environment through waste management (Angelidaki et al., 2007). Additionally, the dairy industries can increase their revenues by selling the electricity produced from their wastes (Wang and Nie, 2001). Future research endeavours focusing on the removal of organic contents from a broad range of dairy effluents and optimising the digestion technology is recommended.

References

Angelidaki I, Alves M, Bolzonella D, Borzacconi L, Campos L, Guwy, A, Jenicek P, Kalyuzhnui S, Van Lier J, 2007. Anaerobic biodegradation, activity and inhibition (abai) task group meeting 9th to 10th October 2006, Prague.

Anon, 2004. National Master Plan for Development of Waste-to-Energy in India. Ministry of Non-conventional Energy Sources. http://www.seas.columbia.edu/earth/wtert/sofos/Natl_Master_ Plan_of_India.pdf

Javedkar P, 2016. Re-categorisation of Industries a landmark decision, new category of white industries will not require environmental clearance, Press Information Bureau, MoEF, GoI.

Anon, 2017. Annual Report 2016-17. Department of Agriculture, Cooperation & Farmers Welfare. http://agricoop.nic.in/sites/default/files/Annual_rpt_201617_E.pdf

Bhadouria BS, Say VS, 2011. Utilisation and treatment of dairy effluent through biogas generation - A case study. *International Journal of Environmental Sciences*, **1(7)**: 1621–30.

Bharati M, Shete S, Shinkar NP, 2013. Comparative study of various treatments for dairy industry wastewater. *IOSR Journal of Engineering*, **3(8)**: 2250–3021.

Birwal P, Deshmukh G, Priyanka, 2017. Advanced technologies for dairy effluent treatment. *Journal of Food, Nutrition and Population Health*, **1(1:7)**: 1–5.

Chavda P, Rana A, 2014. Performance evaluation of effluent treatment plant of dairy industry. *International Journal of Engineering Research and Application*, **4(9)**: 37–40.

Demirel B, Yenigun O, Onay TT, 2005. Anaerobic treatment of dairy wastewaters: A review. *Process Biochemistry*, **40(8)**: 2583–95.

Deshannavar UB, Basavaraj RK, Naik NM, 2012. High rate digestion of dairy industry effluent by upflow anaerobic fixed-bed reactor. *Journal of Chemical and Pharmaceutical Research*, **4(6):** 2895–9.

Gavala HN, Kopsinis H, Skiadas ILG, 1999. Treatment of dairy wastewater using an upflow anaerobic sludge blanket reactor. *Journal of Agricultural Engineering Research*, **73(1)**: 59–63.

Gotmare M, Dhoble RM, Pittule, 2011. Biomethanation of dairy waste water through UASB at mesophilic temperature range. *International Journal of Advanced Engineering Sciences and Technology*, **8(1)**: 1–9.

Gu L, Zhang YX, Wang JZ, Chen G, Battye H, 2016. Where is the future of China's biogas? Review, forecast, and policy implications. *Petroleum Science*, **13(3)**: 604–24.

Hawke RM, Summers S, 2006. Effects of land application of farm dairy effluent on soil properties: A literature review. *New Zealand Journal of Agricultural Research*, **49(3)**: 307–20.

Ince O, 1998. Potential energy production from anaerobic digestion of dairy wastewater. *Journal of Environmental Science & Heal Part A*, **33(6)**: 1219–28.

Islam MM, Anjum S, Modi RJ, Wadhwani KN, 2016. Scenario of livestock and poultry in india and their contribution to national economy. *International Journal of Environmental Science and Technology*, **5(3)**: 956–65.

Jutta Q, Gilbert EM, 2016. Wastewater treatment in the dairy processing industry - recovering energy using anaerobic technology. Technology for Water White Paper. Enviro Chemie. 10.13140/RG.2.1.3875.4965

Karthikeyan VA, Venktesh KR, Arutchelvan V, 2015. Correlation study on physico-chemical characteristics of dairy wastewater. *International Journal of Engineering Science and Technology*, **7(2)**: 89–92.

Krich K, Augenstein D, Benemann J, Rutledge B, 2005. *Biomethane from Dairy Waste – A Sourcebook for the Production and Use of Renewable Natural gasses in California*. West United Dairymen, California.

Kushwaha JP, Srivastava VC, Mall ID, 2011. An overview of various technologies for the treatment of dairy wastewaters. *Critical Revies in Food Science and Nutrition*, **51(5)**: 442–52.

Marek H, 2014. The economics of biogas creating green jobs in the dairy industry in India. International Labour Organisation, Geneva.

Najafpour GD, Hashemiyeh BA, Asadi M, Ghasemi, MB, 2008. Biological treatment of dairy wastewater in an upflow anaerobic sludge-fixed film bioreactor. *American-Eurasian Journal of Agricultural and Environmental Sciences*, **4(2)**: 251–7.

Noorjahan CM, Sharief SD, Dawood N, 2004. Characterisation of dairy effluent. *Journal of Industrial Pollution and Control*, **20(1)**: 131–6.

Plombin C, 2003. *Biogas as Vehicle Fuel: A European Overview*. Environment and Health Adminstration, Stockholm, Sweden.

Ramesh T, Nehru Kumar V, Srinivasan, G, 2012. Kinetic evaluation of fixed film fixed bed anaerobic reactor by using dairy wastewaters. *International Journal of Pharmaceutical and Biological Archieves*, **3(4)**: 835–7.

Rao PV, Baral SS, Dey R, Mutnuri S, 2010. Biogas generation potential by anaerobic digestion for sustainable energy development in India. *Renewable & Sustainable Energy Reviews*, **14(7)**: 2086–94.

Rupera P, 2016. Amul starts bio-CNG generation plant. Times of India, Mumbai. https://timesofindia.indiatimes.com/city/vadodara/Amul-starts-bio-CNG-generation-plant/articleshow/51285225.cms

Sathyamoorthy GL, Saseetharan MK, 2012. Dairy wastewater treatment by anaerobic hybrid reactor – a study on the reactor performance and optimum percentage of inert media fill inside reactor. *Research Jouranl of Chemistry Environment*, **16(1)**: 51–6.

Satpathy P, 2017. Recent developments in biogas technology. *In*: A Vico, N Atemio (Eds) *Biogas: Production, Applications and Global Developments*. Nova Science Publishers, New York: 209–40 pp.

Satpathy P, Biernacki P, Cypionka H, Steinigeweg S, 2016a. Modelling anaerobic digestion in an industrial biogas digester: Application of lactate-including ADM1 model (Part II). Journal of *Journal of Environmental Science and Health Part A*, **51(14)**: 1226–32.

Satpathy P, Biernacki P, Uhlenhut F, Cypionka H, Steinigeweg S, 2016b. Modelling anaerobic digestion in a biogas reactor: ADM1 model development with lactate as an intermediate (Part I). *Journal of Environmental Science and Health Part A,* **51(14)**: 1216–25.

Satpathy P, Steinigeweg S, Cypionka H, Engelen B, 2016c. Different substrates and starter inocula govern microbial community structures in biogas reactors. *Environmental Technology*, **37(11)**: 1441–50.

Satpathy P, Steinigeweg S, Siefert E, Cypionka H, 2017. Effect of lactate and starter inoculum on biogas production from fresh maize and maize silage. *Advances in Microbiology*, **7(5)**: 358–76.

Satpathy P, Steinigeweg S, Uhlenhut F, Siefert E, 2013. Application of anaerobic digestion model 1 (ADM1) for prediction of biogas production. *Interantional Journal of Science Engineering and Research*, **4(12)**: 86–9.

Shete BS, Shanker NP, 2013. Dairy industry wastewater sources, characteristics & its effects on environment. *International Journal of Current Engineering Technology*, **3(5)**: 1611–5.

Tikariha A, Sahu O, 2014. Study of characteristics and treatments of dairy industry waste water. *Journal of Applied Environmental Microbiology*, **2(1)**: 16–22.

US EPA, 2011. Market Opportunities for Biogas Recovery Systems at U.S. Livestock Facilities. Minnesota: Agstar. http://www.epa.gov/agstar/documents/biogas_ recovery_ systems_ screenres.pdf

Velmurugan, B, Narra M, Vyas B, Vahora S, Shah DR, Zala A, Patel B, 2017. Biogas from dairy effluent scum – a case study of Vidya dairy. *Indian Dairyman*, **69(11)**: 44–7.

Vidal G, Carvalho A, Mendez RLJ, 2000. Influence of the content in fats and proteins on the anaerobic biodegradability of dairy waste. *Bioresource Technology*, **74(3)**: 231–239.

Wang H, Nie Y, 2001. Municipal solid waste characteristics and management in China. *Journal of the Air & Waste Management Association*, **51(2)**: 250–63.

9

Valorization of Pineapple Wastes for Biomethane Generation

***Prakash Kumar Sarangi*[a,1] *and Sonil Nanda*[b]**

[a]*Directorate of Research, Central Agricultural University, Imphal, Manipur, India*
[b]*Lassonde School of Engineering, York University, Toronto, Canada, India*
[1]*Corresponding Author: sarangi77@yahoo.co.in*

ABSTRACT

The annual accumulating organic wastes generated through milling, brewing and various other agro-industrial and agri-processing activities in India amount to about 500 MMT. Most of these contain three major structural biopolymers, viz., *cellulose, hemicellulose and lignin, a large proportion of which is carbohydrate and phenolic in nature. Massive accumulations of this kind of recalcitrant biomass not only deteriorate the environment, but also generate huge amount of potentially useful materials. Biological transformation has globally become increasingly popular among researchers and academicians in treating agricultural, industrial, organic and toxic wastes these days. About 60% by weight of the original pineapple fruit in forms of peeled skin, core, crown end, etc. resulting from the pineapple processing remains unutilised being discharged as waste, causing disposal and pollution owes. Of the 10% dry matter content of such waste, about 96% is organic and 4% is inorganic. Being rich in cellulose and hemicellulose, it can be a valuable resource for numerous value-added products by bioconverting it to valuable products, including bioenergy. Biomethane is one such energy resource that can be produced by valorizing pineapple wastes by employing various bioconversion technologies. This chapter focuses on the bioconverstion technologies to produce biomethane from pineapple wastes.*

***Keywords*:** Pineapple waste, bioconversion, bioethanol, biogas

Abbreviations used

Mg : Megagram
NE : North-East (region of India)
TS : Total solids
TSS : Total suspended solids

Introduction

Pineapple (*Ananas comosus*) of Family *Bromeliaceae* is the most important horticultural produce globally, positioning it as an important fruit. The plant develops a cone-shaped juicy and fleshy fruit growing up to a height of 75-150cm with crown at the top (Morton, 1987; Tran, 2006). Major pineapple growing nations are Brazil, Thailand, Philippines, Costa Rica, China and India, the total area under pineapple cultivation in the world being 909.84 thousand ha with around 19412.91 thousand tonnes production. Apart from their use as fleshy raw fruits, these are also processed into food items like juices, concentrates, and jams. Pineapple juice occupies the third most preferred position worldwide, after orange and apple juices (Cabrera *et al.*, 2000).

As discussed, pineapple (*Ananas comosus*) is a commercially important fruit crop of India, abundantly grown in almost the entire North-Eastern (NE) region, and the states of West Bengal, Kerala, Karnataka, Bihar, Goa and Maharashtra. In recent times, there is a sizeable increase in acreage and production of pineapple. In acreage, there is an increase from 87,000 ha in 2006-07 to 89,000 ha in 2010-11. Accordingly, the production increased from 1,362 thousand (2006-07) to 1,415 thousand tonnes (2010-11). Assam has maximum area (14 thousand ha) under pineapple cultivation with medium-scale productivity.

Comprising a total of eight states, the NE region of India has a geographical area of 2.62 million km^2. It has immense climatic diversity, organically fertile soil, and enough rainfall, thereby providing scope for variety of cropping. More than 40% (90-95% of which is organic) of the total pineapple production of the country is from here. Pineapple from NE region is sweeter, has relatively high TSS and less fibre contents. Scientific and academic circles are increasingly exploring the processing and preservation of such an important produce of the region.

Being easily available, the most natural and renewable resources possibly mitigating the future energy crisis are the lignocellulosic materials. Large quantities of such waste biomass are generated through agro-industries and various other agri-practices (Pérez *et al.*, 2002). Because of their renewable nature, bioconversion technologies for lignocellulosic biomass have recently gained increasing interests and special importance among academicians and researchers globally (Asgher *et al.*, 2013; Ofori & Lee, 2013). The availability of huge biomass of such type like the pineapple wastes from the pineapple processing industries can potentially be converted into high-value products including biofuels like biohydrogen and biomethane (Asgher *et al.*, 2013; Iqbal *et al.*, 2013; Irshad *et al.*, 2012; Isroi *et al.*, 2011). Significant developments have been made to harvest energy from renewable resources to provide energy for domestic and industrial usage thereby alleviating the livelihoods, particularly in the rural sectors. Organic biomass from plants and animal wastes converting them to biofuels may be potential sources for the future. In contrast to fossil fuels that contribute to the greenhouse gas emissions and global warming, biofuels are considered carbon-neutral thereby curbing the greenhouse gas emissions. As the CO_2 produced from their combustion is utilised by the plants during photosynthesis, biofuels are carbon-neutral energy sources (Figs. 1 and 2).

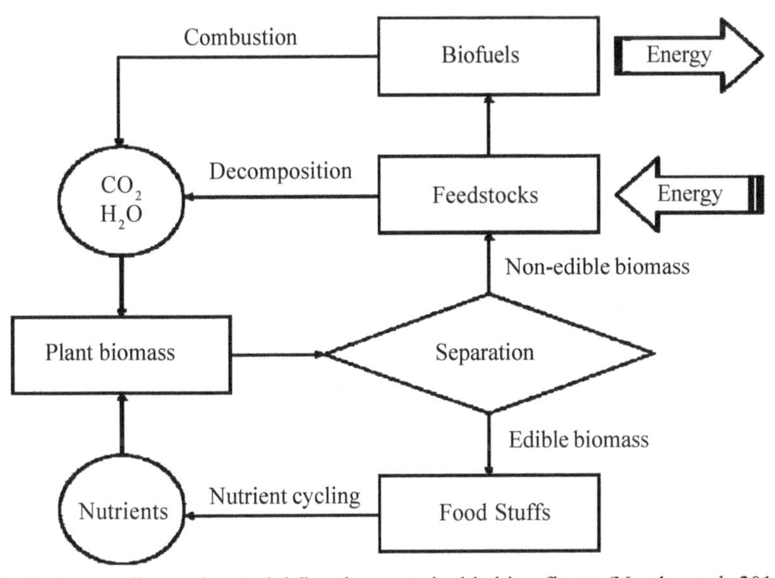

Fig. 1: Carbon cycling and material flow in a sustainable biorefinery (Nanda *et al.*, 2014)

Being non-competitor with food sources, non-food and non-feed residues, food processing wastes and other such organic resources are now considered suitable for biofuel production (Nanda *et al.*, 2015). The second generation biofuel sources, therefore, are the best renewable resource options to mitigate future energy needs. Being far away from the food-versus-fuel debate, these renewable, environmental friendly and easily available resources around the globe are one of the best options. This chapter emphasises on the biomethane production processes from pineapple wastes to meet the energy needs in rural sectors, particularly at the sites of huge production and processing of pineapple.

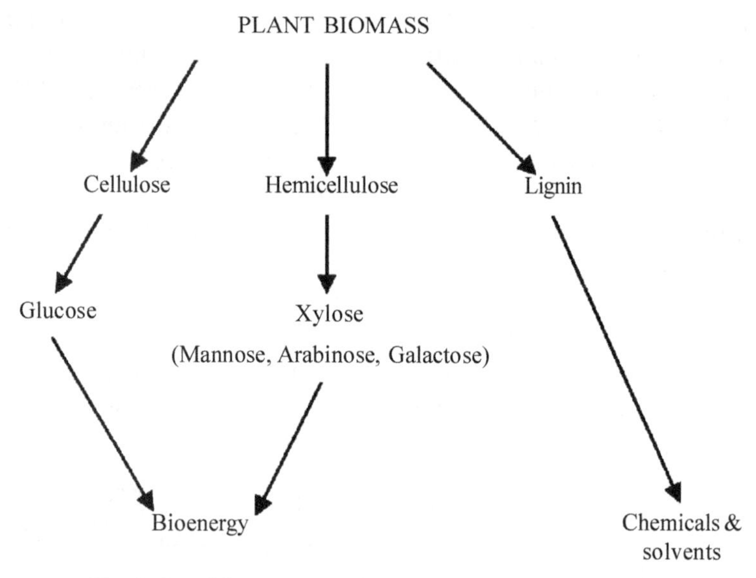

Fig. 2: Possible outcomes of plant biomass bioconversion

Agro-wastes

About 200 billion tonnes of organic matter are produced globally through the photosynthetic process on yearly basis (Zhang, 2008). The majority of this organic matter is not directly edible (or any other direct use) by humans and animals consequently creating environmental concerns. Due to great demands for appropriate nutritional standards, there is rising costs and decreasing availability of raw materials, there is a direct influence on the recovery, recycling and upgradation of these wastes (Laufenberg *et al.*, 2003). This is predominantly valid for the agro-food (primarily, agro-forestry sector and secondarily, food-processing sector) industries which generate large volumes of solid wastes, residues and byproducts, posing serious and continuously increasing environmental threats (Boucque and Fiems, 1988).

Global annual crop residues production is estimated at about 4 billion tonnes, 75% of which are from cereals (Lal, 2008). Cereal straw, corn cob, cotton stalk, grass, sorghum and maize stover, vine prunings, sugarcane and tequila bagasse, coconut and banana residues, corn husk, coffee pulp and husk, cottonseed and sunflower seed hull, peanut shell, rice husk, sunflower seed hull, waste paper, wood sawdust and chip, are examples of organic agri-residues.

The list also includes agri-wastes such as straw, stem, stalk, leaves, husk, shell, peel, lint, seed/stones, pulp, stubble, etc. from cereals (rice, wheat, maize/corn, sorghum, barley, millet), cotton, groundnut, jute, legumes (tomato, bean, soya), coffee, cacao, olive, tea, fruits (banana, mango, coconut, cashew) and palm.

Many of these are widely used as household fuel in developing countries including India. The global estimates of agri-residues production, the annual figures stand at 2.8 billion Mg (Megagram) from cereals, 305 million Mg from legumes, 108 million Mg from oil crops, 373 million Mg from sugar crops and 170 million Mg from tubers (Lal, 2005). The total crop residue production in the world is estimated at 3.8 billion Mg, out of which 74% are from cereals, 8% from legumes, 3% from oil crops, 10% from sugar crops and 5% form tubers. The most useable crop residue, however, is that of cereals.

Fig. 3: Various wastes generated by the processing of pineapple (See colour version after references)

Pineapple wastes

Basically, pineapple byproducts consist of the residual pulp, peels, stem and leaves (Fig. 3). Increased production of processed items from pineapple generates bulk of waste components that are unsuitable for human consumption. These wastes (peel, core, stem, crown and leaves) generally account for 60% (w/w) of the total weight (Sarangi *et al*., 2016; Figs. 4 and 5).

Further, low workmanship, rough handling of fruits and inappropriate transportation and storage could cause up to 55% of product waste (Nunes *et al*., 2009), which are usually prone to microbial spoilage thus limiting further processing as food items. Their drying, storage and shipment is cost intensive and hence efficient, inexpensive and eco-friendly utilisation of these is becoming more and more necessary (Upadhyay *et al*., 2010). Owing to these facts, biotechnological approaches for efficient use of pineapple (being enormously handled in the NE region) byproducts lignocellulosics may be used as cheap biomethanation substrates, thereby also partly catering to the energy needs in a renewable and sustainable manner.

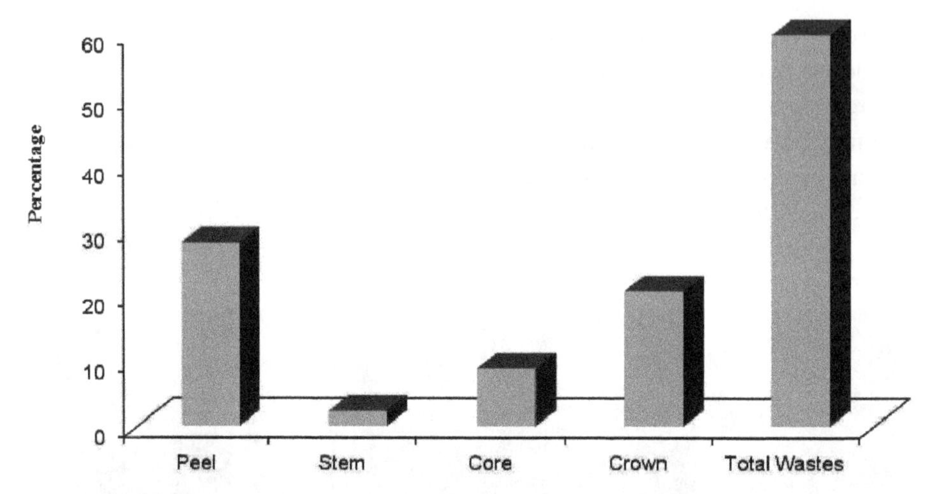

Fig. 4: Mean % waste of pineapple parts during processing (Sarangi *et al.*, 2016)

Fig. 5: Generation of pineapple wastes in rural India during preliminary processing (See colour version after references)

Bioconversion of pineapple wastes

Agricultural residues play an important role in meeting the growing societal energy demands. Lignocellulosic biomass is a renewable resource on earth and it has attracted continued efforts to produce fuels and chemicals. Lignocellulosic agricultural crop residues have vast energy generation potential, which is reportedly underutilised thus far. There is a growing need of new and renewable energy sources due to the inconsiderate consumption of non-renewable energy by the growing urbanisation, and industrial and automobile developments (Reddy, 2011; Örlygsson, 2010). Converting a renewable non-fossil carbon, such as energy crops and lignocellulosic residues (plants, grasses, fruit wastes and algae) to fuel would assure a continual energy supply (Wyman, 1996).

Biofuel fermentation economics are significantly influenced by the cost of the raw materials, which accounts for more than half of the production costs (Classen, 1999). To reduce the production costs, supply of cheap raw material is a necessity (Reddy, 2011). Lignocellulosic material is the most abundant biopolymer on

Earth with an estimated annual production at approximately 50 billion tonnes (Paepatung, 2009). Ethanol and hydrogen is produced by microbial fermentations from such biomass (Koskinen, 2008). Pineapple waste, a byproduct of pineapple processing, is one other such material, which is rich in cellulose, hemicelluloses, sugars and other carbohydrates.

Lignocellulosic biomass contains three major structural polymers: cellulose, hemicellulose and lignin. Many microbial species could degrade these releasing valuable products, which may otherwise remain locked. Specific sugar residues may be released from the cell-wall by the action of carbohydrases. Various possible value-addition of the pineapple byproducts have been explored. Such wastes from the pineapple canneries, which are high on moisture content and full of carbon sources (Table 1), have been used as a substrate for biomethanaion. Biogas has been accepted as one of the best alternative renewable energy since long (Vimal *et al.*, 1976). Organic wastes management with this appropriate technology could also alleviate environmental issues, addressing the global warming.

Table 1: The chemical compositions of pineapple peel solid waste (Vimal *et al.*, 1976)

Compositions	Percentage
Moisture content (%)	79.97
pH	4.6
Total carbon (%)	53.83
Total nitrogen (%)	0.7
C:N (carbon to nitrogen) ratio	76.9

Biomethanation processes

Agro-industrial fruit waste consists of high moisture content and full of carbon source, which is unquestionably the best alternative for renewable energy. Biogas production is an effective solution to manage such wastes. Biogas production mechanism involves acidogenesis and methanogenesis through anaerobic digestion (Fig. 6).

As a preliminary study for value-addition to pineapple peel agro-industrial solid waste, Chulalaksananukul *et al.* (2012) studied the biogas production potential in lab-scale (6 litre capacity) batch experiments at ambient temperature (approx. 30°C), with an initial COD concentration of 20,000mg/l. The conditions such as the type of microorganisms, pH condition, C:N ratio, as well as the organic loading rate affected the process. They reported that, pineapple peel solid waste could generate 48% methane containing biogas successfully in 20 days using indigenous microbes that reduced COD by 61.49% at pH optima of 7.0 and a controlled C:N ratio of 20. Further, higher amount of biogas was produced in fed-batch operating system operated with lower organic loading rate (1kg/m^3/d). Bardiya *et al.* (1996) produced up to 1682ml/day of biogas with a maximum 51% methane content using semi-continuous anaerobic digestion.

Rani and Nanda (2004) reported biogas yields ranging from 0.41-0.67m^3/kg volatile solids from pineapple peels with 41-65% methane content. Lane (1984) and Viswanath *et al.* (1992) used pineapple waste as one of the substrates along with other fruit wastes for biogas generation. Vijayaraghavan *et al.* (2007) used 15% pineapple peel in a mixed fruit peel waste to generate bio-hydrogen at a rate of 0.73m^3/kg volatile solids.

A comparative biogas production efficiency study of various post-harvest pineapple wastes separately from peels, leaves and mixture of peels and leaves with two different inoculum types, *viz.*, Cattle Manure (CM) and Novel microbial Consortia (NC) was conducted which prolonged till the 80th day (cessation of biogas production). Comparing the biogas production potentials among the different substrate-inoculum combinations, NC was more efficient than CM. Pineapple leaves with NC was the best substrate-inoculum combination with maximum (208.28L/kg TS) biogas yield with 72.45% methane, whereas the same substrate with CM as inoculum yielded 35.96L/kg TS biogas with 34.7% methane. Codigestion of peels and leaves yielded 187.19L/kg TS with 56.61% methane with NC, which was 2.2 fold higher compared with CM having a yield of 84.46L/kg TS and 49.2% methane. In all these assays, the total biogas yield with NC was significantly higher than CM inoculum. Also, a high percentage (44.72–61.4%) of volatile solids conversion and methane yield (48.8–72.45%)

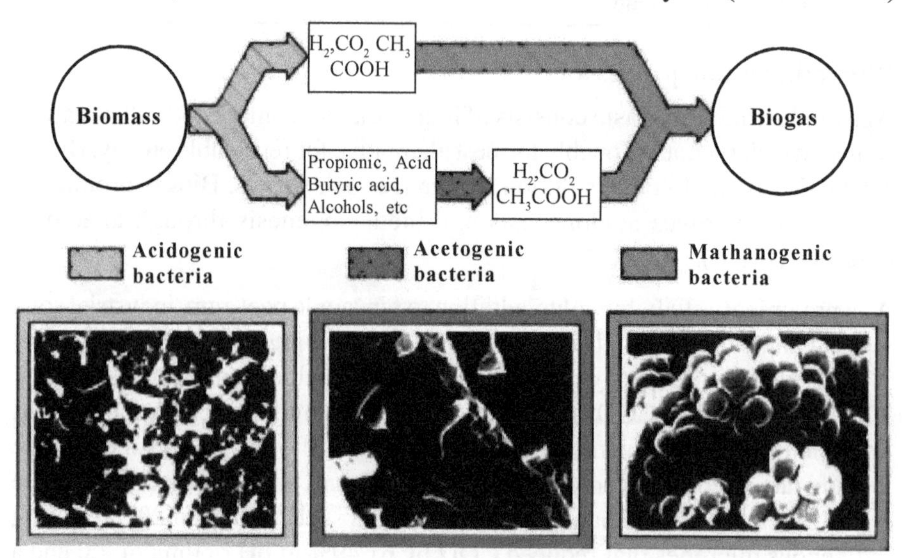

Fig. 6: A basic process flow of the biogas production process and microbial mechanisms (See colour version after references)

using NC were observed. Thus, the efficiency of NC as an inoculum for anaerobic digestion of pineapple wastes instead of cattle manure inoculum is validated for commercial operations.

Conclusion

The energy and environmental crises that the modern world is facing is forcing revaluation of the efficient utilisation or finding alternative uses for natural, renewable resources, using clean technologies. Lignocellulosic biomass holds considerable potential to meet the current energy demand of the modern world, thereby overcoming the excessive dependence on petroleum fuels. The accumulating bulk of agro-waste creates not only ecological issues but also is a loss of huge value-added compounds due to improper bioconversion. There is a global urgency to convert these into value-added products through microbial as well as chemical conversion routes. Advanced biotechnological interventions are crucial to discover and characterise new enzymes, produce homologous or heterologous systems and ultimately lead to low-cost conversion of lignocellulosic biomasses into biofuels and biochemicals. To overcome the current energy problems it is envisaged that lignocellulosic biomass in addition of green biotechnology will be in focus.

From the about 50-60% pineapple processing wastes generated, their high potentials for biofuels and valuable food additives needs to be exploited and calls for emphasis by the academicians and researchers. Research trends are being directed towards lignocellulose biotechnology and genetic engineering for improved processes and products. Biocatalyst having novel actions to utilise pineapple wastes can be explored for sustainable betterment of life. The various unexplored value-added food additives and medicinally important compounds in it are useful in pharmaceutical industries, food and beverage industries, and chemical and fibers industries. Different bioconversion biocatalysts can be explored to source novel compounds of global significance, especially meeting various food and pharmaceutical needs. Improved bioconversion technologies employing recombinant/engineered microbes will definitely help tap important compounds, thereby supplementing into renewable energy sources while addressing other issues.

References

Asgher M, Ahmad Z, Iqbal HMN, 2013. Alkali and enzymatic delignification of sugarcane bagasse to expose cellulose polymers for saccharification and bio-ethanol production. *Industrial Crops and Products*, **44**: 488–95.

Bardiya N, Somayaji D, Khanna S, 1996. Biomethanation of banana peel and pineapple waste. *Bioresource Technology*, **58(1)**: 73–6.

Boucque CHV, Fiems LO, 1988. II. 4. Vegetable by-products of agro-industrial origin. *Livestock Production Science*, **19(1-2)**: 97–135.

Cabrera HAP, Menezes HC, Oliveira JV, Batista RFS, 2000. Evaluation of residual levels of benomyl, methyl parathion, diuron, and vamidothion in pineapple pulp and bagasse (cmooth cayenne). *Journal of Agriculture and Food Chemistry*. **48(11)**: 5750–3.

Chulalaksananukul S, Sinbuathong N, Chulalaksananukul W, 2012. Bioconversion of pineapple solid waste under anaerobic condition through biogas production. *Asia-Pacific Journal of Science and Technology*, **17(5)**: 734–42.

Claassen PAM, Van Lier JB, Lopez Contreras AM, Van Niel EWJ, Sijtsma L, Stams AJM, De Vries SS, Weusthuis RA, 1999. Utilization of the biomass for the supply of energy carriers. *Applied Microbiology and Biotechnology*, **52(6)**: 741–55.

Iqbal HMN, Kyazze G, Keshavarz T, 2013. Advances in valorization of lignocellulosic materials by biotechnology: an overview. *BioResources*, **8(2)**: 3157–76.

Irshad M, Anwar Z, But HI, Afroz A, Ikram N, Rashid U, 2012. The industrial applicability of purified cellulose complex indigenously produced by *Trichoderma viride* through solid-state bio-processing of agro-industrial and municipal paper wastes. *BioResources*, **8(1)**: 145–57.

Isroi, Millati R, Syamsiah S, Niklasson C, Cahyanto MN, Lundquist K, Taherzadeh, MJ, 2011. Biological pretreatment of lignocelluloses with white-rot fungi and its applications: a review. *BioResources*, **6(4)**: 5224–59.

Koskinen PE, Beck SR, Örlygsson J, Puhakka JA, 2008. Ethanol and hydrogen production by two thermophilic, anaerobic bacteria isolated from Icelandic geothermal areas. *Biotechnology and Bioenginering*, **101(4)**: 679–90.

Lal R, 2005. World crop residues production and implications of its use as a biofuel. *Environment International*, **31(4)**: 575–84.

Lal R, 2008. Crop residues as soil amendments and feedstock for bioethanol production. *Waste Management*, **28(4)**: 747–58.

Lane AG, 1984. Laboratory scale anaerobic digestion of fruit and vegetables solid waste. *Biomass*, **5(4)**: 245–59.

Laufenberg G, Kunz B, Nystroem M, 2003. Transformation of vegetable waste into value added products:: (A) the upgrading concept; (B) practical implementations. *Bioresource Technology*, **87(2)**: 167–98.

Morton JF, 1987. *Fruits of Warm Climates*. JF Morton, Miami, USA, ISBN: 0961018410: 517 pp.

Nanda S, Azargohar R, Dalai AK, Kozinski JA, 2015. An assessment on the sustainability of lignocellulosic biomass for biorefining. *Renewable and Sustainable Energy Reviews*, **50**: 925–41.

Nanda S, Mohammad J, Reddy SN, Kozinski, JA, Dalai, AK, 2014. Pathways of lignocellulosic biomass conversion to renewable fuels. *Biomass Conversion and Biorefinery*, **4(2)**: 157–91.

Nunes MCN, Emond JP, Rauth M, Dea S, Chau KV, 2009. Environmental conditions encountered during typical consumer retail display affect fruit and vegetable quality and waste. *Postharvest Biology and Technology*, **51(2)**: 232–41.

Ofori-Boateng C, Lee KT, 2013. Sustainable utilization of oil palm wastes for bioactive phytochemicals for the benefit of the oil palm and nutraceutical industries. *Phytochemistry Reviews*, **12(1)**: 173–90.

Örlygsson J, Sigurbjornsdottir MA, Bakken HE, 2010. Bioprospecting thermophilic ethanol and hydrogen producing bacteria from hot springs in Iceland. *Icelandic Agricultural Sciences*, **23**: 73–85.

Paepatung N, Nopharatana A, Songkasiri W, 2009. Bio-methane potential of biological solid materials and agricultural wastes. *Asian Journal of Energy and Environment*, **10(1)**: 19–27.

Pérez J, Munoz-Dorado J, de la Rubia TDLR, Martinez J, 2002. Biodegradation and biological treatments of cellulose, hemicellulose and lignin: an overview. *International Microbiology*, **5(2)**: 53–63.

Viswanath P, Devi SS, Nand K, 1992. Anaerobic digestion of fruit and vegetable processing wastes for biogas production. *Bioresource Technology*, **40(1)**: 43–8.

Rani DS, Nand K, 2004. Ensilage of pineapple processing waste for methane generation. *Waste Management*, **24(5)**: 523–8.

Reddy LV, Reddy OVS, Wee YJ, 2011. Production of ethanol from mango (*Mangifera indica L.*) peel by *Saccharomyces cerevisiae* CFTRI101. *African Journal of Biotechnology*, **10(20)**: 4183–9.

Sarangi PK, Singh NJ, Singh TA, 2016. Extraction of Bromelain from pineapple wastes. CAU *Research Newsletter*, **7**: 14–15.

Tran VA, 2006. Chemical analysis and pulping study of pineapple crown leaves. *Industrial Crops and Products*, **24(1)**: 66–74.

Upadhyay A, Lama JP, Tawata S, 2010. Utilization of pineapple waste: a review. *Journal of Food Science and Technology Nepal*, **6**: 10-18.

Vijayaraghavan K, Ahmad D, Soning C, 2007. Bio-hydrogen generation from mixed fruit peel waste using anaerobic contact filter. *International Journal of Hydrogen Energy*, **32(18)**: 4754–60.

Vimal OP, Adsule PG, 1976. Utilization of fruit and vegetable wastes. *Research Industry*, **21**: 1–6.

Wyman CE, 1996. Handbook on Bioethanol: Production and Utilization. Taylor & Francis, Bristol, Paris, ISBN: 1-56032-553-4: 420 pp.

Zhang YHP, 2008. Reviving the carbohydrate economy via multi-product lignocellulose biorefineries. *Journal of Industrial Microbiology and Biotechnology*, **35(5)**: 367–75.

10

Optimised Process Considerations for Enhanced Kitchen Refuse Biomethanation

Vijay K. Malesu[1] and Puneet K. Singh

Bioenergy Lab, Biogas Development and Training Centre
School of Biotechnology, KIIT University, Bhubaneswar, Odisha - 751024, India
[1]Corresponding Author: vijay.malesu@gmail.com

ABSTRACT

*Biomethanation is the process of generating energy from anaerobic digestion (AD) of organic refuse. This helps reduce dependency on fossil fuels while reducing CO_2 emission. Organically-rich kitchen refuse (KR) generated daily from the households all over the world could be a good candidate to biogasify, though the quantity and quality of biogas depends on the KR proximate. To enhance biogasification, KR is subjected to physical, chemical and/or biological pretreatment. Physical pretreatments include chopping, grinding, ultrasonication and steam explosion, whereas chemical pretreatments could be by an alkali or a base. Biological treatment involves digestion of the organic matters by individual microbe or its consortium, although it takes place naturally. For biological treatment, standardising the substrate and inoculum ratio is essential. Candidate strains must be mutually compatible to formulate a microbial consortium. Bacillius seems to be a potential candidate as it is a facultative anaerobe exhibiting multiple essential enzyme activities (*viz.*, amylase, cellulase and lipase). Maintaining an optimal pH of 7.0 and temperature of 37°C before and during anaerobic digestion is suggested. Other important factors in digestion include organic loading rate, retention time, C/N ratio, volatile fatty acids (VFA) concentration, etc. VFA content is vital as its accumulation during acidogenesis adversely affects the AD by dropping pH values to unfavourable levels. Acetogens convert VFA to acetate, CO_2 and H_2 in acetogenesis, finally converting it to CH_4 in methanogenesis. For effective methanation,* *methanogen-rich cattle dung or a consortium is supplemented to KR. Biogas from KR is a cheap and useful fuel at household and community levels, as also a successful waste management strategy.*

Keywords: Anaerobic digestion, Biogas, Consortium, Methanation, Volatile fatty acids

Abbreviations used

BMP : biomethanation potential
C/N : carbon to nitrogen ratio
CD : cattle dung
cm^3 : cubic metre
HRT : hydraulic retention time
KC : cattle dung + kitchen refuse
KR : kitchen refuse
OLR : organic loading rate
SSF : solid state fermentation
VFA : volatile fatty acids
% : percent

Introduction

As per the FAO of the United Nations, the volume of food wasted globally is estimated at 1.6 billion tonnes out of which 1.3 billion tonnes was otherwise edible. Public concern regarding the utilisation of resources is promoting the production and consequent use of biogas as a renewable energy resource globally. Anaerobic digestion (AD) of organic refuse can potentially appreciably contribute to replacing fossil fuel with renewable energy. Disposal of kitchen refuse (KR) is a household issue which can be made useful by transforming it biologically to biogas which can be used as fuel for household energy needs. KR is purely organic having high calorific and nutritive values for microbial activity thereby ensuring efficient CH_4 production by several orders of magnitude. One general problem related to such a waste by anaerobic digestion is the inhibition of the digestion process owing to accumulation of various intermediate products and byproducts. The optimum carbon to nitrogen (C/N) ratio for efficient digestion is 20–25 or below, while the C/N ratio in KR is lower, which make it an appropriate substrate for AD. The pH plays a significant role in AD, and the preliminary stages, *i.e.*, hydrolysis to acetogenesis lower the pH. As organic matter in KR is easily hydrolysable to volatile fatty acids (VFAs), it may limit digestion by unfavourably dropping the pH of digester if it builds up rapidly during the course of digestion. The growth of methanogens and, thus, methane production are inhibited at very low pH. The optimum pH for efficient biogas yield in AD is 6.5–7.5.

The biodegradability of waste depends on various factors like the presence of lignocellulosic complex structure comprising of hemicellulose, cellulose and lignin. Cellulose and hemicellulose are composite polysaccharides found in plant cell wall. Hemicellulose is readily hydrolysable owing to its amorphous structure, which is more susceptible than cellulose and lignin structure to enzymatic attack. Cellulose possesses a simple structure and few enzymes can essential digest it. Cellulose solubilisation is reliant on the source of the inoculum, concentration of the biomass and availability of cellulose. Pretreatment is aimed at releasing sugars or enhancing enzymatic contact to carbohydrate polymers. An ideal pretreatment requires minimal energy and inflicts no environmental damage and they are categorised into physical, chemical and biological.

Hydraulic retention time (HRT), the period for which the feedstock stays in a digester, is usually 10-25 days depending on the substrate size and its structural complexity. Average time essential to decompose the organic matter depends on the type and state of feedstock, environmental conditions and the intended use of digested material. Organic substrate containing recalcitrant substrates like cellulose often needs longer HRT than easily degradable ones like the dissolved sugars. Mostly, HRT becomes critical at high loading rates to facilitate degradation. Loading rate is usually more than biomethanation rate which means that not all the material degrades.

Commissioning of an AD system is fragile and an important stage for successful operation. The feed is supplemented with either formulated or natural microbial inocula for this purpose to provide essential biotic environment to initiate digestion. As the rate of substrate degradation and the lag time depends on the quality and quantity of microbial population, the substrate to inoculum ratio is vital.

Stages of biogasification

Scavenging microbes break the complex carbon into smaller substances through a series of steps.

Biogasification involves four metabolic stages as shown in Fig. 1, *viz.*, hydrolysis, acidogenesis, acetogenesis, methanogenesis.

Hydrolysis

The microbial enzymes active in an anaerobic digestion system help hydrolyse insoluble complex organic wastes into simple soluble compounds in the first phase. Thus, lipids degrade to fatty acids by lipase enzymes, carbohydrates degrade to sugars by amylase and cellulase enzymes, and proteins degrade to amino acids by proteases. Thus, a diverse microbial community remains active to degrade various feed stocks in a biodigester. Following equations discuss the breakdown of carbohydrate as a case example. Similar instances are also encountered in other polymeric organics whose equations have been left out to make the discussion precise and more focussed.

$$(C_6H_{10}O_4)_n + 2H_2O \rightarrow C_6H_{12}O_6 + H_2 \text{ (glucose formation)} \quad (1)$$

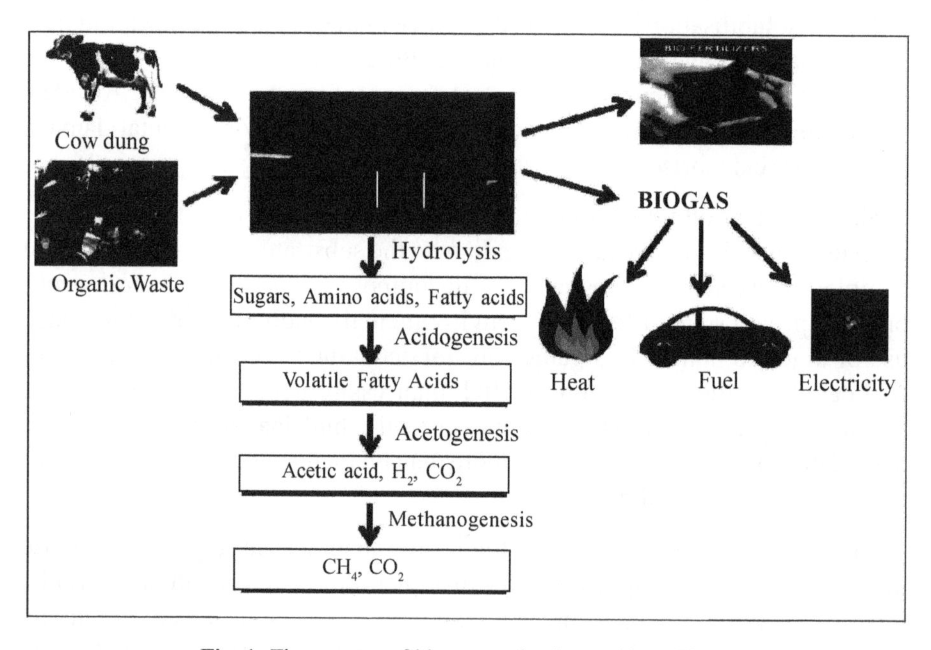

Fig. 1: The process of biogas production and its utilities

Acidogenesis

At the second stage, the acidogenic bacteria transform the end-products of the first stage into short chain volatile acids, ketones, alcohols, hydrogen and CO_2. The most critical end-products are butyric acid ($CH_3CH_2CH_2COOH$), propionic acid (CH_3CH_2COOH), acetic acid (CH_3COOH), lactic acid ($C_3H_6O_3$), formic acid (HCOOH), and methanol (CH_3OH). Equations 2, 3 (Ostrem *et al.*, 2004) and 4 (Bilitewski *et al.*, 1997) represent respectively the three typical acidogenesis reactions converting glucose to ethanol, propionate and acetic acid.

$$C_6H_{12}O_6 \leftrightarrow 2CH_3CH_2OH + 2CO_2 \text{ (reversible reaction)} \quad (2)$$

$$C_6H_{12}O_6 + 2H_2 \leftrightarrow 2CH_3CH_2COOH + 2H_2O \text{ (reversible reaction)} \quad (3)$$

$$C_6H_{12}O_6 \rightarrow 3CH_3COOH \quad (4)$$

Acetogenesis

At this third stage of biomethanation, short chain volatile compounds such as propionate, butyrate and alcohols transform by acetogenic bacteria into acetate, CO_2 and hydrogen. Hydrogen is an important intermediary product; hydrogen concentration of a digester is a sign of its health. Biomethanation takes place only if the hydrogen partial pressure is low enough to permit the transformation of all the acids thermodynamically. For instance, the conversion of propionate to acetate (Eq. 5) is possible only at low hydrogen pressure. Lowering of partial pressure is carried out by various hydrogen scavenging bacteria (Mata-Alvarez, 2003). Glucose (Eq. 6) and ethanol (Eq. 7) and others also convert to acetate during this stage (Ostrem, 2004).

$$CH_3CH_2COO^- + 3H_2O \leftrightarrow CH_3COO^- + H^+ + HCO_3^- + 3H_2 \quad (5)$$

$$C_6H_{12}O_6 + 2H_2O \leftrightarrow 2CH_3COOH + 2CO_2 + 4H_2 \quad (6)$$

$$CH_3CH_2OH + 2H_2O \leftrightarrow CH_3COO^- + 2H_2 + H^+ \quad (7)$$

Methanogenesis

At this fourth and final stage, methanogens convert hydrogen and acetate into methane and carbon-di-oxide. Methanogens are essentially anaerobes from the domain Archaebacteria. Methane production is a prerogative of the methanogenic bacteria (Gerardi, 2003). This process is considered as a rate-limiting step in anaerobic digestion (Gerardi, 2003; Zinder, 1993).

$$CO_2 + 4H_2 \rightarrow CH_4 + 2H_2O \quad (8)$$

$$2C_2H_5OH + CO_2 \rightarrow CH_4 + 2CH_3COOH \quad (9)$$

$$CH_3COOH \rightarrow CH_4 + CO_2 \quad (10)$$

In a single-chamber microbial electrolysis cell, bioconversion of hydrogen to CH_4 take place on the anode or the cathode due to the discharging hydrogen into solution. Methanogenic pathway, which utilises CO_2 and H_2, involves methanogenesis-specific enzymes catalysing unique reactions using novel coenzymes. Methyl-Coenzyme M Reductase (MCR) is an enzyme which is present in Archaea, helps in catalysing the methane formation by uniting the hydrogen donor coenzyme B along with the methyl donor coenzyme M. It contains two active sites and each one is occupied by the nickel containing F_{430} cofactor. MCR presence is considered as a diagnostic indicator of methanogenesis (Lueders *et al.*, 2001; Luton *et al.*, 2002; Narihiro and Sekiguchi 2011). The genomes present in every methanogenic archaea code for at least a copy of the mcrA operon (Thauer, 1998). Composed of two alpha (mcrA), beta (mcrB) and gamma (mcrG) subunits, the mcrA holoenzyme helps catalyse heterodisulphide formation between coenzyme M as well as coenzyme B from methyl coenzyme M and coenzyme B and the following release of methane (Ellermann *et al.*, 1998).

Parameters affecting anaerobic digestion

Parameters of factors affecting anaerobic digestion are pH, temperature, C:N ratio, VFA, organic loading rate, agitation and retention time (Fig. 2).

pH

The pH plays a significant role in the anaerobic digestion. Preliminary stages of anaerobic digestion, *i.e.*, hydrolysis and acidogenesis lower the pH. Growth of methanogenic bacteria and methane production are inhibited in high acidic conditions. Although several Swedish biogas plants reportedly efficiently operate at a pH of 8.0 (Schnürer and Jarvis, 2010), methanogens love a range of 7.0–7.5. Though it is verified that the pH optima for maximum biogas yield in anaerobic digestion is 6.5–7.5, the range fairly vary from plant to plant, and the pH optima vary with substrate and digestion technique (Liu *et al.*, 2008). Excess methanogen population can lead to higher ammonia concentration, thereby raising the pH over 8.0 which inhibit acidogenesis (Lusk, 1999). Sanchez *et al.* (2000) while studying the effect of temperature and pH on the kinetics of methane production from the anaerobic digestion of cattle dung (CD) revealed that the apparent kinetic constants of the biomethanation increased 2.3 times when the initial pH of the influent increased from 7.0 to 7.6 at mesophilic temperature.

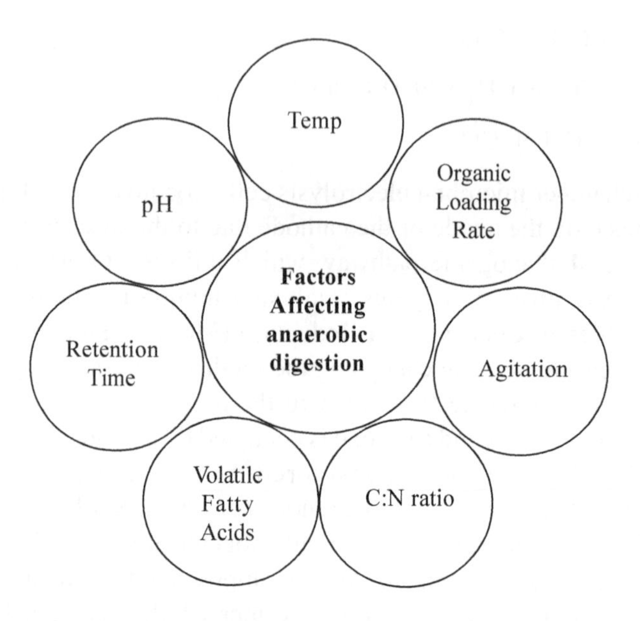

Fig. 2: Parameters affecting the anaerobic digestion

Organic loading rate (OLR)

Biodigestion efficiency is reliant on the loading rate. CH_4 production increased as loading rate reduced (Vartak *et al.*, 1997a). Sreekrishnan *et al.* (2004) reported that an everyday loading rate of about 16 kg VS/m^3 of digester capacity yielded 0.04–0.074 m^3 of gas/kg of dung fed. Another digester which was working at diverse OLRs yielded a highest of 0.36 m^3/kg VS at an OLR of 2.91 kg VS/m^3 per day. As per pilot plant results, maximum yield was noted with a loading rate of 24 kg dung/m^3 though reduction of VS was only two-third of that with low loading rate. An each day loading rate of 16 kg VS/m^3 yielded 0.04–0.074 m^3 of gas/kg feed. Digester operating at diverse OLRs produced a highest yield of 0.36 m^3/kg VS at an OLR of 2.91 kg VS/m^3/day (Sundrarajan *et al.*, 1997).

Smaller particles offer a better surface area for enzymic attack (Hartmann and Ahring, 2006). Hence, the substrate particle size is critical particularly during the preliminary stage in anaerobic digestion. Kim *et al.* (2000) reported a decrease in substrate utilisation rate coefficient in anaerobic digestion with an increase in the average particle size of the substrate. Decreasing the particle size from 100 to 2mm improved fibres degradation and methane yield (Mshandete *et al.*, 2006). Size reduction in recalcitrant fibres between 5–20 mm fibre sizes had no significant difference, and the biogas potential increased by 16% at 2 mm fibre size and 20% at 0.35 mm fibre size (Angelidaki and Ahring, 2000).

Retention time

The HRT (Hydraulic Retention Time), the period for which the feedstock remains in the digestor, is usually 10-25 days. The average time essential for decomposition of organic matter depends upon the type and state of feedstock, environmental conditions and the intended use of the digested material (Ostrem *et al.*, 2004). Organic substrate containing cellulose like recalcitrant often needs a longer HRT than easily degradable materials like the dissolved sugars. Mostly, HRT becomes critical at high loading rates to facilitate better degradation. The loading rate is usually more than the biomethanation rates, which signify that all the substrate is not degraded. Biodigester residue contains non-degraded organic material, inorganic and inert material, salts, and microbial biomass and water.

Temperature

Temperature is another important factor determining the digestor performance; optimal temperature is necessary for the survival and performance of microbial consortia. A minute temperature change (35–30°C and 30–32°C) reduces biogasification rate (Chae *et al.*, 2008). The temperature for anaerobic digesters determines the active methanogens species. Mesophilic digestion reportedly occurs at around 30–38°C, or even at ambient temperatures within a range 20–45°C, where mesophilic microorganisms are primarily active. Similarly, the optimal thermophilic digestion occurs at about 49–57°C (up to 70°C), where thermophiles are active. Hegde and Pullammanappallil (2007) recovered 95% methane post eleven days under thermophilic conditions when compared to 27 days in mesophilic conditions. A reduced biomethanation in thermophilic digestion owing to high metabolic and growth rates, and high microbial death rate here compared to the mesophilic is a major drawback (Duran and Speece, 1997). Also, thermophilic digestion is considered less attractive from the energy input view point (Zaher *et al.*, 2007) as the net energy production from it is less than that of mesophilic process. Although mesophilic process has a longer HRT, its stability makes it trendier in anaerobic digestion (Zaher *et al.*, 2007).

Carbon to Nitrogen (C/N) ratio

Another determinant of the quality of digestion is the substrate C/N ratio. Microbes usually utilise C/N in the ratio of 25-30:1, hence a C/N ratio in case of biodegradable fraction of the C/N within the range of 25–30 is taken as optimum for anaerobic digestion (Kayhanian and Rich, 1995). Anaerobic digestion of banana and plantain peels in equal proportions (Ilori *et al.*, 2007) generated 8,800 ml and 2,409 ml of biogas respectively, acquiring a high volume of 13,356 ml. A low C/N ratio has relatively better nitrogen content, but may lead to toxic ammonia accumulation while a high C/N ratio is unsuitable for microbial metabolism due to substrate recalcitrance and nitrogen deficiency thereby lowering the rate of gas production and solids degradability (Hartmann and Ahring, 2006).

Low nitrogen results in reduced biogas production. High nitrogen content will increase the pH level beyond 8.5 due to the accumulating ammonia which inhibits the microbial activity and consequently gas production. During anaerobic digestion microbes reportedly utilise carbon 25–30 times quicker than nitrogen. Therefore to meet this necessity, microbes require 20–30:1 ratio of C to N with a larger percentage of readily degradable carbon (Bardiya and Gaur, 1997). The C/N is also manipulated by intensity of phosphorus and other trace elements (Speece, 1987).

Agitation

Proper agitation of substrate ensures resourceful transfer of organic material for the actively multiplying microbes, discharge of gas bubbles present in the substrate and check sedimentation of heavy particles. Improved substrate-bacteria interactions demand a certain level of agitation although more of it reduces biogas production. Consistent intermittent agitation for several times a day is determined by the reactor type, agitator type, and total solids content in the feedstock (Burton and Turner, 2003). The employed agitations vary significantly (Karim *et al.*, 2005). Propellers with screw in a central tube help if feedstock is of low viscosity for downward/upward movement depending on the torsion. Recirculation of gas through the reactor base or hydraulic stirringing by recirculation of digestate by a pump helps to adequately mix. When installed internally, agitation systems not only affect digestion but are costly to install, maintain and run. Well-organised mixing system helps increase productivity and reduces the cost. Slow agitation allows better shock loading absorption than high speed (Gomez *et al.*, 2006). Reduced mixing enhances the performance and helps unstable digester stabilise (Stroot *et al.*, 2001).

Volatile fatty acid

The concentration of total and individual volatile fatty acid (VFA) is considered an important controlling parameter in particularly liquid-phase digestion, and failure sign. The VFA, main source for biomethanation, generated in hydrolysis and acidification is a vital intermediate (Buyukkamaci and Filibeli, 2004), though its high concentration decreases the pH and harms methanogens thereby ultimately leading to digester failure. When VFA concentration reaches 13000–21000 mg/l it reportedly inhibits acidogenesis. Bouallagui *et al.* (2004) found butyrate as the major VFA product in anaerobic digestion of fruit and vegetable waste (FVW) and municipal solid waste (MSW), reaching beyond 4000 mg/l. Siegert and Banks (2005) reported differential consequence of VFA on the metabolically diverse hydrolysis, acidogenesis and biogasification phases linked with batch anaerobic digestion with increasing concentrations.

Pretreating the biodigestion substrates

Enhanced degradation and biogasification rates necessitate physical, chemical and biological pretreatments. A few of these methods popular in field applications are discussed below. Table 1 shows some pre-treatment factors and their effect on the net CH_4 yield although the figures may vary owing to deviations in the physicochemical factors in field conditions.

Table 1: Pre-treatment factors and their effect on the net methane yield

Type	Pretreatment	% increase in CH_4 yield	Reference
Physical	Ultrasonication	17-30	Závacký *et al.* (2010)
	Steam explosion	50	Dereix *et al.* (2005)
	Milling	5-25	Delgenes *et al.* (2002)
	High temperature	35–70	Lissens *et al.* (2004)
Chemical	Acid hydrolysis	74	Kivaisi *et al.* (1994); Chen *et al.* (2007)
	Alkaline hydrolysis	2-fold increase	Dar and Tandon (1987); Mouneimne *et al.* (2003)
Biological	*Bacillus*	210	Miah *et al.* (2004)
	B. licheniformis	57	Merrylin *et al.* (2013)
	B. aryabhattai	18	Ojha (2015)
		76	Zhong (2011)

Physical pretreatment

Physical treatment increases the accessible surface area and pore size, and decreases crystallinity and degrees of cellulose polymerisation in substrates. Different physical processing like milling, irradiation and high pressure steaming improve the enzymic hydrolysis or degradability of recalcitrant materials. Mechanical treatment uses numerous approaches for physically disintegrating the cells and partially solubilising the matter. Engelhart *et al.* (2000) worked on the consequence of high-pressure homogeniser on mechanical disintegration on AD of sewage sludge and achieved 25% augment in VS reduction. Hartmann *et al.* (2000) concluded that an enhancement of up to 25% in biogas from pretreated fibres in manure feedstock.

Milling: Milling decreases particle size as well as crystallinity. Particle size reduction leads to increased availability of precise surface area and a reduced degree of polymerisation (Palmowski and Muller, 1999). According to Delgenes *et al.* (2002) milling increase hydrolysis rate and caused 5–25% increased methane yield. Mechanical disintegration to physically disrupt cellular material enhanced the net biogas production (Baier and Schmidheiny, 1997).

Irradiation: Gamma rays, electron beam and microwaves irradiation progress enzymatic break down of recalcitrant substrates. In grouping with the other treatment methods like acid treatment, irradiation expedites enzymatic hydrolysis further, particularly the enzymic degradation of cellulose to glucose.

Steam explosion (moist heat): Steam explosion pretreatment has successfully improved biogasification from various materials, *viz.*, municipal solid wastes (Solheim *et al.*, 2004), municipal wastewater sludge (Dereix *et al.*, 2006).

Ultrasonic treatment: High-frequency sound waves disintegrate the substrates, thereby increasing the available surface area for the enzyme action. Závacký *et al.* (2010) obtained 13–29.5% enhancement in biogas and 16.9–29.5% in methane yield respectively from maize silage by ultrasonic treatment. Wang *et al.* (1999) reported that 30 min ultrasonic pretreatment of activated sludge increased methane production by 64%.

Stirring: Continuous/continual stirring of digester contents enhances chance-meeting of microbes with the substrate, ultimately resulting in improved digestion. However, this needs to be kept to the barest minimum as biodigestion is essentially an anaerobic process. There are various electronically controlled mixing devices available in the market that can be employed for the purpose. Non-mechanical stirring such as through gas recirculation is also a good alternative to mixing leading to increased gas production (Nopharatana *et al.*, 2007).

Chemical pretreatment

The reaction time, temperature and O_2 pressure are critical in wet oxidation. Chemical treatment is exothermic and hence the reaction commences by self-supporting heat. Palonen *et al.* (2004) suggested treatment of biogasifying materials at a high temperature above 120°C for 30 min. Lissens *et al.* (2004) reported 35–70% improved methane yield from lignocellulosics, food and yard refuse, and digested biowastes with a 15 min 185–220°C pretreatment at an oxygen pressure of 0–12 bar. Alkaline solutions, such as, sodium and calcium hydroxides, or ammonia eliminate lignin and a fraction of the hemicelluloses thereby increasing the ease of access of enzyme to the cellulose component of a lignocellulosic biomass. Dar and Tandon (1987) reported that alkali treated (1% NaOH for 7 days) CD-supplemented plant residues improved the microbial digestibility by 31–42% and a two-fold increase in biogasification. Knezevic *et al.* (1995) reported improved gas production with increasing sodium hydroxide dosage. Thermal pretreatment with alkalis like sodium hydroxide, urea and ammonia combined reportedly has shown better results than individual pretreatments.

Studies on alkali pre-treatment of lipid rich wastewaters revealed that addition of alkaline chemicals like the hydroxides of sodium, potassium and calcium to abattoir wastewater plus oil mill effluents enhanced break down of solid fatty residuals (Beccari *et al.*, 1999; Masse *et al.*, 2001; Mouneimne *et al.*, 2003). Further, Masse *et al.* (2001) opined that as it results in an unacceptable increase in biodigestor pH, alkali pretreatment may be kept to the minimum. However, the advantage with such pretreatments is that these can be carried out at ambient temperature thereby saving on the energy input. Cellulose-lignin union is known to be the main limiting issue on long-term AD, particularly in case of agro-residues. Acids pretreatment of lignocellulose at ambient temperature increases hydrolysis rate to solubilise hemicelluloses. Solubilised hemicelluloses (oligomers) produce monomers, and volatile products (Ramos, 2003). This facilitates the accessiblility of enzymes to cellulose. During the process, solubilised lignin condenses and precipitates (Liu and Wyman, 2003). Chen *et al.* (2007) used acetic and nitric acids to pretreat lignin in organic waste for biogas production. Kivaisi and Eliapenda (1994) reported 74% enhanced biogas production in coconut fibers by treating with hydrochloric acid. Solubilisation of hemicellulose and precipitation of solubilised lignin are more distinct in pretreatments with strong acids.

Biological pretreatment

Disposal of huge quantity of biodegradable refuse without proper treatment is an environmental nuisance. Microbial role in biodigestion is significant due to their capacity to utilise a range of nutrient sources, the foundation of decomposition of dead matter in environment. KR is nutrient rich that contains elevated levels of lipids, proteins, carbohydrates and other organic molecules to support abundant microbial population. Biological pretreatment improved biogas and methane production by 120% and 134% respectively in *Agropyron elongatum* 'BAMAR' (Lalak *et al.*, 2016). Zhong (2011) reported an improved biogas and methane yield by 33.07% and 75.57% respectively in biologically pretreated corn straw. The potentially better degradability of KR shows great promise as a biogasifying substrate. Codigestion of waste paper and algal sludge was helpful and offered two benefits, a balanced C/N ratio, optimised range being 20–25/1 for enhanced cellulase activity. A boost in cellulase activity can be useful in the degradation of algal sludge that could supply nutrients, which finally would enhance CH_4 production (Yen *et al.*, 2007). Miah *et al.* (2004) noted a 210% improvement in biogas production in thermophilic digestion (at 65°C) due to activity of protease from the *Geobacillus* sp. strain AT1. KR pretreated with 5.0 U/g amylase of *Bacillus aryabhattai* (KIIT BE-1) enhanced the BMP by 18% (Ojha, 2015). *Bacillus licheniformis* pretreatment of MSW improved biogas production by 57% (Merrylin *et al.*, 2013).

Role of microbial enzymes

Enzymes play a vital role to catalyse a reaction. Since its synthesis and secretion is expensive (some being dogmatic methods), micro-ecological schemes are developed to guarantee their availability for catalysis. Numerous parameters affect the structure of enzymes, for instance the temperature and pH, because of which enzymes tend to evolve with time. Temperature is a critical parameter that roots to enzyme denaturation by raising the molecular quivering responsible for breaking the intermolecular bonds. Similarly, pH is another such parameter. Amylase enzymes are unstable at low pH levels and show highest enzyme stability at 5.0–8.0 (Nerkar *et al.*, 2011; Sajedi *et al.*, 2005).

Amylase

Amylase, a well-known starch degrader, has received immense attention due to their apparent scientific relevance and commercial significance. Out of many amylases, *á*-amylases are prominent; they hydrolyse *á*-1,4 glucosidic linkages in starch and generate smaller dextrin and oligosaccharide subunits. According to the sequence based classification of glucoside hydrolases (GH-13) they are classified into 13 families (Henrissat *et al.*, 1996). Some subunits arrange between themselves and form an active site in amylase molecule, each one of which contain a glucose residue (Talamond *et al.*, 2005). Synthesised by a few selected fungal and bacterial strains, it has numerous industrial applications. Amylases are used in elucidating beverages (beer or fruit juices), enhancing the digestibility of fibrous animal feed (Gavrilescu *et al.*, 2005; Van Der Maarel *et al.*, 2002). Paper industries employ such microbially obtained amylases (Amizyme - PMP fermentation products, Peoria, USA; Termamyl, Fungamyl, BAN - Novozymes, Denmark, and α-amylase G9995 - Enzyme Biosystems, USA). Amylases are used in new-age enzyme-based (containing 90% active enzymes) detergents, and most liquid detergents (Gupta *et al.*, 2003; Hmidet *et al.*, 2009). Among the various amylase-producing bacteria a few promising ones are, *Bacillus* sp. *(subtilis, tequilensis, licheniformis, barbaricus)*, *Aeromonas veroni* and *Stenotrophomonas maltophilia.*

However, the applications of the enzyme are limited when it's derived from such fungi and bacteria (Sivaramakrishnan *et al.*, 2006). Thus, search is on for novel sources to extend their applicability. It was found that, peptone and yeast extract are the favourable media to synthesise amylase by *Bacillus cereus* (Teodoro and Martins, 2000). Another strain capable of producing thermostable alkaline α-amylase was reported by Annamalai *et al.* (2011). Vijayabaskar *et al.* (2012) observed that yeast extract is a better nitrogen source for amylase isolation. Divakaran *et al.* (2011) maximised α-amylase production from *B. licheniformis* strain MTCC 2618 by using starch (3.64 IU/ml/min), rice powder (2.93 IU/ml/min), wheat powder (2.67 IU/ml/min), and ragi powder (2.36 IU/ml/min). *B. subtilis* is a facultative 'anaerobe', with a wide range of enzymatic activity, *e.g.*, amylase, cellulase, lipase, caseinase, gelatinase etc. (Barros *et al.*, 2013). Unakal *et al.* (2012) maximised α-amylase production (24 hrs, pH 7.0, 35°C) from *B. subtilis* by using banana waste as a substrate in solid state fermentation (SSF). The α-amylase produced by *B. licheniformis* has wide use in food, fermentation, textile and paper industries (Pandey *et al.*, 2000). *Manatsirikiat (1999)* used pineapple peels to produce amylase from B. licheniformis which was isolated from biodigestor maintained at mesophilic conditions.

Cellulase

Cellulose in paper, single-cell protein, glucose, and sorbitol can be transformed into energy sources (Coral *et al.*, 2002). Cellulase enzyme is an enzyme complex – the exoglucanase, most importantly, cleaves non-reducing end of a cellulose chain and splits elementary fibrils from the crystalline cellulose, and $\hat{a}$-1,4-glucosidase hydrolyses cellobiose and water soluble cellodextrin to glucose (Gupta *et al.*, 2011). Cellobiohydrolase I (Cel7A) and cellobiohydrolase II (Cel6A) are the chief cellulase proteins with major commercial worth. About 10% of the extracellular extracts are represented by endoglucanases and hemicellulases (Chundawat *et al.*, 2011). Biofuels and biorefinery products can be manufactured by using cellulase (Kubicek, 2012). Some cellulolytics are *Trichonynpha, chlostridium, Bacteroides succinogenes, Butyrivibrio fibrisolvens*, *Actinomycetes*, *Ruminococcus albus* and *Methanobrevibacter ruminntium*. Cellulase enzyme complex was important from the present study point of view owing to the fact that an appreciable quantity of cellulose and cellulose-like recalcitrant polysaccharides are commonly present in a KR.

Lipase

Lipase catalyses free fatty acid production, esterification of oils/fats and synthesis of esters and peptides. Prominent lipase producers are *B. amyloliquifaciens*, *Bacillus subtilis*, *B. licheniformis*, *Serratia marsescens*, *Pseudomonas aeruginosa* and *Staphylococcus aureus*. Many nutritional as well as physicochemical factors influence microbial production of extracellular lipase (Tari *et al*., 2007). Several bacterial species produce lipases with different enzymological properties and specificities (Kashmiri *et al*., 2006). Ghaima *et al*. (2014) produced maximal lipase in *Bacillus cereus* at 35°C with pH 8.0. Devi *et al*. (2012) predicted maximal lipase production by *B. subtilis* (16.627 U/min/ml) in optimised culture conditions (yeast extract - 9.3636 g, $CaCl_2$ - 0.8986 g, 43½ hrs incubation). Yedavalli *et al*. (2013) found that almost 65% of the surface of *Bacillus subtilis* lipase was engaged by the loops, and to improve its stability and activity they subjected all the 91 amino acids to site-saturation mutagenesis as the loops were superficial, substitution-tolerant, and dynamic in the protein.

Caseinase

Caseinase hydrolyses milk protein to casein. Milk casein solubilises against phosphate buffer by dialysing, or by dilution. Milk protein hydrolysates (MPH) are utilised in nutrition usually as useful food ingredient (Abd El-Salam and El-Shibiny, 2017). Various bacilli show good caseinase activity (Bairagi *et al*., 2004; Deshmukh, 2013). *Bacillus subtilis*, *B. licheniformis*, *B. cereus*, *B. mascerans*, *Butyrivibrio fibrisolvens*, *Prevotella ruminicola* and *Eubacterium ruminantium* are few organisms of interest. This enzyme, in combination with other enzymes, could play a critical role as an active protease to degrade kitchen refuse.

Co-substrate biogasification of KR

Various types of biomass which are composed of macronutrients such as carbohydrates, lipids, cellulose, proteins and hemicelluloses are appropriate candidates for biogasification. Theoretically the production of biogas differs mainly due to the existence of three chief macronutrients which are lipids, protein and carbohydrates. Apart from these lignocellulosic wastes were also studied extensively for the process of co-digestion; the chief nutrients present in this waste are cellulose and hemicelluloses. The degradability pattern of these nutrients differs mostly. Lipids are known to be the peak producers of biogas with maximum HRT owing to retarded biodegradability, while proteins and carbohydrates get convert at a faster rate but with low biogas yield.

Co-digestion of food waste with manure

Food division is a massive producer of possible waste for energy revival. Several works on co-digesting food waste with cattle manure exist. Food waste primarily consist of vegetable and fruit waste, KR, fish, poultry and meat waste etc. Further, Zhang *et al.* (2011) also studied the anaerobic co-digestion of CD and food waste so as to determine the main factors responsible for biogas and CH_4 production. They were successful in determining the main factors for both semi continuous as well as batch tests which revealed that methane yield could be increased by digestion of food waste with CD (ratio of 2) and C/N ratio of 15.8. While in batch set-up the methane production was increased by 41.1% corresponding to the methane production of 388 ml/g VS and in semi-continuous mode whereas methane production at organic loading rate of 10g VS food waste enhanced by 55.2% corresponding to methane production of 317 ml/g VS. Sagagi et al used vegetable and fruits waste such as pineapple, pumpkin, orange, spinach as well as CD individually to study their biogas producing capability and found that highest weekly biogas production rate was in the order of CD (1554 ml); pineapple waste (965 ml); orange waste (612 ml); pumpkin waste (373 ml); spinach waste (269 ml) respectively.

De Vrieze *et al.* (2015) reported enhanced methane production by codigesting molasses with anaerobic sludge. Anaerobic codigestion of *Phragmites australis* and KR with addition of clinoptilolite not only increased the production of methane but also enhanced the internal mechanism of nitrogen and organics control for the improved yield of methane and biogas. Lee *et al.* (2012) reported methane production from KR, fat, grease and oil as co-substrates in AD of activated sludge. Gannoun *et al.* (2007) anaerobically digested combined olive mill and abattoir wastewater at 37 and 55°C and concluded that the latter removed COD and yielded biogas better than the former, and the performance remained the same even at a high organic loading.

Biogas production from various wastes

Anaerobic codigestion of diverse organic refuse together can perk up nutrient balance, reduce potentially toxic compounds like sulphur-containing substances, and consequently boost processing aptitude and biogas yield (Mahanty *et al.*, 2013). Tewelde *et al.* (2012) reported 74% organic solids conversion while codigesting brewery waste (BW) and CD, and achieved a maximum 69% methane yield with a CD/BW ratio of 70:30. Alvarez and Liden (2008) reported enhanced methane production at 80% by codigesting slaughter house with various other substrates. Similar observations are reported by Ukpai *et al.* (2012) who attained 76% methane yield by co-digesting CD and cassava peel. Ward (2008) showed 82% conversion of volatile solids while optimising anaerobic digestion of agro wastes with an increased methane yield. According to Malesu *et al.* (2016) the cumulative mean biogas production (ml) levels were 1281, 4448 and 3256 in KR, CD and KC, respectively. Methanation started by the 7th day in CD and KC, and the biogas quality slowly improved with the methane content reaching up to 63.65% and 53% respectively by the 21st day in batch operation. Discernible methanation in KR took longer time and also the quality and quantity remained relatively low. It was observed that the biogasification of KR without a combination of interventions would remain low, and there was a risk of system failure due to the accumulating high level of VFAs. Though challenges were obvious, KR was a promising biogasification candidate thus opening a plethora of opportunity to extend this biogas technology even to urban and periurban locations that are low on cattle resources *albeit* rich in other organic refuse.

Microbes in anaerobic digestion

Microorganisms are ubiquitous and their diversity in the environment is vast. They play crucial roles in geochemical cycling. Their contribution towards the environment is yet to be duly evidenced and appreciated. Multiple catalogues of microbially-mediated carbon and nitrogen cycling includes decomposition of materials that are extensively observed in crop rhizospheres. From the entire environmentally active microbial forms, *Bacillus* is the most common. It has broad industrial applications. The most significant role of these is their capacity to utilise nutrients obtained from decomposing dead organic matters in the environment. *Bacillus* has proved to be a promising strain particularly in environmental and clinical biotechnology, and thus various genetic diversity studies on this challenging genus have been popular amongst the researchers. *Bacillus* is of extraordinary interest for biotechnological processes owing to their remarkable thermo-stability and efficient enzyme expression systems (Prakash *et al.*, 2009). Many bacilli show same morphological and biochemical characteristics and so it is still not easy to differentiate the isolates (Harrel *et al.*, 1995), especially without molecular differentiation.

Microbial consortium as inocula for biogas production

Different AD phases include considerably diverse species of symbiotic microbes that can be generally classified in two classes or groups: methanogens and acidogens. As methanogenesis is supposed to be rate limiting in the overall AD, accurate management of the methanogenic phase has been a key factor in effectively operating most anaerobic processes. Methanogens belong to Archaea which are recognised to have perceptive growth and active behaviour in relation to diverse substrates and operation settings (Calli *et al.*, 2005).

Microbial inocula play an important role in degrading the organic matter, biogasification, and turn-over time in anaerobic digestion (Malik *et al.*, 2012). Selecting an active and appropriate inoculum prevents process failure during the start-up phase (Li *et al.*, 2010). Inocula from diverse sources are employed under similar (mesophilic temperature, neutral pH) operating environment. The source of inoculum not only influences the amount of biogas formed but also the progress of AD. The choice of inocula to activate a full-scale digester is critical and vital in surmounting the acidification stage. Microbial consortia differ in aspects akin to enzyme profile, species profile, biofilm-forming behaviour, nutritional needs, and physiological qualities (Jensen *et al.*, 2009). Microbial performance and biochemical profile of a consortium greatly influence the biogas production performance, particularly in lignocellulosic biomass (Griffin *et al.*, 1998).

Inocula are sourced ideally from anaerobic digesters or the digestate with appropriate and diverse group of effective microbes. Sludge is a good source for aerobic/anaerobic treatment of organic refuse. According to Lopes *et al.* (2004), choosing the refuse and the inocula requires appropriate ratio along with estimated anaerobic biodegradability. In case of solid refuse, highly active inocula lessen the experimental period or decrease the quantity of inocula necessary in full-scale batch digester and thereby the related digester capacity (Obaja *et al.*, 2003). Foster *et al.* (2007) studied the consequence of inoculum on anaerobic thermophilic degradation of individual organic fractions. During a 60-d period, they saw that the reactor removed 44% COD and 43% VS, and superior volumetric biogas generation of 78.9mL/day (35.6mL CH_4/day), with a CH_4 yield of 0.53L CH_4/g VS in stabilisation phase.

Microbial treatment of biological (organic) wastes

This involves anaerobic microbial digestion of the refuse by employing the microbes as multienzyme biofactories. Anaerobic technique is a better option than aerobic treatment (open-air composting) to stabilise and recycle organic refuse. Anaerobic digestion of organic materials generates biogas that comprises 50–70% CH_4, 30–40% CO_2 and traces of CO, H_2S, NH_3, H_2, O_2 and moisture. There are some of the potential *Bacillus* strains which have the potential of degrading the components of KR at a much faster rate. Table 2 shows some latest reports on the enzymatic activities of various *Bacilli*.

Table 2: Compilation of the enzymatic activities of various reported *Bacilli*

Strains	Enzyme Activity	References
B. subtilis	amylase, cellulase, lipase	Barros *et al.*, 2013; Kokab *et al.*, 2003; Patagundi *et al.*, 2014
	α-amylase	
	cellulase	
	extremely acidic lipases, thermo tolerant acidic lipase (TAL) and acidic lipase (AL)	Saranya *et al.*, 2014; Jurdao *et al.*, 2004; Tari *et al.*, 2007; Dhevahi *et al.* (2014)
B. tequilensis	Amylase	Tiwari *et al.*, 2014; Gatson *et al.*, 2006
	Cellulase	Malesu *et al.*, 2014; Gatson *et al.*, 2006
	Lipase	Bonala *et al.*, 2012
B. licheniformis	α-amylase	Divakaran *et al.*, 2011; Alk and Ibrahim, 2011; Pandey *et al.*, 2000
	Lipase	Khyami-Horani, 1996; Xi *et al.*, 2014
B. cereus	Amylase	Teodoro and Martins, 2000; Anto *et al.*, 2006
	Lipase	Ghaima *et al.*, 2014
	Cellulose	Patagundi *et al.*, 2014

B. subtilis: *B. subtilis* is a facultative anaerobe, with a wide range of enzymatic activity, *e.g.*, amylase, cellulase, lipase, caseinase, gelatinase etc. (Barros *et al.*, 2013). Kokab *et al.* (2003) maximised α-amylase production (9.06 IU/mL/min; 24 hrs, pH 7.0, 35°C) from *B. subtilis* through SSF of banana peel. *B. subtilis* has the potential to also produce extremely acidic lipases, thermo-tolerant acidic lipase and acidic lipase (Saranya *et al.*, 2014). Many physicochemical (pH and temperature) and nutritional (carbon, nitrogen and lipid) factors affect microbial production of extracellular lipase (Jurdao *et al.*, 2004). Under defined culture conditions, Dhevahi *et al.* (2014) reported an increased lipase production in *Bacillus subtilis* BDG-8 by optimising metal concentration, pH and temperature. Patagundi *et al.* (2014) could maximise cellulase production in *B. subtilis* (0.357 IU/ml/min).

B. tequilensis: *B. tequilensis* produced good amylase in presence of starch, peptone, and Ca^{2+} ions (Tiwari *et al.*, 2014). Pradhan *et al.* (2013) isolated a halophilic biosurfactant-producing *B. tequilensis* from Chilika Lake. BLAST search revealed that the partial 16S rRNA gene sequence of *B. tequilensis* was very similar to the *B. subtilis* subgroup members (Malesu *et al.*, 2014). According to Gatson *et al.* (2006) *B. tequilensis* grows at 5.5–8.0 pH and 25–50 °C. The spores are located centrally, without swollen sporangia. The colonies after 24 h at 37 °C are yellowish in colour, round, smooth and about 3.8 mm in diameter. Casein, starch and gelatin can be easily hydrolysed by it. *B. tequilensis* has the potential to reduce nitrite to nitrogen gas.

B. licheniformis: The α-amylase produced by *B. licheniformis* has ample use in food industries (Alk and Ibrahim, 2011). Manatsirikiat (1999) produced amylase from B. licheniformis isolated from biodigestor maintained at 35°C using pineapple peels. Both capable to produce extracellular enzymes, B. licheniformis plays dominant role in nutrient cycling (Rey *et al.*, 2004). Lipases from thermotolerant *B. licheniformis* (Khyami-Horani, 1996) have good applications. Sangiliyandi and Gunasekaran (1996) found high lipase production by *B. licheniformis* upon addition of olive oil in the medium. Xi *et al.* (2014) reported maximal lipase production (59.03±0.26 U/ml) in 7 days by *B. licheniformis* A3 with *Jatropha* seed cake as substrate.

B. cereus: Teodoro and Martins (2000) found that, peptone and yeast extract are the favorable media to synthesise amylase by *Bacillus cereus*. *B. cereus* MTCC 1305 can produce *á*-amylase through SSF by utilising wheat bran and rice flake refuse (Anto *et al.*, 2006). Patagundi *et al.* (2014) could maximise cellulase production in *Bacillus cereus*, *B. subtilis* and *B. thuringenesis* by using pods of *Acacia arabica*, *Bauhinia forficata*, *Cassia surattensis* and *Peltophorum pterocarpum*. *Bacillus cereus* produced appreciable amount of lipase when maintained at 37°C with pH 7.0 (Mohan and Palavesam, 2012). Senthilkumar and Selvakumar (2007) reported lipase production by *B. cereus* till 6.2 U/ml from stored food cereals.

Malesu (2016) found *B. subtilis*, *B. tequilensis*, *B. cereus* and *B. licheniformis* as ecologically compatible. These were clearly good candidates to formulate an effective microbial consortium 'ProbSuTeCeLiKIIT'. ProbSuTeCeLiKIIT exhibited better enzymatic activities in a consortium than each individual member. Though 1.0% CD-supplementation helped in KR biomethanation, ProbSuTeCeLiKIIT as inoculum enhanced biomethanation and biogasfication potentials of KR by 13.75% and 11.05% respectively.

Conclusion

Hence, appropriate pretreatment of the substrate is essential to maximise biogas and CH_4 yields. Substrate and inoculum ratio standardisation is necessary for microbial pretreatment. While preparing microbial consortia, it is important that the strains used are mutually compatible with each other. The efficacy of the formulated consortium can be checked by evaluating the biogas and biomethanation potentials of substrates and by studying the various physicochemical parameters before and after digestion in wet-lab, alongwith employing statistical tools.

References

Abd El-Salam MH, El-Shibiny S, 2017. Preparation, properties and uses of enzymatic milk protein hydrolysates. *Critical Reviews in Food Science and Nutrition*, **57(6)**: 1119–32.

Ahmed A, Ibrahim HM, 2011. A potential new isolate for the production of a thermostable extracellular-amylase. *African Journal of Bacteriology Research*, **3(8)**: 129–37.

Alvarez R, Liden G, 2008. Semi-continuous co-digestion of solid slaughterhouse waste, manure, and fruit and vegetable waste. *Renewable Energy*, **33(4)**: 726–34.

Angelidaki I, Ahring BK, 2000. Methods for increasing the biogas potential from the recalcitrant organic matter contained in manure. *Water Science and Technology*, **41(3)**: 189–94.

Annamalai N, Thavasi R, Vijayalakshmi S, Balasubramanian T, 2011. Extraction, purification and characterisation of thermostable, alkaline tolerant á-amylase from *Bacillus cereus*. *Indian Journal of Microbiology*, **51(4)**: 424–9.

Anto H, Trivedi U, Patel K, 2006. á-Amylase production by *Bacillus cereus* MTCC 1305 using solid-state fermentation. *Food Technology and Biotechnology*, **44(2)**: 241–5.

Baier U, Schmidheiny P, 1997. Enhanced anaerobic degradation of mechanically disintegrated sludge. *Water Science and Technology*, **36(11)**: 137–43.

Bairagi A, Sarkar Ghosh K, Sen SK, Ray AK, 2004. Evaluation of the nutritive value of *Leucaena leucocephala* leaf meal, inoculated with fish intestinal bacteria *Bacillus subtilis* and *Bacillus circulans* in formulated diets for rohu, *Labeo rohita* (Hamilton) fingerlings. *Aquaculture Research*, **35(5)**: 436–46.

Bardiya N, Gaur AC, 1997. Effects of carbon and nitrogen ratio on rice straw biomethanation. *Journal of Rural Energy*, **4(1-4)**: 1–16.

Barros FFC, Simiqueli APR, de Andrade CJ, Pastore, GM, 2013. Production of enzymes from agroindustrial wastes by biosurfactant-producing strains of *Bacillus subtilis*. *Biotechnology Research International*, **2013**: 1–9.

Beccari M, Majone M, Riccardi C, Savarese F, Torrisi L, 1999. Integrated treatment of olive oil mill effluents: effect of chemical and physical pretreatment on anaerobic treatability. *Water Science and Technology*, **40(1)**: 345–55.

Bilitewski B, Härdtle G, Marek K, 1997. *Waste Management*. (Original German translated to English by H Boeddicker and A Weissbach). Springer-Verlag, New York.

Bonala KC, Mangamoori LN, 2012. Production and optimization of lipase from *Bacillus tequilensis* NRRL B-41771. *International Journal of Biotechnology Application*, **4(1)**: 134–6.

Bouallagui H, Torrijos M, Godon JJ, Moletta R, Cheikh RB, Touhami Y, Delgenes JP, Hamdi M, 2004. Two-phase anaerobic digestion of fruit and vegetable wastes: bioreactors performance. *Biochemical Engineering Journal*, **21(2)**: 193–7.

Burton CH, Turner C, 2003. Manure management: Treatment strategies for sustainable agriculture. 2nd Edition, Silsoe Research Institute, Silsoe, UK, ISBN: 0-9531282-6-1: 472 pp..

Buyukkamaci N, Filibeli A, 2004. Volatile fatty acid formation in an anaerobic hybrid reactor. *Process Biochemistry*, **39(11)**: 1491–4.

Calli B, Mertoglu B, Inanc B, Yenigun O, 2005. Community changes during start-up in methanogenic bioreactors exposed to increasing levels of ammonia. *Environmental Technology*, **26(1)**: 85–91.

Chae KJ, Jang AM, Yim SK, Kim IS, 2008. The effects of digestion temperature and temperature shock on the biogas yields from the mesophilic anaerobic digestion of swine manure. *Bioresource Technology*, **99(1)**: 1–6.

Chen Y, Jiang S, Yuan H, Zhou Q, Gu G, 2007. Hydrolysis and acidification of waste activated sludge at different pHs. *Water Research*, **41(3)**: 683–9.

Chundawat SPS, Beckham, GT, Himmel ME, Dale BE, 2011. Deconstruction of lignocellulosic biomass to fuels and chemicals. *Annual Review of Chemical and Biomolecular Engineering*, **2:** 121–45.

Coral G, Arikan B, Ünaldi MN, Güvenmez H, 2002. Some properties of crude carboxymethyl cellulase of *Aspergillus niger* Z10 wild-type strain. *Turkish Journal of Biology*, **26(4)**: 209–13.

Dar GH, Tandon SM, 1987. Biogas production from pretreated wheat straw, lantana residue, apple and peach leaf litter with cattle dung. *Biological Wastes*, **21(2)**: 75–83.

De Vrieze J, Saunders AM, He Y, Fang J, Nielsen PH, Verstraete W, Boon N, 2015. Ammonia and temperature determine potential clustering in the anaerobic digestion microbiome. *Water Research*, **75**: 312–23.

Delgenes JP, Penaud V, Moletta R, 2002. Pretreatments for the enhancement of anaerobic digestion of solid wastes. *In*: JM Alvarez (Ed) *Biomethanization of the Organic Fraction of Municipal Solid Wastes*, IWA Publishing, London, 201–228 pp.

Dereix M, Parker W, Kennedy K, 2005. Steam-explosion pre-treatment for enhancing anaerobic digestion of municipal wastewater sludge. *Proceedings of the Water Environment Federation*, **2005(16)**: 53–73.

Dereix M, Parker W, Kennedy K, 2006. Steam-explosion pretreatment for enhancing anaerobic digestion of municipal wastewater sludge. *Water Environment Research*, **78(5)**: 474–85.

Deshmukh HV, 2013. Study of yeasts and molds from distillery waste, dung, ipomoea weed substrate and biogas digester effluent. *Asian Journal of Science and Technology*, **4(10)**: 089–91.

Dhevahi B, Gurusamy R, 2014. Factors influencing production of lipase under metal supplementation by bacterial strain, *Bacillus subtilis* BDG-8. *Journal of Environmental Biology*, **35(6)**: 1151–5.

Divakaran D, Chandran A, Pratap Chandran R, 2011. Comparative study on production of α-amylase from *Bacillus licheniformis* strains. *Brazilian Journal of Microbiology*, **42(4)**: 1397–404.

Duran M, Speece RE, 1997. Temperature-staged anaerobic processes. *Environmental Technology*, **18(7)**: 747–53.

Ellermann J, Hedderich R, Böcher R, Thauer RK, 1988. The final step in methane formation: Investigations with highly purified methyl CoM reductase (component C) from *Methanobacterium thermoautotrophicum* (strain Marburg). *European Journal of Biochemistry*, **172(3)**: 669–77.

Engelhart M, Krüger M, Kopp J, Dichtl N, 2000. Effects of disintegration on anaerobic degradation of sewage excess sludge in downflow stationary fixed film digesters. *Water Science and Technology*, **41(3)**: 171–9.

Gannoun H, Othman NB., Bouallagui H, Moktar H, 2007. Mesophilic and thermophilic anaerobic co-digestion of olive mill wastewaters and abattoir wastewaters in an upflow anaerobic filter. *Industrial & Engineering Chemistry Research*, **46(21)**: 6737–43.

Gatson JW, Benz BF, Chandrasekaran C, Satomi M, Venkateswaran K, Hart ME, 2006. *Bacillus tequilensis* sp. nov., isolated from a 2000-year-old Mexican shaft-tomb, is closely related to *Bacillus subtilis*. *International Journal of Systematic and Evolutionary Microbiology*, **56(7)**: 1475–84.

Gavrilescu M, Chisti Y, 2005. Biotechnology-a sustainable alternative for chemical industry. *Biotechnology Advances*, **23(7-8)**: 471–99.

Gerardi MH (2003) The microbiology of anaerobic digestors. ISBN 0-471-20693-8, John Wiley & Sons, New Jersey: 177 pp.

Ghaima KK, Mohamed AI, Mohamed MM, 2014. Effect of some factors on lipase production by *Bacillus cereus* isolated from diesel fuel polluted soil. *International Journal of Scientific and Research Publications*, **4(8)**: 416–20.

Gomez X, Cuetos MJ, Cara J, Morán A, Garcia AI, 2006. Anaerobic co-digestion of primary sludge and the fruit and vegetable fraction of the municipal solid wastes: conditions for mixing and evaluation of the organic loading rate. *Renewable Energy*, **31(12)**: 2017–24.

Griffin ME, McMahon KD, Mackie RI, Raskin L, 1998. Methanogenic population dynamics during start-up of anaerobic digesters treating municipal solid waste and biosolids. *Biotechnology and Bioengineering*, **57(3)**: 342–55.

Gupta R, Gigras P, Mohapatra H, Goswami VK, Chauhan B, 2003. Microbial *á*-amylases: a biotechnological perspective. *Process Biochemistry*, **38(11)**: 1599–616.

Harrell LJ, Andersen GL, Wilson KH, 1995. Genetic variability of *Bacillus anthracis* and related species. *Journal of Clinical Microbiology*, **33(7)**: 1847–50.

Hartmann H, Ahring BK, 2006. Strategies for the anaerobic digestion of the organic fraction of municipal solid waste: an overview. *Water Science and Technology*, **53(8)**: 7–22.

Hartmann H, Angelidaki I, Ahring BK, 2000. Increase of anaerobic degradation of particulate organic matter in full-scale biogas plants by mechanical maceration. *Water Science and Technology*, **41(3)**: 145–53.

Hegde G, Pullammanappallil P, 2007. Comparison of thermophilic and mesophilic one-stage, batch, high-solids anaerobic digestion. *Environmental Technology*, **28(4)**: 361–9.

Henrissat B, Bairoch A, 1996. Updating the sequence-based classification of glycosyl hydrolases. *Biochemical Journal*, **316(Pt 2)**: 695–6.

Hmidet N, Ali NEH, Haddar A, Kanoun S, Alya SK, Nasri M, 2009. Alkaline proteases and thermostable á-amylase co-produced by *Bacillus licheniformis* NH1: Characterization and potential application as detergent additive. *Biochemical Engineering Journal*, **47 (1-3)**: 71–9.

Ilori MO, Adebusoye SA, Iawal AK, Awotiwon OA, 2007. Production of biogas from banana and plantain peels. *Advances in Environmental Biology*, **1(1)**: 33–9.

Jensen PD, Hardin MT, Clarke WP, 2009. Effect of biomass concentration and inoculum source on the rate of anaerobic cellulose solubilization. *Bioresource Technology*, **100(21)**: 5219–25.

Jurado ECLV, Camacho F, Luzón G, Vicaria JM, 2004. Kinetic models of activity for *β*-galactosidases: influence of pH, ionic concentration and temperature. *Enzyme and Microbial Technology*, **34(1)**: 33–40.

Karim K, Hoffmann R, Klasson T, Al-Dahhan MH, 2005. Anaerobic digestion of animal waste: Waste strength versus impact of mixing. *Bioresource Technology*, **96(16)**: 1771–81.

Kashmiri MA, Adnan A, Butt BW, 2006. Production, purification and partial characterisation of lipase from *Trichoderma viride*. *African Journal of Biotechnology*, **5(10)**: 878–82.

Kayhanian M, Rich D, 1995. Pilot-scale high solids thermophilic anaerobic digestion of municipal solid waste with an emphasis on nutrient requirements. *Biomass and Bioenergy*, **8(6)**: 433–44.

Khyami-Horani H, 1996. Thermotolerant strain of *Bacillus licheniformis* producing lipase. *World Journal of Microbiology and Biotechnology*, **12(4)**: 399–401.

Kim IS, Kim DH, Hyun SH, 2000. Effect of particle size and sodium ion concentration on anaerobic thermophilic food waste digestion. *Water Science and Technology*, **41(3)**: 67.

Kivaisi AK, Eliapenda S, 1994. Pretreatment of bagasse and coconut fibres for enhanced anaerobic degradation by rumen microorganisms. *Renewable Energy*, **5(5-8)**: 791–5.

Knezevic Z, Mavinic DS, Anderson, BC, 1995. Pilot scale evaluation of anaerobic codigestion of primary and pretreated waste activated sludge. *Water Environment Research*, **67(5)**: 835–41.

Kokab S, Asghar M, Rehman K, Asad MJ, Adedyo O, 2003. Bio-processing of banana peel for α-amylase production by *Bacillus subtilis*. *International Journal of Agriculture and Biology*, **5(1)**: 36–9.

Kubicek CP, 2012. Systems biological approaches towards understanding cellulase production by *Trichoderma reesei*. *Journal of Biotechnology*, **163(2)**: 133–42.

Lalak J, Kasprzycka A, Martyniak D, Tys J, 2016. Effect of biological pretreatment of *Agropyron elongatum* 'BAMAR' on biogas production by anaerobic digestion. *Bioresource Technology*, **200**: 194–200.

Lee DW, Koh YS, Kim KJ, Kim BC, Choi HJ, Kim DS, Suhartono MT, Pyun YR, 2012. Isolation and characterisation of a thermophilic lipase from *Bacillus thermoleovorans* ID-1. *FEMS Microbiology Letters*, **179(2)**: 393–400.

Li L, Yang X, Li X, Zheng M, Chen J, Zhang Z, 2010. The influence of inoculum sources on anaerobic biogasification of NaOH-treated corn stover. *Energy Sources, Part A: Recovery, Utilisation, and Environmental Effects*, **33(2)**: 138–44.

Lissens G, Thomsen AB, De Baere L, Verstraete W, Ahring BK, 2004. Thermal wet oxidation improves anaerobic biodegradability of raw and digested biowaste. *Environmental Science and Technology*, **38(12)**: 3418–24.

Liu C, Wyman CE, 2003. The effect of flow rate of compressed hot water on xylan, lignin, and total mass removal from corn stover. *Industrial & Engineering Chemistry Research*, **42(21)**: 5409–16.

Liu CF, Yuan XZ, Zeng GM, Li WW, Li J, 2008. Prediction of methane yield at optimum pH for anaerobic digestion of organic fraction of municipal solid waste. *Bioresource Technology*, **99(4)**: 882–8.

Lopes WS, Leite VD, Prasad S, 2004. Influence of inoculum on performance of anaerobic reactors for treating municipal solid waste. *Bioresource Technology*, **94(3)**: 261–6.

Lueders T, Chin KJ, Conrad R, Friedrich M, 2001. Molecular analyses of methyl coenzyme M reductase *á* subunit (mcrA) genes in rice field soil and enrichment cultures reveal the methanogenic phenotype of a novel archaeal lineage. *Environmental Microbiology*, **3(3)**: 194–204.

Lusk P, 1999. Latest progress in anaerobic digestion. *Biocycle,* **40(7)**: 52–4.

Luton PE, Wayne JM, Sharp RJ, Riley PW, 2002. The mcrA gene as an alternative to 16S rRNA in the phylogenetic analysis of methanogen populations in landfill. *Microbiology*, **148(11)**: 3521–30.

Mahanty B, Zafar M, Park HS, 2013. Characterisation of co-digestion of industrial sludges for biogas production by artificial neural network and statistical regression models. *Environmental Technology*, **34(13-16)**: 2145–53.

Malesu VK, Mishra S, Ojha SK, 2014. *In silico* and *in vitro* study of two novel closely related biogas digestate bacilli strains. *Journal of Advanced Microbiology,* **1(3)**: 163–9.

Malesu VK, Mishra S, Ojha SK, Naik K., Singh PK, Nayak B, 2016. Bioprospecting kitchen refuse as a suitable substrate for biogasification. *Air, Soil and Water Research*, **9**: 63–7.

Malik B, 2012. *Evaluation of Process Parameters and Treatments of Different Raw Materials for Biogas Production.* Lund University, Sweden.

Manatsirikiat S, 1999. *Glucose Degradation by Pure Culture Isolated Microorganisms from the Anaerobic Digestors of Pineapple Peel* (Master's thesis). King Mongkut's University of Technology Thonburi, Bangkok, Thailand.

Masse L, Kennedy KJ, Chou S, 2001. Testing of alkaline and enzymatic hydrolysis pretreatments for fat particles in slaughterhouse wastewater. *Bioresource Technology*, **77(2)**: 145–55.

Mata-Alvarez J, 2003. Anaerobic digestion of the organic fraction of municipal solid waste: a perspective. *In*: Mata†Alvarez, J (Ed) *Biomethanization of the Organic Fraction of Municipal Solid Wastes*. IWA Publishing, London: 1–17.

Merrylin J, Kumar SA, Kaliappan S, Yeom IT, Banu JR, 2013. Biological pretreatment of non-flocculated sludge augments the biogas production in the anaerobic digestion of the pretreated waste activated sludge. *Environmental Technology*, **34(13-14)**: 2113–23.

Miah MS, Tada C, Sawayama S, 2004. Enhancement of biogas production from sewage sludge with the addition of *Geobacillus* sp. strain AT1 culture. *Japanese Journal of Water Treatment Biology*, **40(3)**: 97–104.

Mouneimne AH, Carrere H, Bernet N, Delgenes JP, 2003. Effect of saponification on the anaerobic digestion of solid fatty residues. *Bioresource Technology*, **90(1)**: 89–94.

Mshandete A, Björnsson L, Kivaisi AK, Rubindamayugi MS, Mattiasson B, 2006. Effect of particle size on biogas yield from sisal fibre waste. *Renewable Energy*, **31(14)**: 2385–92.

Narihiro T, Sekiguchi Y, 2011. Oligonucleotide primers, probes and molecular methods for the environmental monitoring of methanogenic archaea. *Microbial Biotechnology*, **4(5)**: 585–602.

Nerkar GS, Chaudhari YA, Khutle NM, 2011. Extraction and characterisation of alpha amylase from *Phaseolus aconitifolius*. *Current Pharma Research*, **1(2)**: 115–22.

Nopharatana A, Pullammanappallil PC, Clarke WP, 2007. Kinetics and dynamic modelling of batch anaerobic digestion of municipal solid waste in a stirred reactor. *Waste Management*, **27(5)**: 595–603.

Obaja D, Mace S, Costa J, Sans C, Mata-Alvarez J, 2003. Nitrification, denitrification and biological phosphorus removal in piggery wastewater using a sequencing batch reactor. *Bioresource Technology*, **87(1)**: 103–11.

Ojha SK, 2015. *Feasibility and Enhancement of Biomethanation of Kitchen Refuse with Special Mention of the Amylolytic Potential of Bacillus aryabhattai KIIT BE-1* (Doctoral thesis). KIIT, Odisha, India.

Ostrem K, 2004. Greening waste: anaerobic digestion for treating the organic fraction of municipal solid waste (Master's thesis). Columbia University, New York, The United State of America.

Ostrem KM, Millrath K, Themelis NJ, 2004. Combining anaerobic digestion and waste-to-energy. *In*: *12th Annual North American Waste-to-Energy Conference*. American Society of Mechanical Engineers: 265–271 pp.

Palmowski L, Muller J, 1999. Influence of comminution of biogenic material on the bioavailability. *Muell Abfall*, **31(6)**: 368–72.

Palonen H, Thomsen AB, Tenkanen M, Schmidt AS, Viikari L, 2004. Evaluation of wet oxidation pretreatment for enzymatic hydrolysis of softwood. *Applied Biochemistry and Biotechnology*, **117(1)**: 1–17.

Pandey A, Soccol CR, Nigam P, Soccol VT, 2000. Biotechnological potential of agro-industrial residues. I: sugarcane bagasse. *Bioresource Technology*, **74(1)**: 69–80.

Patagundi BI, Shivasaran CT, Kaliwal BB, 2014. Isolation and characterisation of cellulase producing bacteria from soil. *International Journal of Current Microbiology and Applied Sciences*, **3(5)**: 59–69.

Pradhan AK, Pradhan N, Mall G, Panda HT, Sukla LB, Panda PK, Mishra BK, 2013. Application of lipopeptide biosurfactant isolated from a halophile: *Bacillus tequilensis* CH for inhibition of biofilm. *Applied Biochemistry and Biotechnology*, **171(6)**: 1362–75.

Prakash S, Veeranagouda Y, Kyoung L, Sreeramulu K, 2009. Xylanase production using inexpensive agricultural wastes and its partial characterisation from a halophilic *Chromohalobacter* sp. TPSV 101. *World Journal of Microbiology and Biotechnology*, **25(2)**: 197–204.

Ramos LP, 2003. The chemistry involved in the steam treatment of lignocellulosic materials. *Química Nova*, **26(6)**: 863–71.

Rey MW, Ramaiya P, Nelson BA, Brody-Karpin SD, Zaretsky EJ, Tang M, de Leon AL, Xiang H, Gusti V, Clausen IG, Olsen PB, Rasmussen MD, Andersen JT, Jørgensen PL, Larsen TS, Sorokin A, Bolotin A, Lapidus A, Galleron N, Ehrlich SD, Berka RM, 2004. Complete genome sequence of the industrial bacterium *Bacillus licheniformis* and comparisons with closely related *Bacillus* species. *Genome Biology*, **5(10)**: r77.

Sajedi RH, Naderi-Manesh H, Khajeh K, Ahmadvand R, Ranjbar B, Asoodeh A, Moradian F, 2005. A Ca-independent á-amylase that is active and stable at low pH from the *Bacillus* sp. KR-8104. *Enzyme and Microbial Technology*, **36(5-6)**: 666–71.

Sanchez E, Borja R, Weiland P, Travieso L, Martin A, 2000. Effect of temperature and pH on the kinetics of methane production, organic nitrogen and phosphorus removal in the batch anaerobic digestion process of cattle manure. *Bioprocess Engineering*, **22(3)**: 247–52.

Sangiliyandi G, Gunasekaran P, 1996. Extracellular lipase producing *Bacillus licheniformis* from an oil mill refinery effluent. *Indian Journal of Microbiology*, **36**: 109–10.

Saranya P, Kumari HS, Jothieswari M, Rao BP, Sekaran G, 2014. Novel extremely acidic lipases produced from *Bacillus* species using oil substrates. *Journal of Industrial Microbiology and Biotechnology*, **41(1)**: 9–15.

Schnürer A, Jarvis A, 2010. *Microbiological Handbook for Biogas Plants*. Swedish Waste Management U2009: 03, Avfall Sverige, Svenskt Gastekniskt Center AB, Swedish Gas Centre Report 207: 74 pp.

Senthilkumar R, Selvakumar G, 2007. A potential lipase producing *Bacillus cereus* from stored food cereals, *Cauvery Research Journal*, **1(1)**: 51-55.

Siegert I, Banks C, 2005. The effect of volatile fatty acid additions on the anaerobic digestion of cellulose and glucose in batch reactors. *Process Biochemistry*, **40(11)**: 3412–8.

Sivaramakrishnan S, Gangadharan D, Nampoothiri KM, Soccol CR, Pandey A, 2006. *á*-amylases from microbial sources–an overview on recent developments. *Food Technology Biotechno*logy, **44(2)**: 173–84.

Solheim OE, 2004. Method and arrangement for continuous hydrolysis of organic material. *US Patent* 0,**168***: 990.*

Speece RE, 1987. Nutrient requirements. *In*: DP Chynoweth, R Isaacson (Eds) *Anaerobic Digestion of Biomass*. *Elsevier Science Publishing Co.*, New York: 129–40.

Sreekrishnan TR, Kohli S, Rana V, 2004. Enhancement of biogas production from solid substrates using difference techniques–a review. *Bioresurce Technology*, **95(1)**: 1–10.

Stroot PG, McMahon KD, Mackie RI, Raskin L, 2001. Anaerobic codigestion of municipal solid waste and biosolids under various mixing conditions-I. Digester performance. *Water Research*, **35(7)**: 1804–16.

Sundrarajan R, Jayanthi A, Elango R, 1997. Anaerobic digestion of organic fractions of municipal solid waste and domestic sewage of Coimbatore. *Indian Journal of Environmental Health*, **39(3)**: 193–6.

Talamond P, Noirot M, de Kochko A, 2005. The mechanism of action of α-amylase from *Lactobacillus fermentum* on maltooligosaccharides. *Journal of Chromatography B*, **834(1-2)**: 42–7.

Tari C, Gogus N, Tokatli F, 2007. Optimization of biomass, pellet size and polygalacturonase production by *Aspergillus sojae* ATCC 20235 using response surface methodology. *Enzyme and Microbial Technology*, **40(5)**: 1108–16.

Teodoro CEDS, Martins MLL, 2000. Culture conditions for the production of thermostable amylase by *Bacillus* sp. *Brazilian Journal of Microbiology*, **31(4)**: 298–302.

Tewelde S, Eyalarasan K, Radhamani R, Karthikeyan K, 2012. Biogas production from co-digestion of brewery wastes and cattle dung. *International Journal of Latest Trends in Agriculture and Food Sciences*, **2(2)**: 90–3.

Thauer RK, 1998. Biochemistry of methanogenesis: a tribute to Marjory Stephenson: 1998 Marjory Stephenson Prize Lecture. *Microbiology*, **144(9)**: 2377–406.

Tiwari S, Shukla N, Mishra P, Gaur R, 2014. Enhanced production and characterization of a solvent stable amylase from solvent tolerant *Bacillus tequilensis* RG-01: thermostable and surfactant resistant. *The Scientific World Journal*, **2014**: 1–11.

Ukpai PA, Nnabuchi MN, 2012. Comparative study of biogas production from cow dung, cow pea and cassava peeling using 45 litres biogas digester. *Advances in Applied Science Research*, **3(3)**: 1864–9.

Unakal C, Kallur RI, Kaliwal BB, 2012. Production of α-amylase using banana waste by *Bacillus subtilis* under solid state fermentation. *European Journal of Experimental Biology*, **2**: 1044–52.

Van Der Maarel MJ, Van Der Veen B, Uitdehaag JC, Leemhuis H, Dijkhuizen L, 2002. Properties and applications of starch-converting enzymes of the α-amylase family. *Journal of Biotechnology*, **94(2)**: 137–55.

Vartak DR, Engler CR, Ricke SC, McFarland MJ, 1997. Organic loading rate and bioaugmentation effects in psychrophilic anaerobic digestion of dairy manure. Paper-American Society of Agricultural Engineers, Michigan (974051).

Vijayabaskar P, Jayalakshmi D, Shankar T, 2012. Amylase production by moderately halophilic *Bacillus cereus* in solid state fermentation. *African Journal of Microbiology Research*, **6(23)**: 4918–26.

Vrieze J, Saunders AM, He Y, Fang J, Nielsen PH, Verstraete W, Boon N, 2015. Ammonia and temperature determine potential clustering in the anaerobic digestion microbiome. *Water Research*, **75:** 312-323.

Wang Q, Kuninobu M, Kakimoto K, Hiroaki I, Kato Y, 1999. Upgrading of anaerobic digestion of waste activated sludge by ultrasonic pretreatment. *Bioresource Technology*, **68(3)**: 309–13.

Ward AJ, Hobbs PJ, Holliman PJ, Jones DL, 2008. Optimisation of the anaerobic digestion of agricultural resources. *Bioresource Technology*, **99(17)**: 7928–40.

Xi Y, Chang Z, Ye X, Xu R, Du J, Chen G, 2014. Methane production from wheat straw with anaerobic sludge by heme supplementation. *Bioresource Technology*, **172**: 91–6.

Yedavalli P, Madhusudhana Rao N, 2013. Engineering the loops in a lipase for stability in DMSO. *Protein Engineering Design and Selection*, **26(4)**: 317–24.

Yen HW, Brune DE, 2007. Anaerobic co-digestion of algal sludge and waste paper to produce methane. *Bioresource Technology*, **98(1)**: 130–4.

Zaher U, Grau P, Benedetti L, Ayesa E, Vanrolleghem PA, 2007. Transformers for interfacing anaerobic digestion models to pre-and post-treatment processes in a plant-wide modelling context. *Environmental Modelling and Software*, **22(1)**: 40–58.

Závacký M, Ditl P, Prell A, Sobotka M, 2010. Increasing biogas production from maize silage by ultrasonic treatment. *Chemical Engineering*, **21**: 439–45.

Zhang B, Zhang LL, Zhang SC, Shi HZ, Cai WM, 2005. The influence of pH on hydrolysis and acidogenesis of kitchen waste in two-phase anaerobic digestion. *Environmental Technology*, **26(3)**: 329–40.

Zhong W, Zhang Z, Luo Y, Sun S, Qiao W, Xiao M, 2011. Effect of biological pretreatments in enhancing corn straw biogas production. *Bioresource Technology*, **102(24)**: 11177–82.

Zinder SH, 1993. Physiological ecology of methanogens. *In*: JG Ferry (Ed) *Methanogenesis*. Chapman and Hall Microbiology Series (Physiology/Ecology/Molecular Biology/ Biotechnology), Springer, Boston, MA: 128–206.

11

Electrofermentation in Aid of Bioenergy and its Industrial Application

Prasun Kumar[a,1], K. Chandrasekhar[b], Archana Kumari[c] and Beom Soo Kim[a]

[a]*Department of Chemical Engineering, Chungbuk National University Cheongju-28644, South Korea*
[b]*Department of Civil Engineering, Yeungnam University, Gyeongsan-38451 South Korea*
[c]*Department of Biotechnology, Bodoland University, Kokrajhar-783370 Assam, India*
[1]*Corresponding Author: prasun@chungbuk.ac.k*

ABSTRACT

The rapid industrialisation and speedy progress towards urbanisation across the world have created the huge demand for energy moreover creating enormous quantity of wastes. The latter is in charge for polluting the environment in an exponential manner, resulting out of the harmful and poisonous compounds released by them. In recent years, there has been a pattern change from 'waste-to-wealth', considering the value of high organic fraction accessible in the wastes. The best-accomplished methods are that of anaerobic digestion; contribute to the production of methane (CH_4). Similar biotransformation has partial net energy harvests. Recently, fermentation industries steering their interest towards the production of value added products, viz., *biohydrogen (H_2), biopolymers, ethanol, vitamins, acetic acid, enzymes, butanediols, etc. from carbon rich organic waste like cellulosic substrates have been envisioned to flourish in a multi-step process, such as the 'Biorefinery'. A contemporary development in fermentation technology holds the key to overcome the constraints for such processes that may lead to a sustainable economy. One related technology is electrofermentation that has attracted significant attention recently owing to its capacity to boost the metabolic activities of microbes by means of extracellular electron transfer. It has been tested on numerous methanogens and acetogens where* *the biogas yield was observed to be about 2-fold enhanced. Therefore, electrofermentation displays its potential to use complex inexpensive carbonaceous substrates for synthesis of high-value biochemicals. In this chapter, we have discussed the principles of electrofermentation along with its application in the area of bioenergy.*

Keywords: Electrofermentation, Hydrogen, Electricity, Methane, Microbial fuel cells

Abbreviations used

BES : bioelectrochemical system
BETS : bioelectrochemical treatment system
COD : chemical oxygen demand
EF : electrofermentation
MDS : microbial desalination system
MEC : microbial electrolysis cell
MES : microbial electrosynthesis
MFC : microbial fuel cell
MRC : microbial remediation cell
MV : methyl viologen
PHA : polyhydroxyalkanoate

Introduction

The growing population and demanding industrialisation have led to pollution and energy crisis alarming for a potential threat to whole humankind in the near future (Deval *et al.*, 2017; Kadier *et al.*, 2016a). The prompt utilisation of fossil fuels based energy sources result in the intensification of greenhouse gas effect by discharging the CO_2 the atmosphere, which damage the environment (Chandrasekhar *et al.*, 2015b). The limited sources of fossil fuel and reduction of the environmental pollution (by reducing the CO_2 emission into the atmosphere) are forcing the investigation aimed at viable ecofriendly power sources (Chandrasekhar and Venkata Mohan, 2014a; Chandrasekhar *et al.*, 2015b; Chandrasekhar and Venkata Mohan, 2014b). In this concern, several researchers are presently engaged in a search to develop and optimise environmental friendly and viable power resources to substitute fossil fuels based energy economy. In competition, innovative fermentation methods, for instance the Electro-fermentation (EF) processes are being industrialised to substitute fossil fuels (Chandrasekhar *et al.*, 2015a). EF is an inimitable bioprocess that treaties with the bio-electrochemical controlling of microbial metabolic reactions. EF is a type of device where the electrochemical and biological process happens for the production of value-added bioproducts with simultaneous electrogenesis. The electrode materials employed for the duration of the EF can serve as both as electron (e^-) source or a sink. This arrangement permits unstable fermentation that has the prospective to alter the redox equilibrium in the feed/electrolyte. However, all these arrangements have a significant impact on microbial metabolic reactions. Initially, most of the EF investigation works turned around the electrogenesis and acknowledged as microbial fuel cells (MFCs). From the last decade, the investigation struggles led to the continuation of such approaches for the production of hydrogen (H_2), methane (CH_4), and carbon dioxide (CO_2), polyhydroxyalkanoates (PHA), 1,3-butanediol etc. It has emerged as a new and ecofriendly approach established in the past decade. It has the to increase its use in numerous research areas of bioengineering, materials science, waste treatment, environmental science, bioenergy over integrated method.

Waste biomass as an indirect opportunity

The rapid industrialisation is posturing a severe influence on the environment. The reliance on fossil fuels creating severe universal difficulties of environmental contamination (Kadier *et al.*, 2015; Kadier *et al.*, 2016b). In this circumstance, usage of carbon rich waste biomass as a carbon source in a biorefinery could be a possible substitute. Waste biomass produced from agriculture or other food wastes, discharges of bio-chemical industries etc. are possible sources of raw materials (Chandrasekhar *et al.*, 2015a). A huge amount of carbon rich organic food waste from food supply chain to agriculture and manufacturing systems is produced yearly. The significance of such wastes material is estimated to grow quickly in the coming 25 years (Galanakis, 2012; ElMekawy *et al.*, 2015). Handling such enormous amounts of wastes material has become a global issue (Venkata Mohan and Chandrasekhar, 2011b; Reddy *et al.*, 2011). In the meantime, they typically comprise of cellulose, lipids, lignin, and proteins, it is an imperious process to use them for the production of value-added chemical substances. It thus provides boundless chances and offers a win-win circumstance by the fact these waste, together with municipal solid waste goes to landfills, leading to CO_2 emission, water pollution, and other hurtful difficulties in the surrounding area (Venkata Mohan and Chandrasekhar, 2011b; Imbert, 2017). Consumptions of carbon rich organic waste have been exposed to harvest ethanol, CH_4, H_2, volatile fatty acids (VFAs), PHAs and numerous additional valuable products. Numerous stochastic mockups having great industrial prospective are offered that could be employed in tandem on a large scale (Venkata Mohan *et al.*, 2016). MFC devices have been extensively used for the bioelectricity generate with simultaneous waste remediation and as well as to extract valuable chemicals (ElMekawy *et al.*, 2015; Chandrasekhar *et al.*, 2015a). Food wastes collected from the canteen have been found to produce 530 mA/m^3 of current density (CD) with concurrent waste removal (44% of COD removal (Goud and Venkata Mohan, 2011).

Limitations of Anaerobic digestion

Anaerobic digestion (AD) is a deep-rooted method aimed at the production of methane/biogas (CH_4). The biogas comprises 60 to 80% of CH_4 and 20 to 40% of CO_2, whose relative amounts largely depends on the nature and type of the substrate employed. The AD process is a multistep procedure comprising a sequence of biological reactions namely hydrolysis, acidogenesis, acetogenesis and methanogenesis. These sequence of biological reactions are mediated by definite groups of microbes functioning together under strict anaerobic circumstances (Villano *et al.*, 2012). The oxidation/breakdown of the carbon rich organic substances is enabled by a cascade of microbial enzymes produced by numerous collections of bacteria for example acidogens, acetogens, methanogens etc. The methanogens are the most common microorganisms involves in the substrate degradation and methane production. The optimal temperature for the AD process differs between 35 and 55°C. Nevertheless, the assessment among total biogas produce and energy input is a strategic factor to decide the choice of AD operating temperature, together with other key parameters.

Electrofermentation: An unconventional scheme of fermentation

The EF is an effective e^- bagging practice by means of specialised electrodes as an electron acceptor for the duration of microbial fermentation. It exploits low-cost platform with electrochemically active microorganisms to oxidise carbon rich organic substrate present in waste water (Chandrasekhar *et al.*, 2015a). In this process, series of biochemical (redox) reactions happens toward transformation of chemical energy into electrical energy or other value added chemicals, in which proton exchange membranes (PEMs) detach the cathode and anode compartments (Chandrasekhar and Venkata Mohan, 2012; Venkata Mohan and Chandrasekhar, 2011a; Velvizhi and Venkata Mohan, 2017). This phenomenon occurs in all most all the MFCs. However, reduction based reactions will take place in the cathode compartment, on the other hand, substrate oxidation will take place in the anode compartment mediated by anodophilic bacteria to produce e^- and protons (H^+). These H^+ transported via PEM and reach cathode compartment. In the meantime, e^- were bagged by an anode electrode and transported them via an electrical circuit (generally copper or titanium wires were used as an electrical circuit) to the cathode electrode. In the cathode chamber, e^- and H^+ get reduced with oxygen (O_2) molecule to form water (H_2O). Here, O_2 will act as the terminal e acceptor. Based on the quantity of compartment and nature of the process, EF could be classified into single chambered (SC), double chambered (DC), tubular, stacked, baffled, up-flow system etc. (Aelterman et al., 2006; Venkata Mohan *et al.*, 2016; Moscoviz *et al.*, 2016; Schievano *et al.*, 2016). Alternatively, with various alterations, EFs can additionally be classified as follows, bio-electrochemical treatment system (BET), microbial desalination system (MDS), microbial electrolysis cell (MEC), and bio-electrochemical system (BES) extending its utility (Venkata Mohan *et al.*, 2016; Choi and Sang, 2016; Nikhil *et al.*, 2017; Wang 2013). In the course of EF, specific microorganisms are employed based on their electrogenic nature and can be collected into exoelectrogens, electricigens, or anode respiring microbes, suggesting that these microorganisms are capable to transfer e^- to the exterior of the cell membranes in a diverse way. Electrochemically active microbes are able to generate bioelectricity at anode electrode over and done with e^- release during the substrate oxidation process, seemingly over their capability of reducing metals. Contrary is the situation with endoelectrogens, they utilise e^- at cathode electrode to through oxidation of metals (Choi and Sang, 2016; Nikhil *et al.*, 2016). These endoelectrogens be able to both openly use membrane structures/membrane proteins, pili, filaments or cytochromes or mediators for e^- transfer. Alternatively, *Shewanella* spp. owns both e mediators such as riboflavin in addition to conductive filaments. Hypothetically, any carbon rich biodegradable substrates are have a habit of bio-transformation into bioenergy and value added chemicals (Logan and Rabaey, 2012). The most extensively exploited carbon rich waste materials include waste biomass, industrial effluents, food waste, inorganic substrates like ammonia, petroleum sludge (petroleum hydrocarbons), acid mine drainage, etc. (Chandrasekhar and Ahn, 2017). Nonetheless, to attain the target co-operation among microorganisms are essential thus, typically mixed microbial population/cultures are used for this purpose (Kumar *et al.*, 2012). The first sets of microorganisms typically had great hydrolytic capabilities to breakdown composite waste substrate into simple molecules. Further, these simple molecules can be consumed by electrogenic microorganisms and breakdown them into intermediate metabolites. This overall substrate oxidation process occurs in the anode compartment. On the other hand, the cathode centered EF process varies from the anodic substrate oxidation reactions since for the reduction of oxidised substances customised systems must be necessitated. Fig. 1 illustrates the numerous methods for e^- transfer from solid electrodes to microbes. Thus far, four common variations of EF are revealed. The basic one is generally recognised as MFC employed for the bioelectricity production from a wide variety of carbon rich substrates. To reduce the cathode potential, MFC can be operated with external power supply, it turns out to be a MEC. In this MEC systems, H_2 and other value-added chemicals are produced in the cathode compartment (Liu and Hu, 2012). Consequently, there are variants of the MECs for specific applications. If the MEC device is employed to treat/remove pollutants (aromatic hydrocarbons, chlorinated compounds, dye compounds, bioreactor effluents, etc.) from the medium, it is designated as microbial remediation cell (MRC) while it is recognised as microbial electrosynthesis (MES) if value-added products are synthesised through bio-catalysis at the cathode compartment (Table 1). In addition, an extra separating compartment can be placed among cathode and anode compartments. It utilises the internal potential for water desalination. The latter is acknowledged as MDC and it does not necessitate external power supply. There are various other variants accessible, which are modified for definite requirements but the working principle same. It makes this area of research so special and objective-oriented.

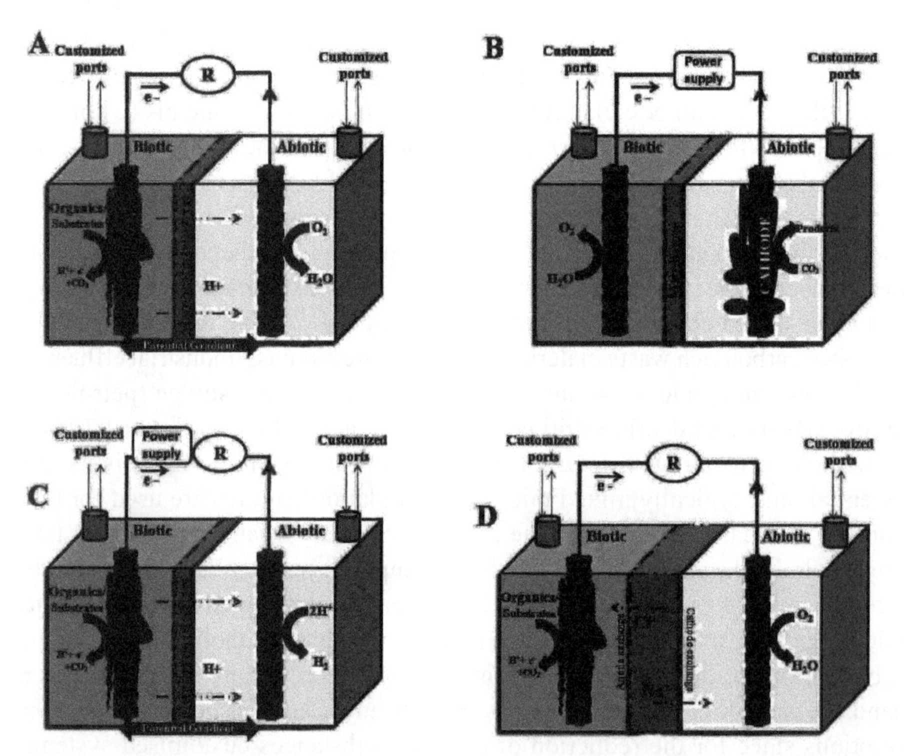

Fig. 1: Schematic presentation of various MES types: A) Microbial fuel cell; B) Microbial electrolysis cell; C) Microbial electrosynthesis; and D) Microbial desalination cells (Kumar *et al.*, 2018a)

Electrofermentation in aid of

i) Hydrogen

The H_2 is deliberated as the light-weight, a high calorific value containing, clean source of energy and has the possibility to substitute fossil fuels (Hu *et al.*, 2008; Liu *et al.*, 2005). In recent years, researchers are showing much interest in H_2 production as a fuel source. It has been noticed that large-scale production of H_2 gas as fuel source through biological routes is inadequate due to several

Table 1: MES (microbial electrosynthesis) variants and their potential applications in aid of bioenergy (Adapted from Kumar *et al.*, 2018a)

MES Type	Substrate(s) used (oxidation at Anode)	Reduction of cathode	Major output	Reference(s)
Microbial Fuel Cells (MFCs)				
Tubular MFC	Glucose, Acetate, wastewater	Kƒ [Fe(CN)†]	Electric Current	Rabaey *et al.* (2005)
Up-flow UMFC	Sucrose	Kƒ [Fe(CN)†], O_2	Electric Current	He *et al.* (2005), He *et al.* (2006)
Up-flow anaerobic sludge blanket–MFC	Glucose	O_2	Electric Current	Zhang *et al.* (2012)
Baffled air-cathode – MFC	Glucose, corn stover hydrolysates	O_2	Electric Current	Feng *et al.* (2010)
Stacked – MFC	NaOAc	Kƒ [Fe(CN)†]	Electric Current	Aelterman *et al.* (2006)
Microbial reverse electrodialysis cell (MRC) Kim and Logan (2011b)	NaOAc	O_2	Electric Current	Cusick *et al.* (2012),
Microbial reverse-electrodialysischemical-production cell (MRCC)	NaOAc	O_2	Electric Current, acid, alkali	Zhu *et al.* (2013)
Microbial electrolysis cells (MECs)				
MEC-based systems for chemical production Microbial electrolysis cells (MECs) — in general	Any biodegradable material	Proton	H_2, H_2O_2, CH_4, NaOH	Cheng *et al.* (2009), Liu *et al.* (2005),
Bioelectrochemically assisted microbialreactor (BEAMR)	Wastewater	Proton	H_2	Ditzig *et al.* (2007)
Solar-powered microbial electrolysis fuel (solar MEC)	Acetate	Proton	H_2	Chae *et al.* (2009)
Microbial reverse-electrodialysiselectrolysis cell (MREC)	Acetate	Proton	H_2	Kim and Logan (2011)
Submersible microbial electrolysis cell (SMEC)	Acetate	Proton	H_2	Zhang and Angelidaki (2012)

Contd.

MES Type	Substrate(s) used (oxidation at Anode)	Reduction of cathode	Major output	Reference(s)
Microbial electrosynthesis (MES)				
MES-based systems for chemical production	Organic, H_2 sulphide, H_2O	Acetate or other organics, CO_2	Ethanol, acetate, 2-oxobutyrate, formate	Gong *et al*. (2013), Kumar *et al*. (2018a) Steinbusch *et al*. (2010)
Microbial desalination cells (MDCs)				
MDC-based systems for water desalination and beneficial reuse	Any biodegradable material	O_2, Kƒ $[Fe(CN)_6]$, organics, or other oxidants	Desalination	Cao *et al*. (2009)
Microbial saline-wastewater electrolysis cell (MSC)	NaOAc	H_2	Treated saline wastewater, electric current	Kim and Logan (2013b)
Osmotic MDC capacitive adsorptioncapability (cMDC)	NaOAc, xylose, wastewater	O_2, Kƒ $[Fe(CN)_6]$, proton	Water desalination, electric current	Zhang and He(2012)
	NaOAc	Kƒ $[Fe(CN)_6]$	Water desalination	Forrestal *et al*. (2012a)
MDC packed with ion-exchangeresin (R-MDC)	NaOAc	O_2	Water desalination, electric current	Morel *et al*. (2012)
Electrolysis-MDC	NaOAc	Proton	H_2, water desalination	Luo *et al*. (2011)
Microbial electrolysis desalination and chemical productioncell (MEDCC)	NaOAc	O_2	Water desalination, NaOH, HCl	Chen *et al*. (2012)
Submerged MDC -denitrificationcell (SMDDC)	NaOAc	Nitrate	Electric current, N_2	Zhang and Angelidaki (2013)
Stacked microbial desalination cell (SMDC)	NaOAc	O_2	Water desalination, electric current	Chen *et al*. (2011)
Upflow microbial desalination cell (UMDC)	NaOAc	O_2	Water desalination, electric current	Jacobson *et al*. (2011)

limiting factors such as low yield, availability, and cost of the substrate and overall process costs (Kumar *et al.*, 2013). The theoretical H_2 yield that is 4 mol of H_2/mol glucose can be attained through dark fermentation (DF) that further confines the up-scaling at a larger scale (Liu *et al.*, 2005). Nevertheless, it was noticed that integrating such bioprocess using MEC could increase overall H_2 production equal to four folds. In this direction, common metabolic reactions during DF through $C_2H_4O_2$ or $C_4H_8O_2$ have been shown in the following reaction:

$$C_6H_{12}O_6 + 2\ H_2O \rightarrow 4\ H_2 + 2\ CO_2 + 2\ C_2H_4O_2 \quad (1)$$

$$C_6H_{12}O_6 \rightarrow 2\ H_2 + 2\ CO_2 + C_4H_8O_2 \quad (2)$$

This confines the attainable H_2 yield to 2 mol/mol and 4 mol/mol with $C_4H_8O_2$ and $C_2H_4O_2$ as an end-product, correspondingly (Liu *et al.*, 2005). Hence, by integrating this process with an MEC system employing electrochemically active microorganisms such as *Shewanella*, *Pseudomonas* etc. leads to form biohydrogen from H^+ and e^- generated by microorganisms mediated substrate oxidation as presented here:

$$\text{Anodic reaction: } C_2H_4O_2 + 2\ H_2O \rightarrow 2\ CO_2 + 8\ e^- + 8\ H^+ \quad (3)$$

$$\text{Cathodic reaction: } 8H^+ + 8e^- \rightarrow 4H_2 \quad (4)$$

The $C_2H_4O_2$ is two carbon containing simple substrate that generate a $H\text{-}_2$ in MEC device. Moreover, $H\text{-}_2$ producing bioreactor effluents contains excessive amount of VFAs, which can be used as a substitute to artificial feedstock for $H\text{-}_2$ generation in MECs (Venkata Mohan *et al.*, 2013). MEC systems could be successfully integrated with $H\text{-}_2$ bioreactor to improve H_2 production rate and overall yield. In MEC devices, H_2 generation at cathode employing $C_2H_4O_2$ and electrochemically active microbes (E_{An}E"0.2 V) is inadequate at the cathode (E_{cell} = -0.414 V, pH=7). It was noticed that H_2 gas production in the cathode at MEC system necessitates very less potential as related to water electrolysis. Moreover, strict anaerobic microenvironment upsurges the coulombic efficiency (CE) of the MEC process. As a result an extreme yield of 11 mol H_2/mol glucose was attained in the MEC system.

The H_2 production rate can be meaningfully increased by operating dual chamber MEC without membranes, which makes the process more economically viable. On the other hand, it also enhances the growth of methanogenic bacteria, where these methanogens utilise produced H_2 as a substrate to produce CH_4. In the course of MEC operation, active methanogenic microorganisms compete with electrochemically active microorganisms for H_2 and consume substrate, thus leads to a reduction in the overall process efficacy (Kaider *et al.*, 2016a). As well, it has been noticed that the purity of H_2 produced in the MEC system is also significantly inclined by hydrogenotropic methanogens (HM). It relics as a foremost obstacle in H_2 production by MEC, where pretreatment of inoculum (such as addition of methanogen inhibitors or O_2 supply for a limited period to create aerobic conditions) may help to a certain level. One more constraint to MEC for H_2 production is the cost of the metals used in the cathode as catalyst material.

In this perspective, single compartment MEC system was operated with bioreactor effluent to produce extra H_2 by supplying external voltage of 600 mV to the MEC system with simultaneous consumption of 68% VFAs (Modestra *et al.*, 2015). Different types of electrode materials (copper, gold, titanium, platinum, graphite, etc.) were employed in MECs systems. Nevertheless, several researchers suggested that extremely conductive electrode materials could be more suitable as an electrode material in MECs (Chandrasekhar *et al.*, 2017).

ii) Biogas

The biogas/CH_4 is typically produced by methanogenic bacteria consuming acetate, formate and H_2 generated from synthetic feed or various types of carbon rich waste in the AD (Wang *et al.*, 2015; Michalopoulos *et al.*, 2017). The utilisation of acetate as feedstock is functioned by acetoclastic methanogens (AM) whereas the utilisation of H_2 as a substrate for CH_4 production is by HM (Wang *et al.*, 2009). The thermodynamically possible reactions of CH_4 production through CO_2 reduction (at 169 mV) had also been suggested.

$$C_2H_4O_2 \rightarrow CH_4 + CO_2 \quad [AM] \tag{5}$$

$$4H_2 + CO_2 \rightarrow CH_4 + 2H_2O \; [HM] \tag{6}$$

$$CO_2 + 8H^+ + 8e^- \rightarrow CH_4 + 2H_2O \tag{7}$$

Unlike AD, it has been noticed that during MEC operation most of the CH_4 resulting from the H_2. Though, the source of CH_4 production (acetate or H_2) during MEC operation is yet to be evidently understood. Methanogens mediated CH_4 production from the H_2 leads to ~15% energy loss due to thermodynamics. Hence, as a viable option for energy generation, MEC systems can be successfully integrated with AD systems, where carbon rich effluents from AD will serve as a feedstock in MECs (Pant *et al.*, 2012). It is well known that the substrate concentration, operation time, ammonia levels and system pH are the limiting factors which influence the overall CH_4 yield. In 2009, Cheng and co-researchers employed *Methanobacterium palustre* as first few microorganisms investigated for CH_4 production with an abiotic anode and biocathode (Cheng *et al.*, 2009). From this experiment, they have concluded that the biotic carbon cathode and abiotic cathodes favor CH_4 and H_2 production respectively. At this juncture, external power supply appears to play a key role, at a power supply of less than 500 mV, H_2 concentration followed decreasing trend whereas CH_4 levels followed an increasing trend (Call and Logan, 2008; Tartakovsky *et al.*, 2008). Alternatively, higher H_2 concentration was noticed when the applied voltage is greater than 700 mV (Tartakovsky *et al.*, 2008). In this case, medium pH and buffering capacity alter the bacterial metabolism from H_2 to CH_4. For example, low carbonate and low pH conditions are more favorable for H_2 production whereas high carbonate and high pH conditions are more favorable for CH_4 production (Wang *et al.*, 2009). Recent reports confirmed that electro-methanogenesis process can utilise bioelectricity produced during the oxidation of carbon rich organic waste (renewable resources) into value-added biofuel with concurrent bagging of CO_2. Due the inevitable methanogenesis phenomenon in MEC, the separation of CH_4 from the produced H_2 turn into relatively costly process that makes overall process economically unviable option (due to the rise in production cost). As a result, researchers started focusing on electro-methanogens mediated CH_4 production. In addition, it is well known that the storage, compression, and transportation of CH_4 is relatively easier than H_2 gas. In recent years, several researchers are working to develop membrane-less MEC systems to reduce the cost of the device. A double-chambered MEC system presents loss of concentration due to the accumulation of H^+ or hydroxyl ions (OH^-). Additionally, CH_4 generation in MEC system offers an added benefit of sludge loss and relieving aeration cost (Pant *et al.*, 2012). From the existing literature, it is evident that in a single compartment MEC system, CH_4 production is certainly associated through the H_2 and current generation instead of acetoclastic methanogenesis (Wang *et al.*, 2009). The work of Clauwaert *et al.* is notorious, when they used two-chamber MEC the CH_4 generation was found to increase up to 0.41 mol/mol acetate (Clauwaert *et al.*, 2008). Future, MEC systems operated without membrane using graphite granule electrodes to produce CH_4 as the leading product (Clauwaert and Verstraete, 2009). Nonetheless, constraints of CH_4 solubility at diverse pH and temperature circumstances were noticed as in the case of conventional fermentation. Few reports are available on the utilisation of carbon rich organic waste as a possible feedstock for CH_4 production through EF technology. In this direction, vegetable waste, garbage slurry, agriculture waste, food wastes and wastewaters have been employed as a potential feedstock for CH_4 production (Lee *et al.*, 2017; Sasaki *et al.*, 2016; Srikanth *et al.*, 2010).

iii) 1,3-Propanediol

1,3-Propanediol ($C_3H_8O_2$) is a three carbon containing organic compound with much industrial importance. It is widely used in cosmetic industry, resins manufacturing industry, solvent production, polymer synthesis, etc. It can also be produced through biological ways by employing glycerol as carbon rich organic feedstock. Rising demand for fossil fuels is triggering release of huge quantities of glycerol-rich effluents from biodiesel industry (Kumar *et al.*, 2015a), whereas several microorganisms has the ability to biotransform the carbon rich glycerol into $C_3H_8O_2$ through a series of metabolic reactions. The innovations in cathodic EFs have led to improved $C_3H_8O_2$ produce (Choi *et al.*, 2014) with the subsequent benefits: (i) straight e^- dissipation into $C_3H_8O_2$ production, (ii) choice of microorganisms amended to reduced circumstance, and (iii) improving $C_3H_8O_2$ production as a single NADH dissipating pathway. Limited articles are available regarding mixed microbial population mediated glycerol EF applying negative potential (-1140 mV) to produce $C_3H_8O_2$ (0.50 mol/mol glycerol). In another study, external power supply to the cathode enhances the $C_3H_8O_2$ synthesis leading to a yield of 0.51 mol $C_3H_8O_2$/mol glycerol. In another study, fed-batch system operated with external power supply (-1340 mV) to the cathode achieved 0.46 mol of $C_3H_8O_2$/mol glycerol. In consequence, the EF strategy has been emerged as a new technology to produce enhanced quantities of $C_3H_8O_2$ from glycerol under applied potential at the cathode clearly outbeats the conventional fermentation process. Nonetheless, the low quantity of H_2 produced due to water at cathode compartment makes it hard to assess the prime outcome of EF. Furthermore, when EF systems were operated with *G. sulfurreducens,* it has been noted that the microorganisms could be influenced by EFs. Related analysis was reported with *C. pasteurianum* that documented 21% shift in NADH-intake to synthesise $C_3H_8O_2$ utilising glycerol as carbon rich feedstock (Choi *et al.*, 2014).

iv) Alcohols

The over-all demand for alcohols particularly C_2H_5OH as an effective energy substitute remains to surge (Mathew *et al.*, 2015). Alcohols have traditionally been produced through numerous fermentation approaches that typically necessitate two to four days duration to attain extreme concentration. As a result of rapid research in the area of ethanol fermentation, high gravity ethanol fermentation can be attained with an initial sugar feed of greater than 180 g/L. In recent times, MFC was adapted to BES which can be used for the reduction reactions at cathode compartment chosen for bioelectricity generation along with valued chemicals namely formic acid, acetic acid, ethanol, butyric acid, aldehydes, etc. (Chandrasekhar *et al.*, 2015a). The EF approach had been implemented to direct the glucose feeding (120 g/L of glucose) in *Saccharomyces cerevisiae* under the applied potential conditions of 1500 mV. Under the applied voltage, these strains grew rapidly due to the e^- supplement over the cathode, which attained 53 g/L ethanol produce after 50 h of incubation (Shin *et al.*, 2002). Likewise, electrochemical-electrostatic technique was performed for immediate ethanol production by *S. cerevisiae* (Song *et al*., 2014). Alternatively, *Zymomonas mobilis* is a renowned microorganism which is widely used for the for industrial ethanol production as it can bear higher ethanol concentrations (Jeon *et al*., 2009). In the electrostatic fermentation, glucose was utilised by *S. cerevisiae* leading to 14% v/v ethanol in 20 h of incubation without any external power supply (Mathew *et al*., 2015). Engineered *Shewanella oneidensis* (type of electrochemically active microorganism) was also employed for the biotransformation of glycerol into bioethanol with the accumulation of reducing counterparts at anode. In the same way, glucose was effectively transformed into further reduced compounds such as C_2H_5OH and C_4H_9OH by *Clostridium acetobutylicum*. Under applied potential (-1.5 V) conditions, *C. thermocellum* also showed rapid biotransformation of substrate into intermediate metabolites, which lead to 61% improvement in the total ethanol produce (Shin *et al*., 2002). Also, electrochemical system operated with *C. pasteurianum* showed changes in metabolic reactions to achieve higher butanol yield (Choi *et al*., 2014). Organic compounds like acetic acid, glycerol, CO_2, butyric acid etc. have also been widely used for alcohol production (Steinbusch *et al*., 2009; Choi *et al*., 2014). Methyl viologen (MV) mediated biotransformation of acetate to ethanol has been achieved in cathode compartment of EF system, here MV reduce/inhibit methanogenesis. Same reactor setup can be performed for the production of C_4H_9OH from butyric acid (Pant *et al*., 2012). Recently, Torella and co researchers achieved 216 mg/L isopropanol production from CO_2 and H_2 obtained from water splitting in bio-electrochemical system, by employing engineered strain of *Ralstonia eutropha* (Torella *et al*., 2015).

v) Polyhydroxyalkanoates (PHAs)

The PHAs are the only polymers (polyesters) of biological origin that has an encouraging feature to substitute fossil fuel based plastics. PHAs are produced by bacteria in the form of lipid granules under nutrition imbalance circumstances. PHAs acts as a reserve carbon source which are very useful at nutrient depletion conditions (Kumar *et al*., 2015b; Singh *et al*., 2015). These PHAs are biodegradable and biocompatible in nature, also comprising quite similar thermomechanical properties related to synthetic plastics. However, PHA generation at industrial scale remains uneconomical and needs focused research towards minimisation of overall production budget by exploring inexpensive carbon substrates and green solvents for extraction. As an alternative, co-production of other high-value biochemicals during PHA production seems to be an attractive option. Some of the reported examples include PHA-H_2, PHA-EPS, PHA-rhamnolipids etc. were recently reviewed (Kumar *et al*., 2016; Kumar et al., 2018b; Kumar and Kim, 2018; Pantazaki *et al*., 2011). With the progress in BES it was found that biocathode might be used for efficient electrogenesis with simultaneous PHA accumulation at cathode (Kumar et al., 2018a; Srikanth *et al.*, 2012). Using synthetic water supplemented with glucose, 512 mV and 19% w/w of electricity and PHA accumulation was recorded, respectively. In low oxygen conditions at the cathode, PHA producers can pose as a terminal e^- acceptor. Such system can provide triple benefits i.e. production of electricity and PHA with simultaneous treatment of wastewater in a single reactor (Srikanth *et al.*, 2012). Nevertheless, more research work has to be carried out in this area for enhanced PHA yield in such reactors, since low PHA yield will increase the cost of downstream processes.

Integrative approach using EF

Biomass especially those consisting of lignocellulose have become quite promising raw materials for energy production, although it is limited by the efficient pretreatments strategies being applied during the process. The composition of these biomass varies widely, for instance, corn stover and straw generally contains 30-40% cellulose, 20-30% hemicellulose and 15-20% lignin (Thygesen *et al.*, 2010; Kumar et al., 2018c). Depending upon the pretreatment method, many different mono- di- or oligosaccharides gets released from the lignocellulosic substrates (*viz.*, glucose, cellobiose, xylose, arabinose etc.) that can be exploited for the production of H_2, CH_4, PHA, ethanol, biodiesel, etc. through different approaches (Kumar *et al.*, 2016; Thygesen *et al.*, 2010). An integration of fermentation with MEC while using wheat straw hydrolysate (from ethanol production) showed that toxic chemicals could be removed and/ or effectively transformed the remaining sugars into H_2. In this integrative system about 96% hemicellulose, 100% VFAs and 50% phenolic compounds is consumed (biologically converted or degraded). This could permit re-use of water in the biorefinery and thereby reduce the fresh water usage (Thygesen *et al.*, 2010). Likewise, electrodialysis method was explored to separate the organic acids generated from H_2 reactor by *Escherichia coli* grown on glucose. These organic acids were subsequently utilised as a nitrogen-free carbon substrate for H_2 production by *Rhodobacter sphaeroides* (Redwood *et al.*, 2012a). This method was also carried out using actual food waste hydrolysates (Redwood *et al.*, 2012b). These integrative approaches and extractive reactors may sustain indefinitely, nevertheless, membrane fouling, need of persistent ionic composition, pH imbalance etc. are the important issues to be addressed.

Challenges and opportunity for electrofermentation

The advancement in EFs over the past few years has been carried out in two fronts i.e. power output and efficiency. Nevertheless, consent is still needed for the mechanism of e^- transfer across the microbial cells and the fermenter membrane, few indicative studied are already done in this regard. In a common electrochemical cell, electrolytes are a prime constituent that allows a path for proton from the anode to cathode. For this, the most common electrolyte used is Nafion, which is quite expensive. Moreover, usage of costly separating membrane obstructs the upscaling of any electrochemical cell. Other concerns are low power densities, pH fluctuations, loss of ohmic voltage, ill transport across the membrane, overpotential, oxidant, proton flux towards bio-cathode etc. Quite a few alternatives are being studied to circumvent these limitations. For instance, Nafion could be substituted by clay-based electrolytes. Besides that, mediators could be replenished during BES (Steinbusch *et al.*, 2009). A single chamber EF method had also been developed that does not require any costly membranes between the two chambers (Call and Logan, 2008). The prime benefit provided by any EF is the on-site use of power for biosynthesis, also it does not require any agricultural plots. However, any real large-scale operation would need stacks of EFs, but such arrangements could be highly susceptible to cell reversal (Pant *et al.*, 2012). In EF reactor, the power density declines with increment in the anodic surface area. Therefore, reactor design, chamber separation and sustaining low internal resistance are the major hurdles in the upscaling of EF-based systems (Fornero *et al.*, 2010; Pant *et al.*, 2012). It was proposed that for an EF to be commercially economical and sustainable over AD, the EF-based reactors must surpass 500W/m^3. Recent progressions and efforts are focused on developing a sustainable bioelectrofermentation considerably and hold the promise to get this system available for common-use soon.

Future perspectives

The latest progress in electrofermentation to produce energy, electricity and other biochemicals by utilising biowaste has unveiled a plethora of opportunities. However, few concerns like pH, high ohmic resistance, overpotential etc. still requires sincere attention in order to implement this technique successfully. In this quest, investigation of novel materials and new microbial isolates are quite important for enhanced reactor performance and competitive operation budget. Besides, integrative operation of EF in association with other bioremediation methods might lead to the development of a sustainable biorefinery model.

References

Aelterman P, Rabaey K, Pham HT, Boon N, Verstraete W, 2006. Continuous electricity generation at high voltages and currents using stacked microbial fuel cells. *Environmental Science and Technology*, **40:** 3388–94.

Call D, Logan BE, 2008. Hydrogen production in a single chamber microbial electrolysis cell lacking a membrane. *Environmental Science and Technology*, **42(9)**: 3401–6.

Chae K-J, Choi M-J, Kim K-Y, Ajayi FF, Chang I-S, Kim IS, 2009. A solar-powered microbial electrolysis cell with a platinum catalyst-free cathode to produce hydrogen. *Environmental Science and Technology*, 43, 9525–30.

Cao X, Huang X, Liang P, Xiao K, Zhou Y, Zhang X, Logan BE, 2009. A new method for water desalination using microbial desalination cells. *Environmental Science and Technology*, **43,** 7148–52.

Chandrasekhar K, Ahn YH, 2017. Effectiveness of piggery waste treatment using microbial fuel cells coupled with elutriated-phased acid fermentation. *Bioresource Technology*, **244**: 650–7.

Chandrasekhar K, Amulya K, Mohan SV, 2015a. Solid phase bio-electrofermentation of food waste to harvest value-added products associated with waste remediation. *Waste Management*, **45**: 57–65.

Chandrasekhar K, Lee YJ, Lee DW, 2015b. Biohydrogen production: strategies to improve process efficiency through microbial routes. *International Journal of Molecular Sciences*, **16(4)**: 8266–93.

Chandrasekhar K, Pandit S, Kadier A, Dasagrandhi C, Velpuri J, 2017. Biohydrogen production: integrated approaches to improve the process efficiency. *In*: VC Kalia, P Kumar (Eds) *Microbial Applications Volume 1*, Springer International Publishing AG, Switzerland: 189–210.

Chandrasekhar K, Venkata Mohan S, 2012. Bio-electrochemical remediation of real field petroleum sludge as an electron donor with simultaneous power generation facilitates biotransformation of PAH: Effect of substrate concentration. *Bioresource Technology*, **110**: 517–25.

Chandrasekhar K, Venkata Mohan S, 2014a. Induced catabolic bio-electrohydrolysis of complex food waste by regulating external resistance for enhancing acidogenic biohydrogen production. *Bioresource Technology*, **165**: 372–82.

Chandrasekhar K, Venkata Mohan S, 2014b. Bio-electrohydrolysis as a pretreatment strategy to catabolize complex food waste in closed circuitry: Function of electron flux to enhance acidogenic biohydrogen production. *International Journal of Hydrogen Energy*, **39(22)**: 11411–22.

Chen S, Liu G, Zhang R, Qin B, Luo Y, 2012. Development of the microbial electrolysis desalination and chemical-production cell for desalination as well as acid and alkali productions. *Environmental Science and Technology*, 46, 2467–72.

Chen X, Xia X, Liang P, Cao X, Sun H, Huang X, 2011. Stacked microbial desalination cells to enhance water desalination efficiency. *Environmental Science and Technology*, **45,** 2465–70.

Cheng S, Xing D, Call DF, Logan BE, 2009. Direct biological conversion of electrical current into methane by electromethanogenesis. *Environmental Science and Technology*, **43(10)**: 3953–8.

Choi O, Kim T, Woo HM, Um Y, 2014. Electricity-driven metabolic shift through direct electron uptake by electroactive heterotroph *Clostridium pasteurianum*. *Scientific Reports*, **4**: 1–10.

Choi O, Sang BI, 2016. Extracellular electron transfer from cathode to microbes: application for biofuel production. *Biotechnology for Biofuels*, **9(1)**: 11.

Clauwaert P, Toledo R, van der Ha D, Crab R, Verstraete W, Hu H, Udert KM, Rabaey K, 2008. Combining biocatalyzed electrolysis with anaerobic digestion. *Water Science and Technology*, **57(4)**: 575–9.

Clauwaert P, Verstraete W, 2009. Methanogenesis in membraneless microbial electrolysis cells. *Applied Microbiology and Biotechnology*, **82(5)**: 829–36.

Cusick RD, Kim Y, Logan BE, 2012. Energy capture from thermolytic solutions in microbial reverse-electrodialysis cells. *Science*, **335:** 1474–7.

Deval AS, Parikh HA, Kadier A, Chandrasekhar K, Bhagwat AM, Dikshit AK, 2017. Sequential microbial activities mediated bioelectricity production from distillery wastewater using bio-electrochemical system with simultaneous waste remediation. *International Journal of Hydrogen Energy*, **42(2)**: 1130–41.

Ditzig J, Liu H, Logan BE, 2007. Production of hydrogen from domestic wastewater using a bioelectrochemically assisted microbial reactor (BEAMR). *International Journal of Hydrogen Energy*, **32:** 2296–304.

ElMekawy A, Srikanth S, Bajracharya S, Hegab HM, Nigam PS, Singh A, Venkata Mohan S, Pant D, 2015. Food and agricultural wastes as substrates for bioelectrochemical system (BES): the synchronized recovery of sustainable energy and waste treatment. *Food Research International*, **73**: 213–25.

Feng Y, Lee H, Wang X, Liu Y, He W, 2010. Continuous electricity generation by a graphite granule baffled air–cathode microbial fuel cell. *Bioresource Technology*, **101**, 632–8.

Fornero JJ, Rosenbaum M, Angenent LT, 2010. Electric power generation from municipal, food, and animal wastewaters using microbial fuel cells. *Electroanalysis*, **22(7-8)**: 832–43.

Forrestal C, Xu P, Jenkins PE, Ren Z, 2012. Microbial desalination cell with capacitive adsorption for ion migration control. *Bioresource Technology*, **120**, 332–6.

Galanakis CM, 2012. Recovery of high added-value components from food wastes: conventional, emerging technologies and commercialized applications. *Trends in Food Science and Technology*, **26(2)**: 68–87.

Goud RK, Venkata Mohan S, 2011. Pre-fermentation of waste as a strategy to enhance the performance of single chambered microbial fuel cell (MFC). *International Journal of Hydrogen Energy*, **36(21)**: 13753–62.

He Z, Minteer SD, Angenent LT, 2005. Electricity generation from artificial wastewater using an upflow microbial fuel cell. *Environmental Science and Technology*, **39:** 5262–7.

He Z, Wagner N, Minteer SD, Angenent LT, 2006. An upflow microbial fuel cell with an interior cathode: Assessment of the internal resistance by impedance spectroscopy. *Environmental Science and Technology,* **40:** 5212–7.

Hu H, Fan Y, Liu H, 2008. Hydrogen production using single chamber membrane-free microbial electrolysis cells. *Water Research*, **42(15)**: 4172–8.

Imbert E, 2017. Food waste valorization options: opportunities from the bioeconomy. *Open Agriculture*, **2(1)**: 195–204.

Jacobson KS, Drew DM, He Z, 2011. Use of a liter-scale microbial desalination cell as a platform to study bioelectrochemical desalination with salt solution or artificial seawater. *Environmental Science and Technology*, **45:** 4652–7.

Jeon BY, Hwang TS, Park DH, 2009. Electrochemical and biochemical analysis of ethanol fermentation of *Zymomonas mobilis* KCCM11336. *Journal of Microbiology and Biotechnology*, **19(7)**: 666–74.

Kadier A, Kalil MS, Abdeshahian P, Chandrasekhar K, Mohamed A, Azman NF, Logrono W, Simayi Y, Hamid AA, 2016a. Recent advances and emerging challenges in microbial electrolysis cells (MECs) for microbial production of hydrogen and value-added chemicals. *Renewable and Sustainable Energy Reviews*, **61**: 501–25.

Kadier A, Simayi Y, Abdeshahian P, Azman NF, Chandrasekhar K, Kalil MS, 2016b. A comprehensive review of microbial electrolysis cells (MEC) reactor designs and configurations for sustainable hydrogen gas production. *Alexandria Engineering Journal*, **55(1)**: 427–43.

Kadier A, Simayi Y, Chandrasekhar K, Ismail M, Kalil MS, 2015. Hydrogen gas production with an electroformed Ni mesh cathode catalysts in a single-chamber microbial electrolysis cell (MEC). *International Journal of Hydrogen Energy*, **40(41)**: 14095–103.

Kim Y, Logan BE, 2011. Hydrogen production from inexhaustible supplies of fresh and salt water using microbial reverse-electrodialysis electrolysis cells. *Proceedings of the National Academy of Sciences*, USA, **108:** 16176–81.

Kim Y, Logan BE, 2013. Simultaneous removal of organic matter and salt ions from saline wastewater in bioelectrochemical systems. *Desalination*, **308:** 115–21.

Kumar AK, Reddy MV, Chandrasekhar K, Srikanth S, Venkata Mohan S, 2012. Endocrine disruptive estrogens role in electron transfer: bio-electrochemical remediation with microbial mediated electrogenesis. *Bioresource Technology*, **104**: 547–56.

Kumar P, Chandrasekhar K, Kumari A, Sathiyamoorthi E, Kim BS, 2018a. Electro-fermentation in aid of bioenergy and biopolymers. *Energies*, **11(2):** 343.

Kumar P, Maharajan A, Jun HB, Kim BS, 2018c. Bioconversion of lignin and its derivatives into polyhydroxyalkanoates: Challenges and opportunities. Biotechnology and Applied Biochemistry, https://doi.org/10.1002/bab.1720

Kumar P, Jun HB, Kim BS, 2018b. Co-production of polyhydroxyalkanoates and carotenoids through bioconversion of glycerol by *Paracoccus* sp. strain LL1. *International Journal of Biological Macromolecules*, **107**:2552-2558.

Kumar P, Kim BS, 2018. Valorization of polyhydroxyalkanoates production process by co-synthesis of value-added products. *Bioresource Technology*, **69**:544-556.

Kumar P, Mehariya S, Ray S, Mishra R, Kalia VC, 2015a. Biodiesel industry waste: A potential source of bioenergy and biopolymers. *Indian Journal of Microbiology*, **55(1)**: 1–7.

Kumar P, Patel SKS, Lee JK, Kalia VC, 2013. Extending the limits of *Bacillus* for novel biotechnological applications. *Biotechnology Advances*, **31(8)**: 1543–61.

Kumar P, Ray S, Kalia VC, 2016. Production of co-polymers of polyhydroxyalkanoates by regulating the hydrolysis of biowastes. *Bioresource Technology*, **200**: 413–9.

Kumar P, Ray S, Patel SK, Lee JK, Kalia VC, 2015b. Bioconversion of crude glycerol to polyhydroxyalkanoate by *Bacillus thuringiensis* under non-limiting nitrogen conditions. *International Journal of Biological Macromolecules*, **78**: 9–16.

Lee B, Park JG, Shin WB, Tian DJ, Jun HB, 2017. Microbial communities change in an anaerobic digestion after application of microbial electrolysis cells. *Bioresource Technology*, **234**: 273–80.

Liu H, Grot S, Logan BE, 2005. Electrochemically assisted microbial production of hydrogen from acetate. *Environmental Science and Technology*, **39(11)**: 4317–20.

Liu H, Hu H, 2012. Microbial electrolysis: novel biotechnology for hydrogen production from biomass. *In*: PC Hallenback (Eds) *Microbial Technologies in Advanced Biofuels Production*, Springer, *United States of America*: 93–105.

Luo H, Jenkins PE, Ren Z, 2011. Concurrent desalination and hydrogen generation using microbial electrolysis and desalination cells. *Environmental Science and Technology*, **45**: 340–4.

Logan BE, Rabaey K, 2012. Conversion of wastes into bioelectricity and chemicals by using microbial electrochemical technologies. *Science*, **337(6095)**: 686–90.

Mathew AS, Wang J, Luo J, Yau ST, 2015. Enhanced ethanol production via electrostatically accelerated fermentation of glucose using *Saccharomyces cerevisiae*. *Scientific Reports*, **5**: 15713.

Michalopoulos I, Chatzikonstantinou D, Mathioudakis D, Vaiopoulos I, Tremouli A, Georgiopoulou M, Papadopoulou K, Lyberatos G, 2017. Valorization of the liquid fraction of a mixture of livestock waste and cheese whey for biogas production through high-rate anaerobic co-digestion and for electricity production in a microbial fuel cell (MFC). *Waste and Biomass Valorization* **8(5)**: 1759–69.

Modestra JA, Babu ML, Venkata Mohan S, 2015. Electrofermentation of real-field acidogenic spent wash effluents for additional biohydrogen production with simultaneous treatment in a microbial electrolysis cell. *Separation and Purification Technology*, **150**: 308–15.

Morel A, Zuo K, Xia X, Wei J, Luo X, Liang P, Huang X, 2012. Microbial desalination cells packed with ion-exchange resin to enhance water desalination rate. *Bioresource Technology*, **118:** 43–8.

Moscoviz R, Toledo-Alarcón J, Trably E, Bernet N, 2016. Electrofermentation: how to drive fermentation using electrochemical systems. *Trends in Biotechnology*, **34(11)**: 856–65.

Nikhil GN, Suman P, Venkata Mohan S, Swamy YV, 2017. Energy-positive nitrogen removal of pharmaceutical wastewater by coupling heterotrophic nitrification and electrotrophic denitrification. *Chemical Engineering Journal*, **326**: 715–20.

Nikhil GN, Yeruva DK, Venkata Mohan S, Swamy YV, 2016. Assessing potential cathodes for resource recovery through wastewater treatment and salinity removal using non-buffered microbial electrochemical systems. *Bioresource Technology*, **215**: 247–53.

Pant D, Singh A, Van Bogaert G, Olsen SI, Nigam PS, Diels L, Vanbroekhoven K, 2012. Bioelectrochemical systems (BES) for sustainable energy production and product recovery from organic wastes and industrial wastewaters. *RSC Advances*, **2(4)**: 1248–63.

Pantazaki AA, Papaneophytou CP, Lambropoulou DA, 2011. Simultaneous polyhydroxyalkanoates and rhamnolipids production by *Thermus thermophilus* HB8. *AMB Express*, **1(17)**: 1–13.

Rabaey K, Clauwaert P, Aelterman P, Verstraete W, 2005. Tubular microbial fuel cells for efficient electricity generation. *Environmental Science and Technology*, **39:** 8077–8082.

Reddy MV, Chandrasekhar K, Venkata Mohan S, 2011. Influence of carbohydrates and proteins concentration on fermentative hydrogen production using canteen based waste under acidophilic microenvironment. *Journal of Biotechnology*, **155(4)**: 387–95.

Redwood MD, Orozco RL, Majewski AJ, Macaskie LE, 2012a. Electro-extractive fermentation for efficient biohydrogen production. *Bioresource Technology*, **107**: 166–74.

Redwood MD, Orozco RL, Majewski AJ, Macaskie LE, 2012b. An integrated biohydrogen refinery: synergy of photofermentation, extractive fermentation and hydrothermal hydrolysis of food wastes. *Bioresource Technology*, **119**: 384–92.

Sasaki K, Sasaki D, Morita M, Hirano SI, Matsumoto N, Ohmura N, Igarashi Y, 2010. Bioelectrochemical system stabilizes methane fermentation from garbage slurry. *Bioresource Technology*, **101(10)**: 3415–22.

Schievano A, Sciarria TP, Vanbroekhoven K, De Wever H, Puig S, Andersen SJ, Rabaey K, Pant D, 2016. Electrofermentation–merging electrochemistry with fermentation in industrial applications. *Trends in Biotechnology*, **34(11)**: 866–78.

Shin HS, Zeikus JG, Jain MK, 2002. Electrically enhanced ethanol fermentation by *Clostridium thermocellum* and *Saccharomyces cerevisiae*. *Applied Microbiology and Biotechnology*, **58(4)**: 476–81.

Singh M, Kumar P, Ray S, Kalia VC, 2015. Challenges and opportunities for customizing polyhydroxyalkanoates. *Indian Journal of Microbiology*, **55(3)**: 235–49.

Song Y, Wang J, Yau ST, 2014. Controlled glucose consumption in yeast using a transistor-like device. *Scientific Reports*, **4**: 5429.

Srikanth S, Kumar M, Singh MP, Das BP, 2016. Bioelectro chemical systems: a sustainable and potential platform for treating waste. *Procedia Environmental Sciences*, **35**: 853–9.

Srikanth S, Reddy MV, Venkata Mohan S, 2012. Microaerophilic microenvironment at biocathode enhances electrogenesis with simultaneous synthesis of polyhydroxyalkanoates (PHA) in bioelectrochemical system (BES). *Bioresource Technology*, **125**: 291–9.

Steinbusch KJ, Hamelers HV, Schaap JD, Kampman C, Buisman CJ, 2009. Bioelectrochemical ethanol production through mediated acetate reduction by mixed cultures. *Environmental Science and Technology*, **44(1)**: 513–7.

Tartakovsky B, Manuel MF, Neburchilov V, Wang H, Guiot SR, 2008. Biocatalyzed hydrogen production in a continuous flow microbial fuel cell with a gas phase cathode. *Journal of Power Sources*, **182(1)**: 291–7.

Thygesen A, Thomsen AB, Possemiers S, Verstraete W, 2010. Integration of microbial electrolysis cells (MECs) in the biorefinery for production of ethanol, H_2 and phenolics. *Waste and Biomass Valorization*, **1(1)**: 9–20.

Torella JP, Gagliardi CJ, Chen JS, Bediako DK, Colon B, Way JC, Silver PA, Nocera DG, 2015. Efficient solar-to-fuels production from a hybrid microbial–water-splitting catalyst system. *Proceedings of the National Academy of Sciences*, **112(8)**: 2337–42.

Velvizhi G, Venkata Mohan S, 2017. Multi-Electrode bioelectrochemical system for the treatment of high dissolved solids bearing chemical based wastewater. *Bioresource Technology*, **242**: 77–86.

Venkata Mohan S, Chandrasekhar K, 2011a. Self-induced bio-potential and graphite electron accepting conditions enhances petroleum sludge degradation in bio-electrochemical system with simultaneous power generation. *Bioresource Technology* **102(20)**: 9532–41.

Venkata Mohan S, Chandrasekhar K, 2011b. Solid phase microbial fuel cell (SMFC) for harnessing bioelectricity from composite food waste fermentation: Influence of electrode assembly and buffering capacity. *Bioresource Technology*, **102(14)**: 7077–85.

Venkata Mohan S, Chandrasekhar K, Chiranjeevi P, Babu PS, 2013. Biohydrogen Production from Wastewater. *In*: A Pandey, JS Chang, PC Halenbecka, C Larroche (Eds) *Biohydrogen 1st Edition*, Elsevier, *Amsterdam*: 223–57.

Venkata Mohan S, Nikhil GN, Chiranjeevi P, Reddy CN, Rohit MV, Kumar AN, Sarkar O, 2016. Waste biorefinery models towards sustainable circular bioeconomy: critical review and future perspectives. *Bioresource Technology*, **215**: 2–12.

Villano M, Aulenta F, Majone M, 2012. Perspectives of biofuels production from renewable resources with bioelectrochemical systems. *Asia-Pacific Journal of Chemical Engineering*, **7(3)**: S263–S274.

Wang A, Liu W, Cheng S, Xing D, Zhou J, Logan BE, 2009. Source of methane and methods to control its formation in single chamber microbial electrolysis cells. *International Journal of Hydrogen Energy*, **34(9)**: 3653–8.

Wang H, Fotidis IA, Angelidaki I, 2015. Ammonia effect on hydrogenotrophic methanogens and syntrophic acetate-oxidizing bacteria. *FEMS Microbiology Ecology*, **91(11)**: fiv130.

Wang H, Ren ZJ, 2013. A comprehensive review of microbial electrochemical systems as a platform technology. *Biotechnology Advances*, **31(8)**: 1796–807.

Zhang B, Zhang J, Yang Q, Feng C, Zhu Y, Ye Z, Ni J, 2012. Investigation and optimization of the novel UASB–MFC integrated system for sulfate removal and bioelectricity generation using the response surface methodology (RSM). *Bioresource Technology*, **124:** 1–7.

Zhang B, He Z, 2012. Integrated salinity reduction and water recovery in an osmotic microbial desalination cell. *RSC Advances*, **2:** 3265–9.

Zhang Y, Angelidaki I, 2012. Innovative self-powered submersible microbial electrolysis cell (SMEC) for biohydrogen production from anaerobic reactors. *Water Research*, **46,** 2727–36.

Zhang Y, Angelidaki I, 2013. A new method for in situ nitrate removal from groundwater using submerged microbial desalination–denitrification cell (SMDDC). *Water Research*, **47:** 1827–36.

Zhu X, Hatzell MC, Cusick RD, Logan BE, 2013. Microbial reverse-electrodialysis chemical-production cell for acid and alkali production. *Electrochemistry Communications*, **31:** 52–55.

12

Industrial Applications of Anaerobic Digestion

M.K. Mohanty[a] and D.M. Das[b]

[a]Professor, College of Agricultural Engineering & Technology, OUAT, Bhubaneswar
[b]Scientist, Agricultural Engineering, KVK, OUAT, India
[a]Corresponding Author: mohanty65@gmail.com

ABSTRACT

*The global energy demand is increasing day by day. Now-a-days, the supply of global energy is highly dependent on fossil fuels (*viz.*, crude oil, lignite, hard coal, natural gas). But, on one hand the source of fossil fuel is decreasing at a rapid rate, and on the other hand the burning of fossil fuel is also degrading the environment due to huge Greenhouse Gas (GHG) emission (FAO/CMS, 2011). Hence, there is an urgent need of an alternate clean energy source that can sustain the future development undeterred while still protecting the environment. Biomass is considered as a sustainable source of clean energy that is useful in steam, fuel and chemicals production. It can be digested in controlled environment to produce biogas. The biogas is a clean, non-polluting and smoke-free fuel (Weiland, 2010) which can be used for applications that include cooking, space cooling/refrigeration, heating, electricity generation and gaseous fuel (CNG) for vehicles (Weiland, 2003). This paper analyses possible alternatives to process and utilise biogas for energy at industrial scale as a petroleum fuels replacement.*

Keywords: Petroleum fuel, Biomass, Biogas, Green House Gas (GHS)

Abbreviations

AD : anaerobicdigestion
CNG : compressed natural gas
GHG : greenhouse gas

HRT : hydraulic retention time
SRT : solid retention time
LR : loading rate
MSW : municipal solid waste
PSA : pressure swing adsorption
CHP : combined heat and power
BTTP : block type thermal power plant
PING : pilot injection natural gas

Introduction

In contrast to the growing dependency on traditional fossil fuels, *i.e.*, natural gas, coal, and petroleum products, these are depleting day-by-day (Abraham *et. al.*, 2007) while its international pricing is increasing. Further, climate change due to heavy carbon dioxide (greenhouse gas) emission from fossil fuels has led to serious environmental thinking and legislations (Gallucci *et al.*, 2010; White and Sulkowski, 2010). In the present scenario, alternative fuel is the need of the hour. India and the world as a whole are pursuing for alternative fuels to reduce the dependency on fossil fuels. Using renewable energy resources instead of fossil energy is a consensually accepted essential way to reduce greenhouse gas emissions (United Nations, 1997; Hoffert *et al.*, 2002; Bates *et al.*, 2008). Nature produces about 200 billion tonnes of biomass annually through photosynthesis, out of which only 3-4% is utilisable (Jenck *et al.*, 2004). Biomass is considered as a sustainable energy source to be used as a (Berndes *et al.*, 2003). Due to its carbon sequestration characteristic, bioenergy chain is climate neutral and is easily recyclable (Reijnders, 2006).

Despite the technical feasibility and multiple benefits, bioenergy has not been optimally popular in India so far. Biomass material is used directly for combustion in rural India, an inefficient, smoky, and can-easy-to-control process (Goldemberg and Teixeira, 2004). This traditional use of wood for fuel is considered unsustainable as only its heat component is utilised. Instead of direct burning, these can be used for biogas generation, which can cater to both the cooking and lighting requirements of the village. Hence, cogeneration of electricity and heat to utilise the full potential of such raw materials is preferable, besides their potential use as a transportation (vehicular) fuel.

Biomass is converted in to energy through various means and its form of utilisation is application dependent (Demirbas, 2005; McKendry, 2002). In a zero-waste biorefinery approach, biogas is the major product in biological processing. Having proper amount of moisture content, biomass can be converted to usable energy, *i.e.*, biogas through anaerobic fermentation (Pimentel, 2001). Biogas primarily contains microbially-generated methane (CH_4) and carbon dioxide (CO_2). Anaerobic fermentation is also useful in waste management and energy recovery from sewage, wastewater and industrial sludge (Green and Byrne, 2004).

Biogas is generally referred as *gobar gas* as the conventional and main feedstock for anaerobic digestion is cattle dung. In India, biogas technology has been well recognised to provide an alternative and efficient fuel and manure rich in plant nutrients. Such technology helps to reduce deforestation, contribute to smokeless cooking and ensure non-polluting environment (Murphy and Power, 2009). Promoting biogas technology in a large-scale helps to generate employment for village technicians, masons and unskilled workers. In view of the energy crisis and its impact on national economy, the Fuel Policy Committee setup by Government of India in 1974 strongly recommended the popularisation of biogas as an alternate energy source. The promotion of biogas technology has been included in Prime Minister's twenty point programme since 1982. As one of the major future fuels, the aim of this chapter is to revisit the various utilisations and industrial applications of biogas.

Anaerobic digestion

Biogas is produced from organic feedstock by multistage microbial processes in anaerobic digestion (devoid of oxygen). The feedstock influences the reactor configurations, *i.e.*, reactor operation, design and the microbial physiology (Fig. 1; Gallucci *et al.*, 2010).

For instance, a completely different design is required for high-solid and polymeric compound wastes than relatively quickly biodegradable wastewaters (Hossain, 2011). Steffen *et al.* (1998) reported that cellulose digestion takes several weeks whereas hemicelluloses, protein and fat take few weeks for their complete degradation. In contrast, very few hours are needed to degrade alcohol, low molecular sugars, and volatile fatty acids. The desired composition, fluid dynamics, homogeneity and biodegradability, therefore, are necessary to qualify a feedstock for anaerobic digestion.

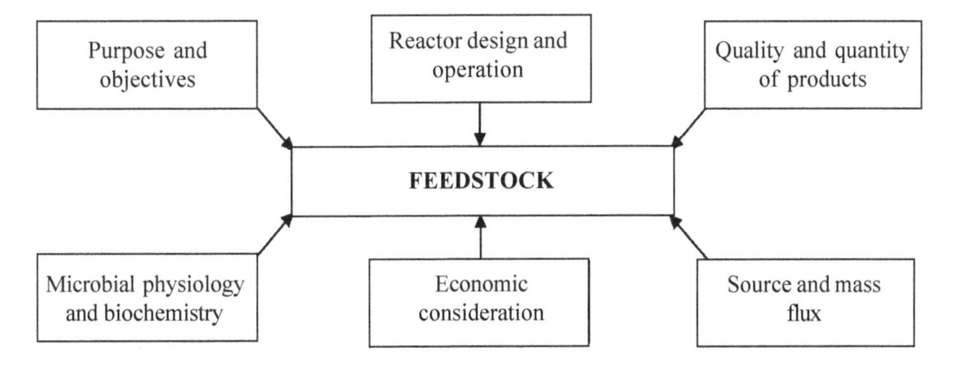

Fig. 1. The role of feedstock in decision-making in aspects of anaerobic digestion

Operational sequence of anaerobic decomposition/digestion

The multistage microbial process of anaerobic digestion (devoid of oxygen) of organic feedstock is dependent on various parameters. Bacterial anaerobic decomposition/digestion of biomass in the absence of oxygen generally occurs in three phases as mentioned by Leggett *et al.*, (2006; Fig. 2). In the 1st (liquefaction) stage, insoluble fibrous materials including proteins, fats and carbohydrates convert to soluble substances by the action of liquefying bacteria. During the 2nd and 3rd steps, almost all liquid soluble compounds are converted into biogas by the action of acid and methane forming bacteria. In the 2nd stage of anaerobic process, volatile acids are formed from soluble organic matter by the action of acidifying bacteria that causes malodour in liquid manure (Smyth *et al.*, 2010). Finally, in the 3rd stage, the volatile acids are converted into biogas by methane-forming bacteria, the quality of which depends upon many factors as described later.

Reactions

Barnett *et al.* (1978) described the following reactions to obtain the theoretical yield of gas from biomass fermentation/anaerobic digestion:

$$C_cH_hO_oN_nS_s + \gamma H_2O \rightarrow xCH_4 + (c - x)\ CO_2 + nNH_3 + sH_2S \qquad \text{(i)}$$

where, $x = 1/8(4_c+h-2_o-3n-2s)$, and $y = \frac{1}{4}(4c+3_n+2_s-h-2_o)$

For example, for cellulose h = 10, c = 6, o = 5, s = 0, n = 0.

Varying the composition of carbohydrate, protein and lipid contents in substrate, can significantly affect the process rates. Though lipid produces the highest gaseous fuel, it requires a relatively high retention time due to its poor biodegradability.

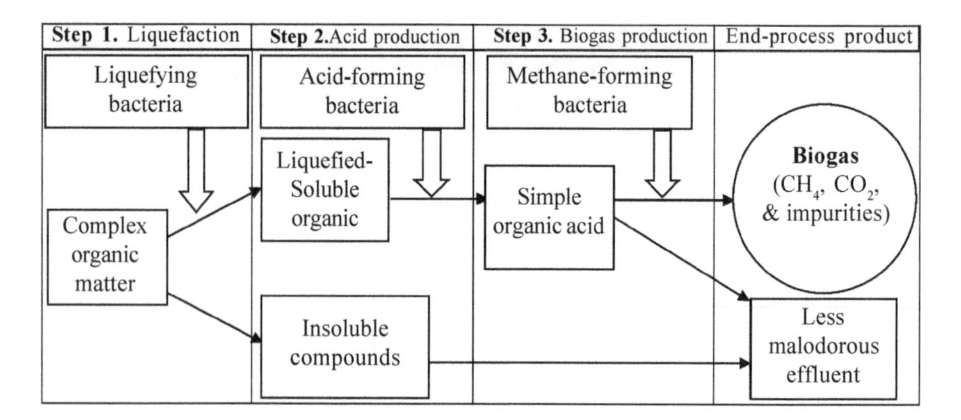

Fig. 2: The three stages of biogas production

Factors affecting anaerobic digestion

Anaerobic fermentation should proceed with minimal external control to increase biogas production. This happens, when the critical parameters lie within the desired range. External control becomes necessary only when they fall beyond the desired limit. All the affecting factors may be divided into two clusters, a) Environmental factors, and b) Operational factors (Hamad, 1983; Myles, 1987; Hansen, 2004; Balsam and Ryan, 2006; Anon., 2011).

Environmental factors

Some environmental factors regulate anaerobic digestion critically when they deviate from optimum levels significantly. The three most critical factors are temperature, substrate pH, and substrate nutritional composition.

Effect of temperature

Temperature is an important factor affecting biogasification in a biogas plant, affecting biogas yield. Anaerobic digestion operates within three different temperature ranges as psychrophilic (10–25°C), mesophilic (25–35°C), and thermophilic (49–60°C). The optimum temperature of digester slurry is 35°C and 55°C in mesophillic and thermophilic zones, respectively. Choices between the mesophilic and thermophilic conditions are influenced by the prevailing climatic condition where the plant is installed. It is reported that no anaerobic psychrophilic bacterial group remains active below 25°C. Wellinger *et al.* (1984) observed that mesophiles very slowly adapt to low temperatures.

Most household biogas plants in rural areas operate under ambient temperature. Thus, biogas production is affected by seasonal variations and it is more dependent on ground temperature (geothermal profile) than the atmospheric temperature. Gas production reduces by 50% in winters, the reactions completely ceasing at slurry temperature below 10°C and fortunately, revives with the increase in temperature. For this, biogas plants should be constructed facing the sun and protected from cold winds. Covering the gasholder with plastic sheets in winter is a way to control the fall in gasfier temperature. Sun-warmed water can be used for slurry preparation to maintain digester temperature.

Effect of pH Value

The pH of the digester contents influences the biogas yield. During anaerobic fermentation, the microbial groups active in a digester require a neutral to mild alkaline environment for efficient biogas production. The optimum pH for efficient biogas production is between 7.0 and 8.5. More acidic or more alkaline nature of the contents retards microbial activity. Limewater is added when the pH becomes very low (high acidity), and hydrochloric acid is added when it becomes high (high alkalinity).

Effect of nutrient composition

The natural degradation of any organic substrate is very dependent on the C:N ratio. Feeding material of a digester may be divided into two types depending on its nitrogen content, viz., nitrogen-rich, and nitrogen-poor. It is necessary to maintain proper C:N ratio in the feedstock within specified limits, ideally below 30.

Both carbon and nitrogen provide the requisite nutrients for efficient microbial functioning during anaerobic digestion (Moller and Muller, 2012). During fermentation, microbes use carbon 25–30 times faster than nitrogen. So the feedstock should ideally have a C:N ratio of 25 to 30:1 with a dry matter concentration of 7–10% (Lapp *et al.*, 1975). Powdered leaves, wheat straw, animal urine are added to the slurry to nutritionally enrich and naturally buffer the digester content, which in turn would stimulate microbial activities thereby enhancing biogas production.

Operational factors

The three operational factors affecting anaerobic digestion the most are retention period, slurry concentration, and mixing and stirring.

Retention period

It is the time duration during which the input is retained inside the digester. When feedstock is relatively liquid, this period is termed as hydrolytic retention time (HRT). If the feedstock is in relatively solid form, it is termed as solid retention time (SRT). The retention period is directly proportional to the reactor temperature. The retention period is 45–55 days at 25–30°C, 20 days at 35–37°C and 6–10 days at 55°C. In the tropical countries like India, the RT is 40–50 days.

Slurry concentration

Feedstock consists of solid matter mixed with equivalent quantity of water, particularly in case of cattle dung. The solid matter contains both non-volatile as well as volatile organic matter. During anaerobiosis, only the volatile solids undergo digestion and the non-volatiles remain unaffected. For optimal biogas production through anaerobic digestion, only 8–10% of total solid in the feedstock is enough. The total solids, loading rate and HRT are related as:

$$LR = k\frac{TS}{HRT} \tag{ii}$$

where, LR is Kg volatile solid/day.m^3, TS is % total solid, HRT is in days, and the constant k has an approximate value of 7 (TERI, 1987).

Mixing and stirring

Proper slurry mixing greatly helps in the intimate contact between the slurry and the resident microbes, ensuring improved fermentation efficiency. However, it should be done at slower speed and infrequently more frequent contact between the microbes and the content occurs if the slurry is fed daily instead of periodically with desired mixing effect. Stirring helps in proper heat distribution throughout the usually underground digester to achieve uniform temperature. It helps in the release of biogas bubbles trapped in the substrate. Stirring also prevents the formation of the undesirable scum layer on the top of the digester content which reduces the effective volume of the digester and restricts the upward gas flow to the storage chamber.

Applications of biogas

Biogas production is one of the best options for wise utilisation of livestock manure. The major objective of this is to produce renewable energy and high quality organic fertiliser. These days, sewage sludge, wastes from food and fermentation industries and municipal solid wastes are mostly treated by anaerobic process to produce biogas thereby reducing their adverse impact on environment (Korhonen and Snakin, 2005).

Agricultural biogas plant

Such biogas plants mostly use agri-residues as feedstock, such as, crop residues, dung and urine, vegetable residues and byproducts, residues from food and fish processing industries, and dedicated energy crops (DEC), etc. Depending on the national frameworks (legislation and energy policies), climatic conditions, energy availability and economy of the country, the design of such biogas plants vary among the nations.

Family-scale biogas plants

Millions of family-scale biogas plants are operational in China, India, Nepal, and they operate on simple technologies; there is no need of control instruments and heating (psychrophilic or mesophilic operation temperatures) as these countries have warmer climate, and the HRT is long. These biogas plants are simple underground reactors. In these biogas plants a floating dome collects biogas and the effluent is collected at the reactor bottom. Fig. 3 provides a schematic presentation of the Indian and Chinese type family-scale plants.

Feedstock for these plants is from households and micro-level farming activities. The produced biogas is usually meant for cooking and lighting purpose at individual households. The operation is in semi-continuous mode where substrates are added once (or twice) in a day and equal amount of spent slurry is removed.

These reactors are often left unstirred, the sedimented solid is removed 2-3 times in a year.

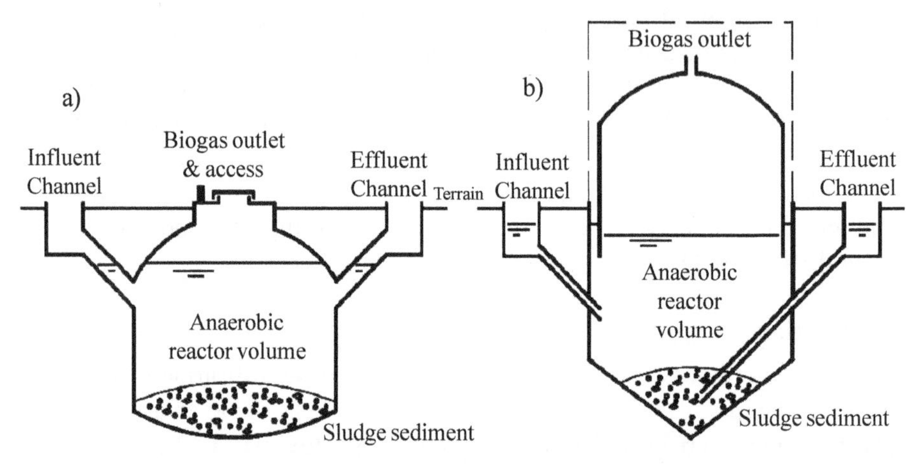

Fig. 3: Family-scale biogas reactors: a) Chinese type, b) Indian type (*Source:* Angelidaki, 2004)

Farm-scale biogas plants

Farm-scale biogas plants use feedstock produced from a particular farm. Such plants have specific specialised technologies, designs, and sizes. Some are small and technologically simple, while others are rather large and technologically complex. In these plants, the feedstock is collected in a pre-storage tank and then pumped into the digester. Digesters are of concrete or steel and insulated to maintain a constant temperature during digestion.

Digesters are generally vertical or horizontal (Fig. 4) with stirring systems for mixing and homogenising the substrate and reducing the risk of sedimentation and floating-layers. Depending on of substrate type and digester temperature the average HRT varies from 20–40 days.

Biogas produced from these plants is used in gas engine for heat production and electricity generation. About 10–30% of the produced energy is used for operating the biogas plant and fulfilling the domestic requirements of the farmer, the surplus is then sold to neighbouring consumers or to power companies. The major limitations of these biogas plants are continuous feedstock availability, operational cost, and availability of a suitable land.

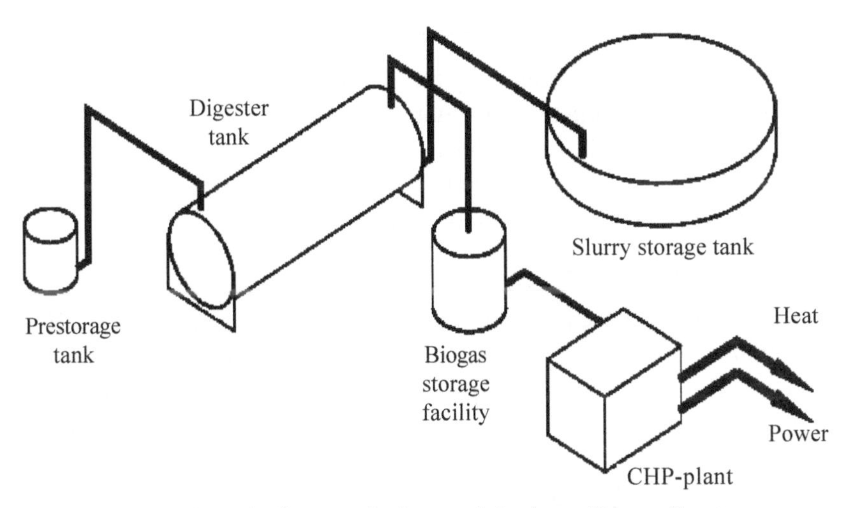

Fig. 4: Schematic diagram of a farm-scale horizontal biogas digester (*Source:* Hjort-Gregersen, 1998)

Co-digestion biogas plants

Centralised co-digestion is based on digesting animal manure and slurries along with suitable co-substrates, like agricultural residues, food and fish waste, household wastes, sewage sludge etc. (Fig. 5). Before feeding, the manure and slurry are mixed with other substrates, homogenised and then pumped inside the digester.

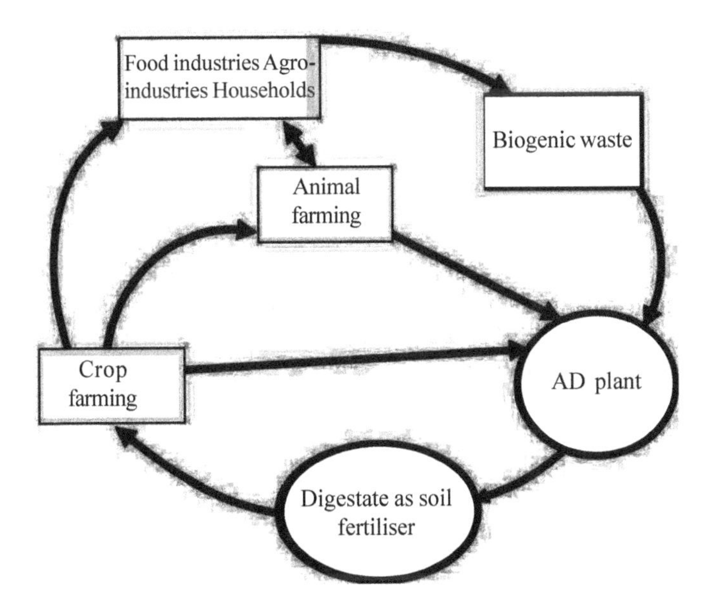

Fig. 5: Schematic diagram of the closed cycle of centralised AD (*Source*:Al Seadi,2001)

Municipal wastewater treatment plant

Municipal wastewater is organically rich with high BOD and COD. It needs to be treated before releasing to open environment. Anaerobic digestion is applied to treat sludge produced from such wastewaters in the treatment plants. This is used to reduce the volume and stabilise the sludge. The solid byproduct thus obtained can be further used as manure in agriculture.

Municipal solid waste (MSW) treatment plant

In many countries, municipal solid waste is generally disposed at landfill sites. This pollutes the environment, and also the utilisable nutrients present in the organic solid waste are lost in this process. Depending upon the type and origin of these solid wastes, biogas and also the nutrient-rich manures can be accordingly produced through AD process. If organic waste is segregated, they can be fed as bulk substrate to operate commercial-size biogas plants.

Industrial biogas plants

Anaerobic digestion is now a standard technology to treat industrial wastes from agro-, food-processing, and pharmaceutical industries (Cote and Cohen, 1998). Due to increased environmental concerns management of industrial solid wastes is controlled by environmental legislations (Roberts, 2004). Industrial biogas plants have several benefits for the environment and the society like, value addition of waste through nutrient recycling, economic benefit through reduction of disposal cost, process energy supplement through the produced biogas in the same plant, and building the image of the industry among the public.

Landfill gas recovery plants

Landfill can act as a large anaerobic plant (Fig. 6). Decomposition process in landfill is discontinuous and depends on the age of the landfill (Themelis and Ulloa, 2007). Biogas recovery from landfills controls methane gas emission into the environment. Also, this is a cheap source of energy and generates revenues (Murphy *et al.*, 2004). The gas is utilised as vehicular fuel and for space heating and cooking. Landfill gas can also be used to generate electricity as the landfill sites are remotely located from the city.

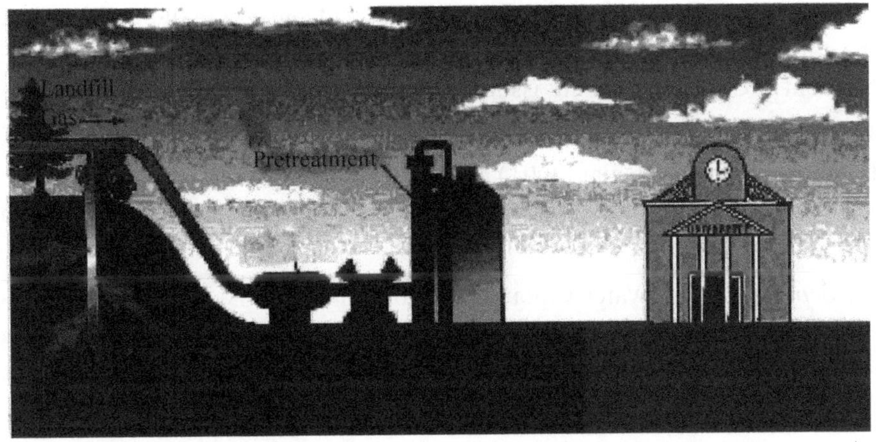

Fig. 6: Landfill gas recovery system (*Source*: Al Seadi & Holm, 2004)

Properties of biogas

Biogas typically contains 40–80% methane and 20–40% carbon dioxide. Other components present in small quantities are nitrogen (N_2) and hydrogen sulphide (H_2S). Among all of these constituents, methane is inflammable. Pure methane has lower heating value of 34300kJ/m^3 at standard temperature and pressure, and biogas has a heating value of 13720–27400kJ/m^3. Biogas has a high specific gravity of 0.85 and very low flame speed of around 0.3m/sec (Sastry, 1985). So during burning, modification of burner is required to reduce gas pressure at the burner nozzle to prevent pressure flame blow off while using large gas jet. Other important property is the flame temperature because the heat transfer is directly proportional to flame temperature. Jensen and Jensen (2000) have reported that the adiabatic flame temperature is 1911°C for 65% methane and 35% carbon dioxide gas composition. With the increase in the concentration of non-combustibles (such as, CO_2 and water vapour) the flame temperature of biogas decreases (Walsh, 1989).

Biogas upgradation

In biogas upgradation process, the unwanted carbon dioxide and other gases are removed various advantages of upgrading the biogas are, reduction of gas volume, thereby reduction of storage capacity and energy needed for compression; getting homogeneous biogas, compatible with fossil fuel standards; and reduced emissions effect of pollutants.

Removal of water

Water is removed by commonly separating the condensed water from gas by absorption or adsorption. Together with water, other impurities such as foam, dust and some ammonia can also be removed during the process.

Drying by cooling: Biogas is cooled in heat exchanger (usually a chiller) and the condensed water is separated out. The condensation point can be lowered to about 1°C due to freezing issue on the heat exchangers surface.

Absorption drying: Water is absorbed by glycol, tri-ethylene-glycol or hygroscopic salts (Hagen, 2001). The drier consists of an absorption vessel filled with granules of hygroscopic salt. As wet biogas is fed from the bottom the salt absorbs water and dissolves (Hagen, 2001).

Adsorption drying: Water vapour reversibly binds on the surface of a drying agent and is adsorbed. Various adsorption materials are, like silica gel, activated carbon, molecular sieves, aluminium oxide or magnesium oxide (Hagen, 2001; AD-Nett, 2001).

Removal of other contaminants

Particles: Particulate materials are common in biogas that can be removed together during water separation. For higher purity as required for appropriate applications, dust cyclones or particle filters are used.

Halogenated hydrocarbons: Halogenated hydrocarbons, particularly chloro- and fluoro-compounds, can be removed with impregnated activated carbon.

Ammonia: Ammonia is removed with activated carbon units. Some upgraded processes such as adsorption and water scrubbing can also remove ammonia (Hagen, 2001). However, some ammonia will already be removed together with water condensate.

Organic silicon compounds: Such compound causes severe damage (silicate fouling) to gas-driven engine. It is removed by absorption, adsorption or condensation. Absorption technique is usually applied as counter-flow columns with or without washing liquid regeneration. Activated carbon is employed to remove silicon compounds by adsorption. Biogas is heated or dried by condensing before treatment to prevent water saturation of the activated carbon.

***Removal of carbon dioxide* (CO_2)**

Membrane separation: It is based on the principle of selective permeability of membranes for different fluid components. Gaseous components in a mixture has characteristic permeation rates through a given membrane material. The permeation rate depends on the ability of gas to dissolve in and diffuse through a characteristic polymeric membrane (Eriksen *et al*., 1998).

Pressure swing adsorption (PSA): PSA is a method to separate carbon dioxide from methane by adsorption/desorption of carbon dioxide on molecular sieves (zeolites, activated carbon) at different pressure levels (Fig. 8). Molecular sieves effectively separate the numerous gaseous compounds in a gas. Contaminating gases like CO_2, O_2, N_2, and H_2S including water vapour can be adsorbed by these materials. The difference in attraction force of contaminating gases to the surface of the fixed material is utilised to separate the biogas components.

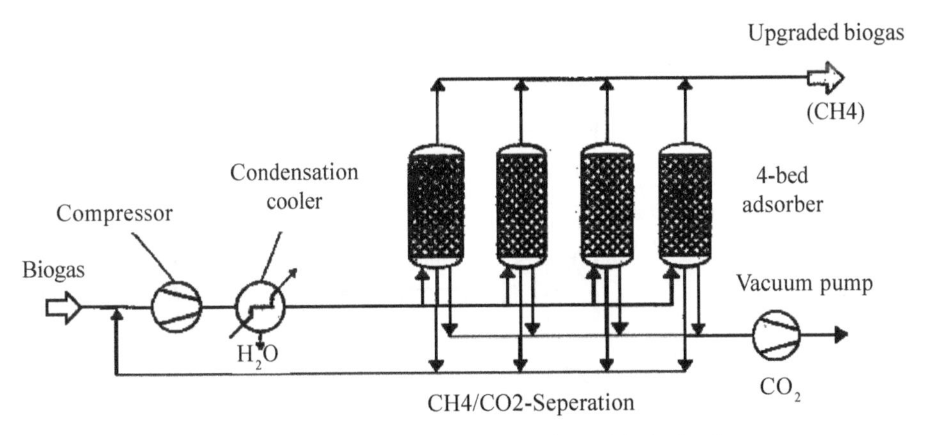

Fig. 8: Flow-chart of PSA process using a 4-bed adsorber (*Source:* Wellinger and Lindberg, 1999)

Removing CO_2 by water scrubbing: Upgrading biogas by this can either be operated in a closed system by regenerating the washing water for reuse, or through a single pass flow without regeneration. Irrespective of the mode, there are many common characteristics for both (Fig. 9). Cleaned biogas is compressed to about 10 bars and fed into the bottom of the absorption column equipped with random packed material to give a large surface for gas-liquid contact and equipment for collection and redistribution of the water (Hagen, 2001).

Water counter-flows from the top of the column. During the flushing of biogas, CO_2 and H_2S are primarily absorbed and the CH_4 content of the gas increases till it reaches the top of the column. The purified gas leaving the vessel has to be dried as it is saturated with water.

Methane is partly soluble in pressurised water and therefore some methane will be washed out with the water. To minimise methane losses, the water is depressurised in two steps: After leaving the absorption column, the water is depressurised in a flash tank to approximately up to 4 bars. The released gas, which is rich in methane, is recycled to the compressor inlet.

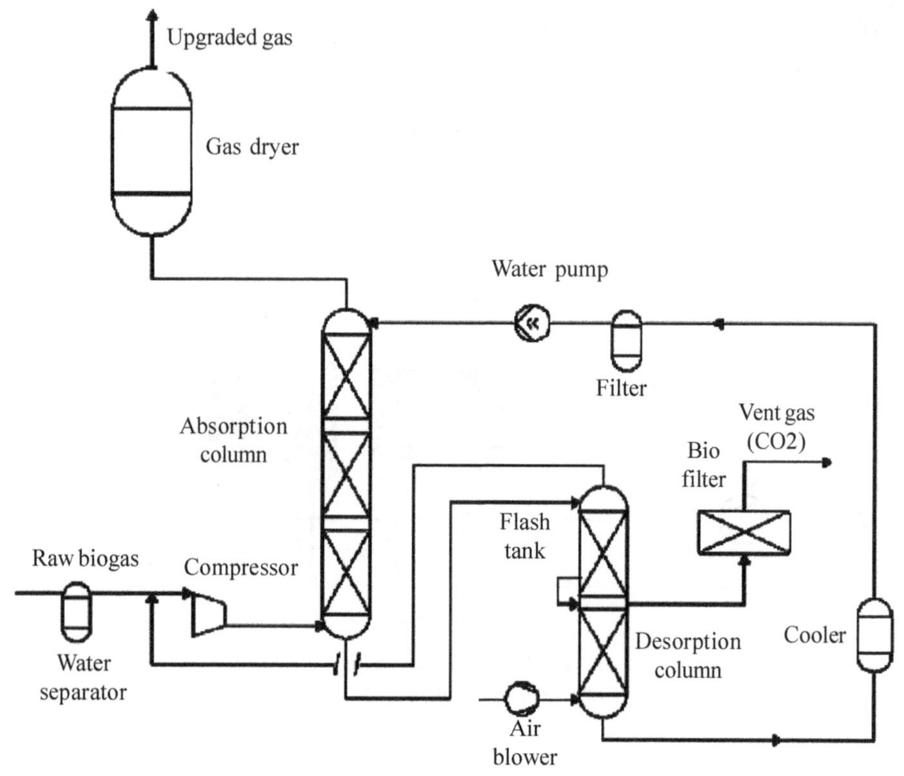

Fig. 9: Flow-chart of a water scrubbing process with wash-water regeneration (*Source*: Hagen, 2001)

Removing CO_2 by polyethylene glycol scrubbing: The principle of such scrubbing is the same as water-scrubbing process with wash-water regeneration. The difference between the two processes is that CO_2 and H_2S are more soluble in polyethylene glycol (PEG), which means a lower solvent demand and reduced pumping. Furthermore, PEG is advantageous as it can absorb water and halogenated hydrocarbons (Wellinger and Lindberg, 1999). The purified gas has low water content and thus, normally does not need to be dried.

Cryogenic removal of CO_2: Raw biogas is compressed in multi stage with inter-cooling to about 80 bars and then dried to avoid freezing in different cooling step. The biogas is cooled with chillers and heat exchangers to -45°C. Condensed CO_2 is removed in a separator and is further processed to recover dissolved CH_4, which is recycled to the gas inlet.

Utilisation of biogas

Generally, biogas can be used for heat production by direct combustion, electricity production by fuel cells or micro-turbines, CHP generation or as vehicle fuel (Fig. 10).

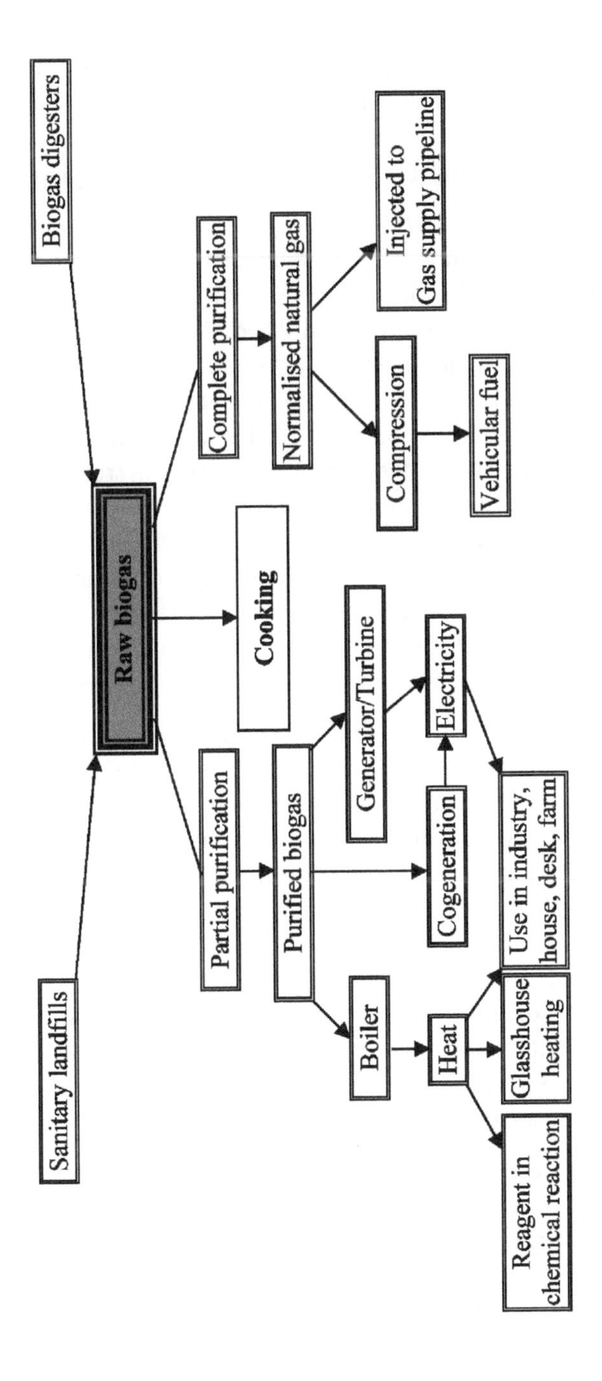

Fig. 10: Biogas utilisation flow chart

Direct combustion and heat utilisation: The simplest way to utilise biogas is direct burning in boilers or burners, a common use for small family digesters (Demirbas, 2005). Biogas is burnt for heat either on site, or transported to the end user by pipeline. Purifying biogas for such use is not needed as the contaminants do not interfere much as in other applications, though its condensation, particulate removal, compression, cooling and drying may be useful.

Combined heat and power (CHP) generation: CHP generation is a standard mode to utilise biogas in countries with developed biogas sector. It is considered an efficient strategy to utilise biogas, for electricity and heat production. Before CHP conversion, biogas is drained and dried. An engine based CHP power plant has an efficiency of up to 90% and produces 35% electricity and 65% heat. The most common types of CHP plants are block type thermal power plants (BTTP) with combustion motors that are coupled to a generator. Generators usually have a constant rotation of 1500rpm (rotations per minute) in order to be compatible with the grid frequency. Motors can be Gas-Otto, Gas-Diesel or Gas-Pilot injection engines. Both, Gas-Otto and Gas-diesel engines operate without ignition oil, on Otto principle. The difference between these engines is only the compression. Alternatives in recent years to the above mentioned BTTPs for CHP generation, such as micro-gas turbines and Stirling motors and fuel-cells, are at important developmental stages.

The produced electricity from biogas can be used as process energy for electrical equipment such as pumps, control systems and stirrers. In many countries with high feed-in tariffs for renewable electricity, all the produced electricity is sold to the grid and the process electricity is bought from the same national electricity grid. Biogas heat can be used by industry processes, agricultural activities or for space heating. Heat quality (temperature) is an important issue for industrial applications. Biogas heat can also be used for drying crops, wood chips or for separation and further treatment of digestate. Finally, the heat can also be used in 'power-heat-cooling-coupling'-systems. This process, known from refrigerators, is used to cool food or for air-conditioning. The input energy is heat converted to cooling through a sorption process, whereby a differentiation is made between adsorption and absorption cooling process.

Gas-Otto motors: Gas-Otto motors are developed specifically to use biogas on Otto principle. In order to minimise carbon monoxide emissions the engines operate with air surplus. This leads to lower gas consumption and reduced motor performance, compensated by using an exhaust turbo charger. Gas-Otto motors require biogas with minimum 45% CH_4 content. This is useful during the start-up of the biogas plant when the heat is used to heat-up the digester.

Pilot-injection gas motors: The Pilot injection engine is based on the diesel engine principle, operated with high air surplus. It is also referred as pilot injection natural gas engine (PING) or dual fuel engine. These engines are often used in tractors and heavy-duty vehicles. The biogas is mixed with the combustion air in a gas mixer and the mixture passes through an injection system in the combustion chamber where it is ignited by the injected ignition oil. Usually up to 10% ignition oil is automatically injected and combusted.

Stirling motors: This motor operates based on the principle that temperature changes result in volume changes in gas. The necessary heat is provided without internal combustion from sources like a gas burner running on biogas. Due to external combustion, biogas with lower CH_4 content can also be used. Electrical efficiency (24-28%) of Stirling engine is lower than Gas-Otto engines. The exhaust temperatures are from 250–300°C, the capacity of such motor is usually below 50kW. Such engine can be used in block type thermal power plants.

Biogas micro-turbines: In biogas micro-turbine, air passes into a combustion chamber at high pressure to mix with biogas. The air-biogas mixture burns causing a temperature increase and the gas mixture expand. The hot gas is released through a turbine connected to the electricity generator (Fig. 11). The electric capacity of micro-turbines is typically below 200kW.

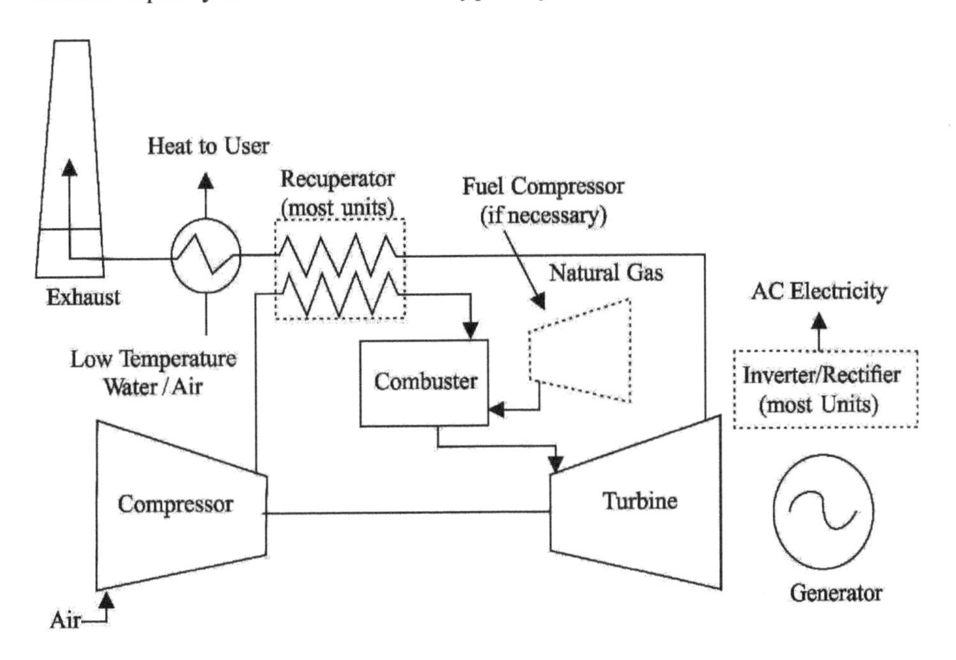

Fig. 11: Micro-turbine structure
(*Source:* www.energysolutionscenter.org)

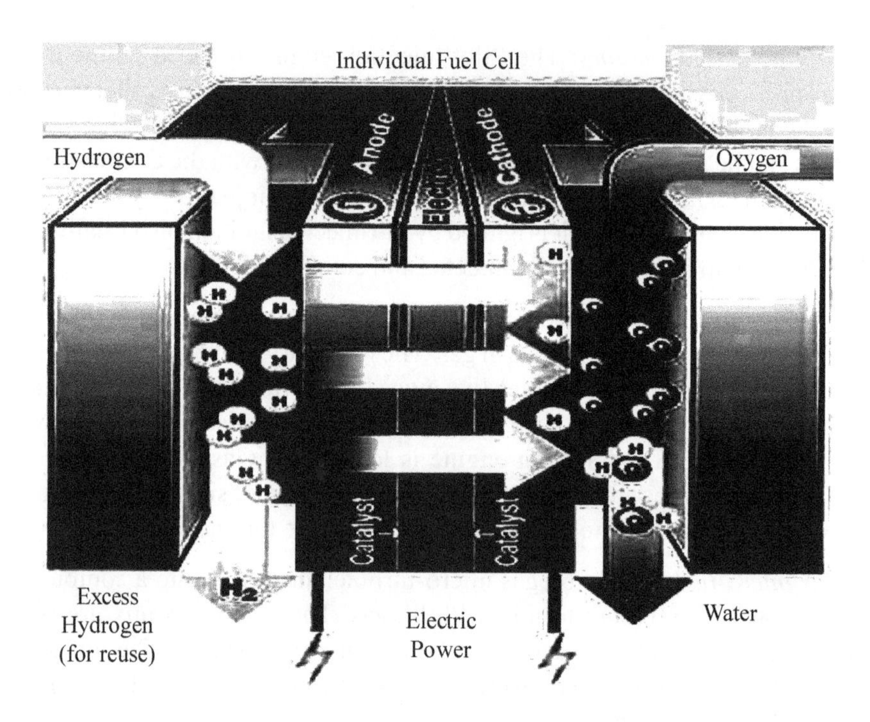

Fig. 12: Simplified schematic diagram of a fuel cell
(*Source:* Emerging environmental issues, 2005)

Fuel cells: The fuel cells are electrochemical devices that convert biochemical energy directly into electrical energy. The basic physical structure (building block) of a fuel cell consists of an electrolyte layer in contact with porous anode and cathode on both sides (Fig. 12).

The gaseous fuel (biogas) is fed continuously to the anode (the negative electrode) compartment and an oxidant (*i.e.*, oxygen in air) is fed continuously to the cathode (the positive electrode) compartment. The electrochemical reaction at the electrodes produces electric current. Research and development work in this area is targeting competitive costs for future models.

Biogas as a raw material in H_2S production plant: Generally H_2 is produced from propane (C_3H_8) which is further used to produce H_2S by using stream reforming process. In this process, using hydrocarbon feedstock other than C_3H_8 in the reactor is possible. Changing from heavier to lighter hydrocarbon, *e.g.*, from propane to methane, in the reformer does not need significant modifications. However, minor issues such as coking of catalyst may occur if process parameters are not altered to a new feedstock (Holladay *et al.*, 2009; Turpeinen *et al.*, 2008; Shu-Ren, 1998). The conversion process is:

$$C_3H_8 + 3H_2O \rightarrow 3CO + 7\ H_2 \quad \text{(iii)}$$

$$CO + H_2O \rightarrow CO_2 + H_2 \quad \text{(iv)}$$

Using methane as a raw material in reforming would result in the main reaction as:

$$CH_4 + H_2O \rightarrow CO + 3H_2 \quad \text{(v)}$$

Fig. 13 illustrates the existing hydrogen production system (the one without shading) and a biogas alternative (the one shaded in grey).

Biogas as a raw material in formic acid and hydrogen peroxide plant: CO and H_2 are two major raw materials needed for formic acid and H_2S preparation. Conventionally these two gases are produced by gasifying hydrocarbon fuels as feedstock, such as natural gas, refinery gas, bunker C-oil, vacuum residue, vacuum-flashed cracked residue, asphalt and liquid waste. Gas is treated in such a way that CO_2 and S-compounds are separated from the gas and CO and H_2 are separated with a membrane system. As conventional feedstock is non-renewable, biogas is a good alternative. Syn-gas is produced without having to purify CO_2 from biogas:

$$CH_4 + {}^1/_2O_2 + CO_2 \rightarrow 2CO + H_2 + H_2O \quad \text{(vi)}$$

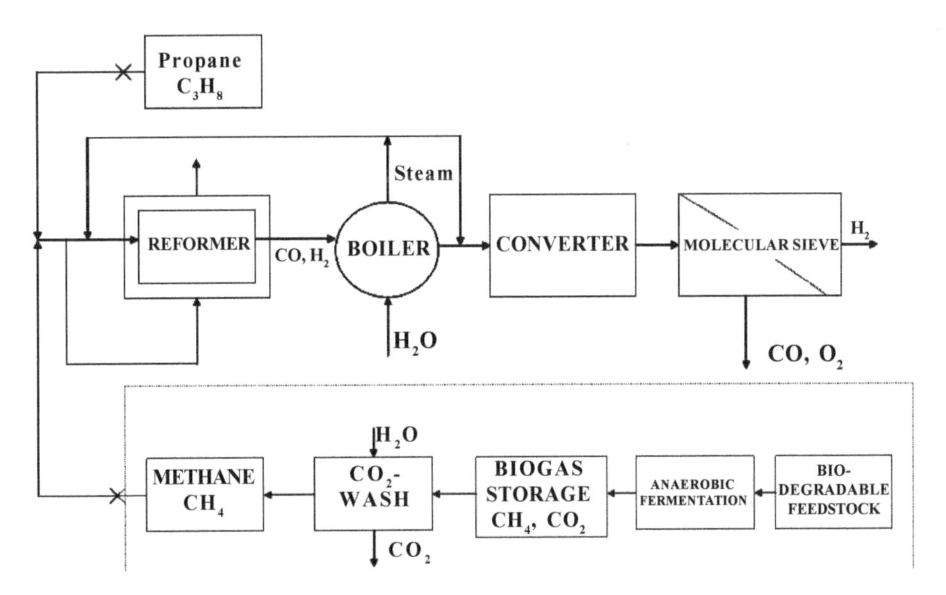

Fig. 13: Alternate use of biogas replacing propane to produce H_2
(*Source:* Arvola *et al.*, 2012)

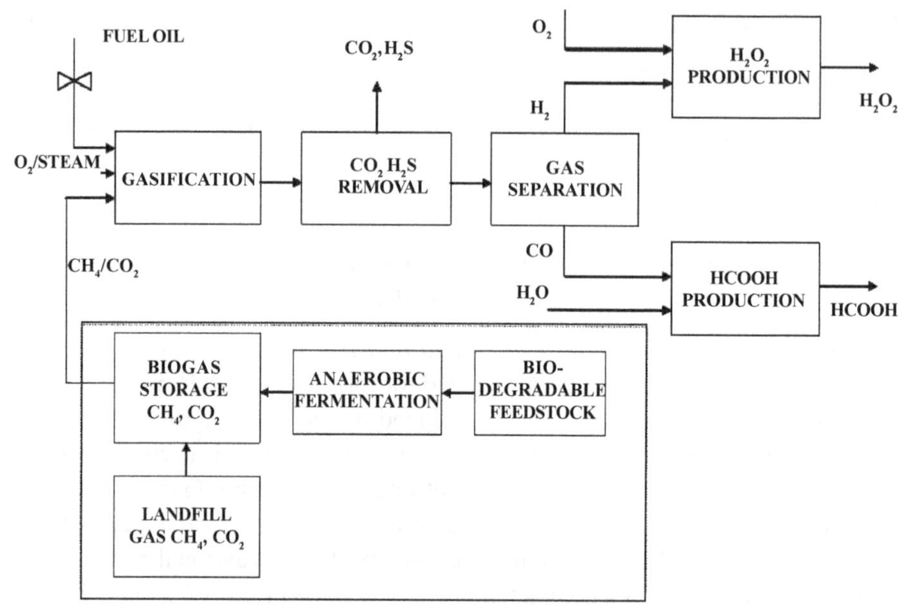

Fig. 14: Alternate use of Biogas in place of Hydrocarbon oil (*Source*: Arvola *et. al.*, 2012)

CO_2 in biogas increases the desired CO formation and creates an additional environmental gain by capturing a part of CO_2 in the end-product instead of releasing it to atmosphere.

The reaction may not reach equilibrium, but a good CO:H_2 ratio is reached as per thermodynamic calculations (Turpeinen *et al.*, 2008). More CO is obtained with higher CO_2 concentrations. The net reaction equations for formic acid and hydrogen peroxide are as follows:

$$CO + H_2O \rightarrow HCOOH \quad \text{(vii)}$$

$$H_2 + O_2 \rightarrow H_2O_2 \quad \text{(viii)}$$

Conclusions

Considering the ever-growing energy demand, limited available limited fossil fuel and environmental degradation, renewable energy sources are the only option for the future. Anaerobic digestion is the easiest and simplest way of converting biomass to green fuel, *i.e.*, biogas. Biogas has a wide application, from household to big industries. Clean or upgraded biogas can potentially replace natural gas in future. Biogas use can be promoted in existing industrial sites currently using fossil-based gas as raw material after techno-economic feasibility study. More research could include studying a larger number of site-specific cases and analysing different biogas production technologies. Additionally, simultaneous use of multiple raw materials such as municipal waste and sewage, industrial waste and other blended biomass in different climatic conditions is worth researching. Investment supports for biogas production and subsidies are needed similar to food production to encourage it as an economical alternative. However, extending subsidies depends on political decisions.

References

FAO/CMS, 1997. A system approach to biogas technology. *In*: *Biogas Technology: A Training Manual for Extension* (FAO/CMS, 1996) http://www.wcasfmra.org/biogas_docs/SustainableBiogas1997.pdf.

Abraham ER, Ramachandran S, Ramalingam V, 2007. Biogas: Can it be an important source of energy? *Environmental Science and Pollution Research*, **14(1)**: 67–71.

AD-Nett, 2000. Anaerobic digestion of agro-industrial wastes: information networks. Technical summary on gas treatment. http://www.ad-nett.org/NL-TS1299.pdf

Al Seadi T, 2001. *Good Practice in Quality Management of AD Residues from Biogas Production.* Task 24-energy from biological conversion of organic waste. IEA Bioenergy, United Kingdom: 32 pp.

Al Seadi T, Holm-Nielsen JB, 2004. Utilisation of waste from food and agriculture (Chap. VI. 1). *In*: Twardowska, Allen, Kettrup, Lacy (Eds) *Waste Management Series 4*. Solid Waste Assessment, Monitoring and Remediation Technologies for Solid Wastes. Elsevier, Netherlands: ISBN 0080443214: 735–756 pp.

Arvola J, Belt P, Harkonen J, Kess P, Imppola R, 2012. Biogas as an option for industrial applications. *International Journal of Sustainable Economy*, **4(1)**: 71–88.

Balsam J, Ryan D, 2006. Anaerobic digestion of animal wastes: factors to consider. Farm energy technical note. *National Sustainable Agriculture Information Service*: 1–12.

Barnett A, Pyle L, Subramanian SK, 1978. *Biogas Technology in the Third World: A Multidisciplinary Review*. The International Development Research Centre, Ottawa, Canada: 132 pp.

Bates BC, Kundzewicz ZW, Wu S, Palutikof JP, 2008. Climate Change and Water. Technical paper of the Intergovernmental Panel on Climate Change. IPCC Secretariate, Geneva. *Climate Change Policy with a Renewed Environmental Ethic*, **21**: 85-101:

Berndes G, Hoogwijk M, Van den Broek R, 2003. The contribution of biomass in the future global energy supply: a review of 17 studies. *Biomass and Bioenergy*, **25(1)**: 1–28.

Cote RP, Cohen-Rosenthal E, 1998. Designing eco-industrial parks: a synthesis of some experiences. *Journal of Cleaner Production*, **6(3-4)**: 181–8.

Demirbas A, 2005. Potential applications of renewable energy sources, biomass combustion problems in boiler power systems and combustion related environmental issues. *Progress in Energy and Combustion Science*, **31(2)**: 171–92.

Eriksen K, Jensby T, Norddahl B. 1998. The upgrading of biogas to be distributed through the natural gas network: Environmental benefits, technology and economy. *Proceedings International Gas Research Conference*, Government Institutes Inc., **1**: 441–50.

Gallucci T, Lagioia G, Dimitrova V, 2010. Opportunities for biofuel sustainable development in Bulgaria. *International Journal of Sustainable Economy*, **2(3)**: 241–57.

Goldemberg J, Teixeira CS, 2004. Renewable energy-traditional biomass *vs* modern biomass. *Energy Policy*, **32(6)**: 711–4.

Green C, Byrne KA, 2004. Biomass: Impact on carbon cycle and greenhouse gas emissions. *Encyclopedia of Energy*, **1**: 223–36.

Hagen M, 2001. Adding gas from biomass to the gas grid. Final report, No.: VII/4.1030/Z/99-412. http://uk.dgc.dk/pdf/altener.pdf.

Hamad MA, 1983. Evaluation of performance of two rural biogas units of Indian and Chinese design. *Energy in Agriculture*, **1**: 235–50.

Hansen RW, 2004. *Methane Generation from Livestock Wastes*. Colorado State University Cooperative Extension Bulletin #5.002. http://www.ext.colostate.edu/pubs/farmmgt/05002.html.

Hjort-Gregersen K, 1998. Danish farm scale biogas concepts-at the point of commercial break through. Proceedings of the International Conference Würzburg, Germany: Biomass for Energy and Industry,8-11 June 1998 Danish Institute of Agricultural and Fisheries Economics: 641–3.

Hoffert MI, Caldeira K, Benford G, Criswell DR, Green C, Herzog H, Jain AK, Kheshgi HS, Lackner KS, Lewis JS, 2002. Advanced technology paths to global climate stability: energy for a greenhouse planet. *Science*, **298(5595)**: 981–7.

Holladay JD, Hu J, King DL, Wang Y, 2009. An overview of hydrogen production technologies. *Catalysis Today*, **139(4)**: 244–60.

Hossain MK, Strezov V, Chan KY, Ziolkowski A, Nelson PF, 2011. Influence of pyrolysis temperature on production and nutrient properties of wastewater sludge biochar. *Journal of Environmental Management*, **92(1)**: 223–8.

Jenck JF, Agterberg F, Droescher MJ, 2004. Products and processes for a sustainable chemical industry: a review of achievements and prospects. *Green Chemistry*, **6(11)**: 544–56.

Korhonen J, Snakin JP, 2005. Analysing the evolution of industrial ecosystems: concepts and application. *Ecological Economics*, **52(2)**: 169–86.

Lapp HM, Schulte DD, Sparling AB, Buchana LC, 1975. Methane production from animal wastes. 1. Fundamental considerations. *Canadian Agricultural Engineering*, **17(2)**: 97–102.

Leggett J, Graves RE, Lanyon LE, 2006. *Anaerobic Digestion: Biogas Production and Odor Reduction from Manure*. Pennsylvania State University College of Agricultural Science Cooperative Extension, http://www.age.psu.edu/extension/factsheets/g G: 77.

McKendry P, 2002. Energy production from biomass II: conversion technologies. *Bioresource Technology*, **83(1)**: 47–54.

Moller K, Muller T, 2012. Effects of anaerobic digestion on digestate nutrient availability and crop growth: a review. *Engineering in Life Sciences*, **12(3)**: 242–57.

Murphy JD, Power N, 2009. Technical and economic analysis of biogas production in Ireland utilising three different crop rotations. *Applied Energy*, **86(1)**: 25–36.

Murphy JD, McKeogh E, Kiely G, 2004. Technical/economic/environmental analysis of biogas utilisation. *Applied Energy*, **77(4)**: 407–27.

Myles RM, Dhussa A, 1987. *AFPRO's Field Guide to Biogas Technology*. Action for Food Production, New Delhi: 57 pp.

Pimentel D, 2001. Biomass utilization, limits of. *In*: RA Meyers (Ed) *Encyclopedia of Physical Science and Technology Vol. 2*, 3rd Edition. Academic, San Diego: 159–71.

Reijnders L, 2006. Conditions for the sustainability of biomass based fuel use. *Energy Policy*, **34(7)**: 863–876.

Roberts BH, 2004. The application of industrial ecology principles and planning guidelines for the development of eco-industrial parks: an Australian case study. *Journal of Cleaner Production*, **12(8-10)**: 997–1010.

Sastry CA, 1985. Factors influencing biogas generation and operation and maintenance of biogas digesters. *Journal of Energy Heat and Mass Transfer*, **7(1)**: 237–250.

Shu-Ren H, 1998. Hydrocarbon steam-reforming process: Feedstock and catalysts for hydrogen production in China. *International Journal of Hydrogen Energy*, **23(5)**: 315–319.

Smyth BM, Smyth H, Murphy JD, 2010. Can grass bio-methane be an economically viable biofuel for the farmer and the consumer? *Biofuels, Bioproducts and Biorefining*, **4(5)**: 519–37.

Steffen R, Szolar O, Braun R, 1998. *Feedstocks for Anaerobic Digestion*. Institute of Agrobiotechnology Tulin, University of Agricultural Sciences, Ad-NETT, *Vienna*: 1-29.

Kishore VVN, Raman P, Rao VR, 1987. Fixed dome biogas plants: A design, construction and operation manual. Tata Energy Research Institute, New Delhi: 44 pp.

Themelis NJ, Ulloa PA, 2007. Methane generation in landfills. *Renewable Energy*, **32**: 1243–57.

Turpeinen E, Raudaskoski R, Pongrácz E, Keiski RL, 2008. Thermodynamic analysis of conversion of alternative hydrocarbon-based feedstocks to hydrogen. *International Journal of Hydrogen Energy*, **33(22)**: 6635–6643.

United Nations, 1997. Kyoto protocol to the United Nations framework convention on climate change. http://unfccc.int/kyoto_protocol/items/2830.php.

Walsh JL, Ross CC, Smith MS, Harper SR, 1989. Utilization of biogas. *Biomass*, **20(3-4)**: 277–90.

Weiland P, 2003. Production and energetic use of biogas from energy crops and wastes in Germany. *Applied Biochemistry and Biotechnology*, **109(1-3)**: 263–74.

Weiland P, 2010. Biogas production: current state and perspectives. *Applied Microbiology and Biotechnology*, **85(4)**: 849–60.

Wellinger A, 1984. Anaerobic digestion: A review comparison with two types of aeration systems for manure treatment and energy production on the small farm. *Agricultural Wastes*, **10(2)**: 117–33.

Wellinger A, Lindberg A, 1999. Biogas upgrading and utilisation. IEA Bioenergy Task 24: Energy from biological conversion of organic waste. 20 pp.

White DS, Sulkowski AJ, 2010. Relative ecological footprints based on resource usage efficiency per capita: macro-level segmentation of 121 countries. *International Journal of Sustainable Economy*, **2(2)**: 224–40.

13

Biomethanation under Biphasic Conditions: Success Story of Nisargruna Biogas Plant

S.T. Mehetre[1] and S.P. Kale

Nuclear Agriculture and Biotechnology Division
Bhabha Atomic Research Centre, Mumbai – 400 085, India
[1]Corresponding Author: smehetre@barc.gov.in

ABSTRACT

Nisargruna biogas technology has been developed in BARC for processing different types of biodegradable waste. This technology is based on biphasic separation of aerobic and anaerobic stages. Introduction of aerobic predigester has increased the efficiency of the process in terms of enhancing the quality and quantity of methane and also reducing the retention time of the process. Process parameters of the plant were studied and found that the modified design has proved very effective in faster digestion of complex waste. The technology has been commercialised and has spread across different parts of country for processing variety of wastes including kitchen waste, municipal waste, agriculture waste, textile waste and poultry waste. Microbial population from different stages of the plant was also studied and found their ability to convert waste to biogas, containing mainly methane.

Keywords: biogas, biphasic digestion, green energy, methane, microbial fermentation, Nisargruna, waste management

Abbreviations used

BARC : Bhabha Atomic Research Centre
BOD : Biochemcial oxygen demand
C/N : Carbon to nitrogen ratio
COD : Chemical oxygen demand

HPLC : High performance liquid chromatography
HRT : Hydraulic retention time
KVIC : Khadi and village industries commission
MPN : Most probable number
TFS : Total fixed solids
TVS : Total volatile solids

Introduction

Biogas plant for biomethanation is a popular global technology to manage the increasing menace of organic wastes. Cattle dung based biogas technology has been demonstrably successful in rural India in last few decades. In India this technology was popularised in the 1970s, and the KVIC and the Janta models were the two most popular designs (Rajabapaiah *et al.*, 1979; Mahkijani and Poole, 1975). These models were primarily processed cattle dung as sole substrate. Biogas plant for biomethanation has various advantages, such as, the odour of the dung is reduced, the pathogenic bacterial population is reduced/ eliminated (treatment temperature dependent), and the digested slurry thus obtained is useful as an organic supplement for agricultural soil conditioning.

This technology for energy production is yet to be fully exploited. In most industrialised nations, this has suffered due to operational difficulties as well as higher input costs (Vijay *et al.*, 1996; Singh and Sooch, 2004). Coupled with the readily available inexpensive non-commercial fuels, the needed urgency seems to be lacking in developing countries.

Considering these, this anaerobic microbial technology promises more economical treatment of sewage waste, agri-wastes, industrial wastes and municipal solid wastes thereby providing large scopes to expand. This gives alternative renewable energy as well as the digested slurry that is useful as manure, as substrate for mushroom cultivation, and to grow *Azolla*, etc.

Biogas plants face constant operational challenges like chocking, unavailability of input material and less biogas output. The main reason for these issues was the lack of concerted research on functional dynamics of the plant. Out of 1.8 million plants installed in India only 60% are functional, the non-functionality being attributed to either chocking of plants or unavailability of feed (Tomar, 1994). Scientific studies to enhance and broaden its scope, improving the process and increasing the efficiency are less addressed thus far. Biogasifiers developed initially were of single digestor, which have undergone quality structural modifications over time with the dual objectives of enhancing biomethanation performance and to expedite the process.

Considering these facts, attempts were made for both structural and as well as functional modification to increase the scope and potential of a biogasifier. The modified plant, called 'Nisargruna Biogas Plant', was developed at Bhabha Atomic Research Centre (BARC), Mumbai, India for comprehensive solid waste management (Kale and Mehetre, 2002). The name Nisargruna literally means 'nature's loan or debt'. The plant aims at repaying the nature for the generated waste (primarily anthropogenic) after processing in appropriate ways.

Aerobic hydrolysis coupled with anaerobic digestion has not been tried on a larger scale for degradation of solid biodegradable waste so far. The Nisargruna plant developed at BARC is (aerobic and anaerobic) biphasic (Fig. 1). There are several advantages of using aerobic and anaerobic digestion serially. The kitchen and vegetable market solid wastes are expected to have complex organic constitution that need to be broken down in size to further facilitate easy degradation by aerobic microorganisms. These, in turn, are passed on to an anaerobic digester for conversion into methane, and microbially highly-fortified slurry.

Biphasic anaerobic digestion process was first reported as more effective than single-phase by Pohland and Ghosh (1971). Thermophilic acidifications in dairy waste (Yu and Fang, 2000) and municipal sludge (Ghosh, 1991) were reported to be successful. Samani *et al.* (2002) reported digestion of 67% of the volatile solids in garden waste (grass) in six months in pilot-scale biphasic biofermentation with 71% CH_4 containing 0.15m^3 biogas/kg waste.

Singh *et al.* (1983) reported increased acetate production by predigesting cattle slurry at 30–35°C for a couple of days. This predigestion enhanced the biogas production by 17–19%, and methane content from 68–75% to 75–86%. Christof *et al.* (2002) has demonstrated two-step 200L capacity continuously stirred tank system to treat the effluent from the second stage composting. Goel *et al.* (2001) studied biphasic (hydrolysis and acidification) digestion for high organic strength liquid production. While reducing the COD by 93%, the system produced 0.48 m^3/kg biogas containing 73% methane.

Ghaly (1996) studied two-stage anaerobic 155L capacity digester system and found it to be more effective than the single-stage system. Subramanyam (1989) analysed the effect of temperature on biogas production and showed a linear increase in gas yield with temperature. Oles *et al.* (1997) reported the best results with an initial high load thermophilic (50–55°C) stage followed by a second mesophilic (35–37°C) condition in a two-stage digestion process. Ros and Zupancic (2004) also studied high-rate two-stage thermophilic aerobic–anaerobic sludge digestion and found that the best process was 3+12 days, which showed a VSS removal of 61.8% and COD removal of 57.4% in 15 days HRT.

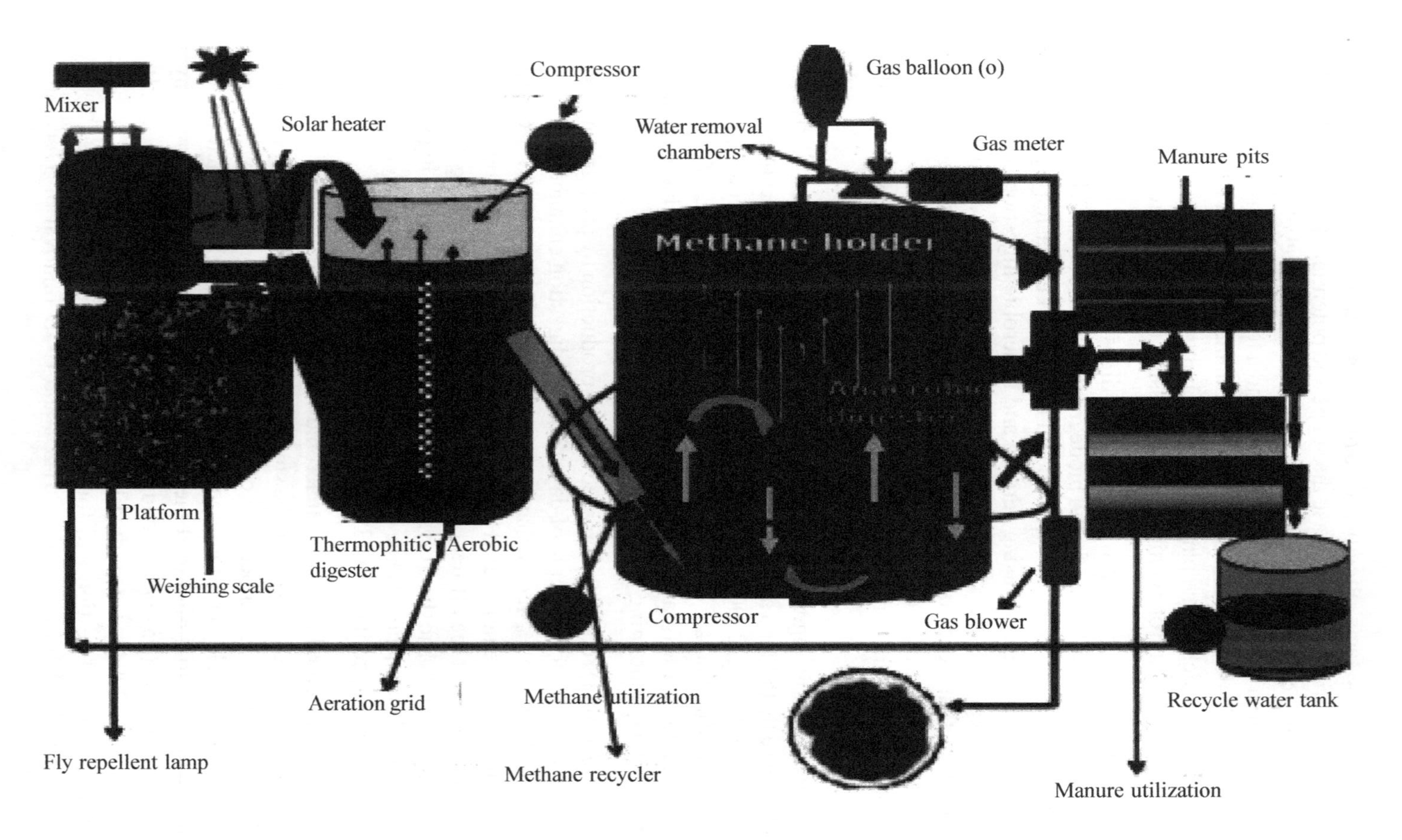

Fig. 1: Different components of a Nisargruna biogas plant

Materials and methods

An array of factors affects the digestion rate of input waste and biogas production output. The findings on different parameters over the period of time of the studies are discussed below.

Waste characterisation

The waste was characterised for carbon and nitrogen ratio (C/N). The input contains stale, cooked food, vegetable refuse and fruit peels. An aliquot of this material was homogenised in a mixer. An equal quantity of water was added while making the slurry and passed through a No. 1 filter paper. The retentate was estimated for C/N contents. The Nitrogen was estimated by Microkjeldhal method (Kjeldahl, 1883) while the organic carbon was estimated by Walkley and Black (1934) method. Other elemental analyses were performed by atomic absorption spectrometer (Model 932 B+, GBC, Australia).

Chemical characteristics of the waste

Raw slurry was used for pH measurement using a pH meter and it was filtered through filter paper and used for further chemical analysis of carbohydrates by Anthrone method (Hedge and Hofreiter, 1962), total fats by Soxhlet extraction method, and proteins by Lowry method (Lowry *et al.*, 1932). To estimate total solids, raw slurry was transferred into a previously weighed crucible and heated in oven at 110°C until constant weight was reached. The difference between the initial and final weight was recorded as the total solids (APHA, 1985). After TS estimation, the dried slurry was further ignited in a muffle furnace at 550°C for 4 hours to estimate the total volatile solids (TVS) and fixed solids (TFS).

Predigested slurry characterisation

The slurry was characterised for temperature, different organic acids by HPLC, and the chemical nature as per procedure described earlier for raw slurry.

Digested slurry characterisation

The chemical nature of the digested slurry with respect to carbohydrates, proteins, fats, total solids etc. was determined by the procedure as described for raw slurry above. It was also characterised for BOD and COD contents by standard methods. The presence of coliforms was studied by the most probable number (MPN) technique following standard method.

Efficiency of the Nisargruna process

Material balance of Nisargruna biogas plant was studied over the period. The daily input of kitchen waste and other biodegradable materials was recorded on kg wet weight basis. The materials were added to the plant considering the C/N ratio suitable for smooth digestion. The outputs in the form of kg biogas generated and kg dry weight manure produced were measured periodically. The quality of the biogas was analysed by quantifying the methane content by gas chromatography (Shimadzu, Japan).

Microbial parameters of Nisargruna biogas plant

Different microbes were enumerated and isolated from the plant's predigestor considering their important role in expediting the bioconversion process. They were isolated following the standard procedure. They were also characterised by biochemical and molecular means.

Production of high quality manure

Manure is an important byproduct of biogasification. This was characterised by studying its composition and performance in yard (pot culture) and field conditions.

Results and discussion

Waste characterisation

The C/N ratio is an important parameter that decides easy and effective processing of the material. The kitchen waste used for processing in the Nisargruna biogas plant was analysed for chemical composition. Its C/N ratio was recorded to be between 20:1 and 25:1 (Table 1).

Table 1: Analysis of the feed (substrate) processed in Nisargruna biogas plant

Sl.No.	Proximate	Measured value	Sl.No.	Proximate	Measured value
1.	Carbon	31.2 ±1.28	7.	Iron	0.78 ±0.02
2.	Nitrogen	1.5 ±0.05	8.	Magnesium	0.02 ±0.002
3.	Hydrogen	4.3 ±0.16	9.	Manganese	0.0072 ±1.47
4.	Calcium	0.55 ±0.03	10.	Phosphorus	0.02 ±0.001
5.	Cadmium	ND	11.	Zinc	0.0061 ± 0.2
6.	Copper	0.0015 ±2.23			

ND: Non-detectable; ±: Standard deviation

Chemical nature of the substrate

It's observed that raw slurry had 6-7 pH. Total solids (TS) content in it ranged from 20-25% while the total volatile solids (TVS) content ranged between 90 and 92%. The carbohydrates were 54.68%, total oil contents were 18.65%, and proteins were 24.2% of the total solids.

Predigested slurry characteristics

The average temperature recorded in the predigestion stage ranged between 52 and 56°C during the summer months and dropped to between 30 and 35°C during winter. Manually-heated water through gas generated at the plant was added during winter months to optimise the heat inputs. This practice helped in increasing the temperature of the predigester to 45°C. Organic acid profile studied by HPLC showed the predominance of acetic acid in the predigested slurry. Other acids formed to a relatively smaller extent were propionic acid, butyric acid and formic acids. Data analyses revealed that out of the total acids generated in the predigester, 62.85% was acetic acid, 11.83% propionic acid, 16.8% butyric acid and 8-9% formic acid. The other unknown or unidentified acids formed were up to 4-5%.

The total solids content dropped from 20-25% to 10-12% during predigestion by thermophilic bacteria. In the predigested slurry, the total carbohydrates were 46%-48%, oil contents were 7.25-8%, and protein contents were 23-24%.

Digested slurry characterisation

The digested slurry was chemically characterised. The colour of the slurry was jet black and its pH was 7.5-8. The mean total carbohydrates was 2.3%, total oil content 1.69% and protein contents were negligible. The TS in the digested slurry was 6-7%. Mostly complex molecules like lignin and hemicelluloses dominated the digested slurry. There BOD of the kitchen waste reduced by 90-92% after processing through different stages of the plant. The COD values at different stages in the biogas plant digester were: raw slurry 5,040mg/L, pre-digestion stage 4,760mg/L and digested slurry stage 840mg/L. Thus, 80-82% COD reduction was recorded during the entire degradation process. The digested slurry was devoid of coliforms. At no stage, faecal coliforms were detected.

Efficiency of Nisargruna biogas plant

The details of the waste processed, gas and manure generated for time to time showed that, an average of 80m^3 gas was produced daily for every single tonne waste processed. Depending on the type of waste, the methane content up to 75% was observed in the biogas. The manure obtained was approximately 70-80kg per tonne waste processed.

Microbial parameters of the Nisargruna biogas plant

It was found that most of the isolated bacteria had very high potential of enzyme production, surviving at elevated temperature and degrading complex substances.

One bacterium *Bacillus licheniformis* showed antifungal activity and was demonstrated to control crop disease (Mehetre and Kale, 2008).

Production of high quality manure

Manure is an important byproduct of biogasification. This increased crop yield and also enhanced the microbial and enzymatic properties of soil (Mehetre *et al.*, 2008).

Conclusion

Studies on process parameters of Nisargruna biogas plant helped in better understanding of overall process and optimisation of the parameters for effective processing of biodegradable waste for maximum output in terms of biogas as well as manure. Material balance study, waste characterisation, predigested as well as completely digested slurry characterisation and biogas analysis regularly. Amongst the different chemical components of the waste, the C/N ratio played a significant role in the functioning of the Nisargruna process. It was found that the C/N ratio of 25:1 or even 30:1 was optimum for processing. This ratio was decreased to 12:1 at the end of process again proving its effectiveness for the application as high quality manure in the field. Analysis of biogas mixture for its composition gave guidelines for optimum parameters for getting maximum gas production at the same time in getting better methane quantity. There was reduction in the gas output during winter as well as rainy season attributable to the needed threshold biogasification temperature. Addition of boiling water made by the generated biogas helped in increasing the temperature of predigestor to desired level and their by maintaining methane quality. Other studies including analysis of organic acids, BOD, COD also helped for better understanding the process of biodegradation of waste. Microbial population showed predominance of *Bacillus* spp. at predigestor stage of the plant. The digested slurry obtained also showed very good manure effect on different crops.

References

APHA, 1985. Standard Methods for the Examination of Water and Wastewater. Washington DC: 16th edition. American Public Health Association.

Christof H, Martin W, Karl-Heinz R, Georg MG, 2002. Two-stage anaerobic fermentation of organic waste in CSTR and UFAF-reactors. *Bioresource Technology*. **81(1)**: 19–24.

Ghaly AE, 1996. A comparative study of anaerobic digestion of acid cheese whey and dairy manure in a two-stage reactor. *Bioresource Technology*, **58(1)**: 61–72.

Ghosh S, 1991. Pilot scale demonstration of two-phase anaerobic digestion of activated sludge. *Water Science and Technology,* **23(7-9)**: 1179–88.

Goel B, Pant DC, Kishore VVN, 2001. Two-phase anaerobic digestion of spent tea leaves for biogas and manure generation. *Bioresource Technology*, **80(2)**: 153–6.

Hedge JE, Hofreiter BT, Whistler RL, BeMiller JN, 1962. *In*: RL Whistler, JN BeMiller (Eds) *Carbohydrate Chemistry 17th Edition.* Academic Press, New York: 34–40.

Kale SP, Mehetre ST, 2002. Biogas plant based on kitchen waste. *BARC Newsletter*, **216**: 8–12.

Kjeldahl JZ, 1883. A new method for the determination of nitrogen in organic bodies. *Analytical Chemistry*, **22**: 366–82.

Lowry OH, Rosebrough NJ, Farr AL, Randall RJ, 1951. Protein measurement with the Folin reagent. *Journal of Biological Chemistry*, **193(1)**: 265–75.

Makhijani A, Poole A, 1975. *Energy and Agriculture in the Third World*. Ballinger Publishing Co., Cambridge.

Mehetre S, Shrivastava M, Kale S, 2008. High quality organic manure generation from solid biodegradable waste using Nisargruna technology. *Journal of Solid Waste Management Technology*, USA: 410–21.

Mehetre S, Kale SP, 2011. Comparative efficacy of thermophilic bacterium, *Bacillus licheniformis* and antagonistic fungi, *Trichoderma harzianum* to control *Pythium* spp. induced damping off in chilli (*Capsicum annuum* L.). *Archives of Phytopathology and Plant Protection*, **44(11)**: 1068–74.

Oles J, Dichtl N, Niehoff H, 1997. Full scale experience of two stage thermophilic/mesophilic sludge digestion. *Water Science and Technology*. **36(6-7)**: 449–56.

Pohland FG, Ghosh S, 1971. Development in anaerobic stabilization of organic waste, the two-phase concept. *Environmental Technology Letters* **1(4)**: 255–66.

Rajabapaiah P, Ramanayya KV, Mohan SR, Reddy AK, 1979. Studies in biogas technology Part 1. Performance of a convential biogas plant. *Proceedings of the Indian Academy of Science*. **2(3)**: 357–63.

Ros M, Zupancic GD, 2004. Two-stage thermophilic anaerobic–aerobic digestion of waste-activated sludge. *Environmental Engineering Science*, **21(5)**: 617–26.

Samani Z, Yu WH, Hanson A, 2001. Energy production from segregated municipal waste and agricultural waste using bi-phasic anaerobic digestion. *In*: AFM VanVelsen, WH Verstraete (Eds) *Anaerobic Conversion for Sustainability. Proc. 9th World Congress Anaerobic Digestion*, Antwerpen, Belgium: 195–7.

Singh R, Jain MK, Tauro P, 1983. Pre-digestion to improve production of biogas from cattle waste. *Agricultural Wastes*, **6(3)**: 167–74.

Singhand JK, Sooch SS, 2004. Comparative study of economics of different models of family size biogas plants for state of Punjab. India. *Energy Conservation Management*, **45(9-10)**: 1329–41.

Subramanyam S, 1989. Use of solar heat to upgrade biogas plant performance. *Energy Conversion and Management*, **29(1)**: 73–5.

Tomar SS, 1994. Status of biogas plant in India. *Renewable Energy*. **5(5-8)**: 829–31.

Vijay VK, Prasad R, Singh JP, Sorayan VPS, 1996. A case for biogas energy application for rural industries in India. *Renewable Energy*, **9(1-4)**: 993–6.

Walkley AJ, Black IA, 1934. An examination of the different method for determining soil organic matter and a proposed modification of the chronic acid titration method. *Soil Science*, **37(1)**: 29–38.

Yu HQ, Fang HHP, 2000. Thermophilic acidification of dairy wastewater. *Applied Microbiology Biotechnology*, **54(3)**: 439–44.

14

Improved Biogas Plants and Related Technologies: A Status Report

Sarbjit Singh Sooch[1]

School of Renewable Energy Engineering, Punjab Agricultural University Ludhiana–141004, Punjab, India
[1]Corresponding Author: sssooch@rediffmail.com

ABSTRACT

The 'steel-drum type' biogas plant installation programme in Punjab was initiated in the early seventies, whose construction works were entrusted on the Department of Agriculture. Because of their high cost and high failure rates then, only the affluent farmers could afford to install. In the meantime, a cheaper Chinese model, also known as Janta *type (dome-type replacing the drum-type) plant, became popular but was still out of the reach of ordinary farmers. With dual objectives of cost reduction and efficient biogas production to make it affordable to small farmers, a coordinated research was initiated. Detailed extension surveys were carried out in Ludhiana district which inferred that the farmers did not install the gas-holding steel drum though they purchased it. A significantly low cost* Katcha-Pucca *plant model of 15m*^{3}*/d capacity was developed and installed in Sangrur district. Similar digester was tried for* Janta *model in which the inlet, outlet and dome (gas holder) were masonry works but the digester was a* Katcha *pit, the dome of the plant being modified to a semi-circular dome. More than 80 such plants are functional in a single village, and more than 40–50% of the plants installed in Punjab and Haryana are these PAU designs. The chapter deals in greater details on the design aspects and the extensive field works carried out to achieve these successes, along with some relevant facts and figures. The write-up is written as a commentary, rather than a typical chapter, for the benefit and interest to the readers.*

Keywords: Biodigester, Biogas, Biogas technology, Entrepreneurship, Economics, Methanation, Rice straw

Abbreviations used

AC (pipe): asbestos cement
AFPRO : action for food production
GI : galvanised iron
HDPE : high-density polyethylene
ICAR : Indian Council of Agricultural Research
KVIC : Khadi and Village Industries Commission
MMT : million metric tonne
PAU : Punjab Agriculture University
RM : running metre
PVC : polyvinyl chloride

Introduction

Biogas plant installation programme in Punjab was initiated in the early seventies adopting few designs under the guidance of Late Shri Jash Bhai Patel, the then Advisor to the Khadi and Village Industries Commission (KVIC), Mumbai. The construction works of the biogas plants was through the department of agriculture, Punjab. People were then reluctant to install such plants because of the high cost, and thus only the affluent farmers were interested. The most popular adopted designs of these plants were steel drum type. The life of these plants was about ten years and the failure rate was high, as trained masons were not available. It was important to remove the shortcomings in the design reducing the plant cost while maintaining or extending its life, and making it biogas production efficient. In the meantime, a cheaper Chinese model, also known as Janta type (dome-type replacing the drum-type) plant, became popular but was still out of the reach of ordinary farmers.

The Indian Council of Agricultural Research (ICAR), New Delhi sanctioned an all India coordinated research project on biogas technology for agriculture and agro-based industries in the year 1981. The aspect of designing of the biogas plant was entrusted to the department of civil engineering, Punjab Agriculture University with the major objective to reduce the cost of plant so as to make it affordable to the small farmers. Before taking up the laboratory study to reduce the costing, a detailed extension survey was carried out in the villages in Ludhiana district. It was found that the farmers had purchased the steel drum (gas holder) but did not install it on the plant, rather they need not feel the need to. Keeping this in view, a *Katcha-Pucca* biogas plant model was developed, significantly reducing the cost. The first *Katcha-Pucca* drum-type biogas plant was installed in Sangrur district with a farmer who held a gas holder since long but did not install it due to the high cost. He was persuaded to install the *Katcha-Pucca* plant of 15m^3/d capacity. The satisfied farmer recommended the design to many other farmers in the area.

Similar digester was tried for Janta model biogas plant in which the inlet, outlet and dome (gas holder) were masonry works but the digester was a *Katcha* pit. The dome of the Janta biogas plant was modified to a semi-circular dome (similar to the *Deenbandhu* biogas plant) to facilitate construction and the digester was the ordinary *Katcha* pit. The first such plant was installed in 1984 again in Sangrur district. The owner farmer's satisfaction led to his advising to other farmers in the contiguous population. Thus, more than 80 such plants are functional in just a single village. Presently, more than 40–50% of the total biogas plants installed in the States of Punjab and Haryana are these PAU designs.

Categories of biogas plants

Biogas plants are typically categorised into two major forms based on the capacity, the smaller individual (household-level), and the large capacity (community-level and institution-level) plants. The capacity of the individual (household) level plants is limited to $6m^3/d$. A community plant for, of and by a group of people in a Village/Mohalla/Town/City, installed by a Village Panchayat/ Municipal, has a capacity more than $15m^3/d$. An institutional plant is installed by an institution, *viz.*, religious (Gurudwara/Mandir/Gowshala), Educational (school/ college) and Industry etc., for their in-house use, having a capacity of more than $15m^3/d$.

Based on another categorisation, related to the feeding strategies, the biogas plants are two types, batch (periodically fed) and semi-continuous (daily fed) plants. In a batch operation, the raw material (substrate) which mostly consists of the farm waste is loaded in the digester, preferably after 'chaffing'. It has a fixed dome and the gas produced is stored in a separate gasholder kept floating over water. Substrate is fed once as the batch system is a constant volume process, and there is no provision for fresh additions or withdrawal of the contents while undergoing anaerobic digestion before the completion of the retention period. The digester is emptied at the end of the completion of the process and recharged again. Contrarily, in daily fed plants, mainly animal wastes like the cattle dung etc. is used as the feeding material. These plants are also called continuous plants. Such plant consists of a mixing tank, a digester or the fermentation well, an inlet pipe/chamber for input and an outlet displacement chamber for the digested slurry. For gas collection, either a 'floating' drum or a 'fixed' dome is used which acts as a gas holder from which gas is supplied to the point of use. The size of these semi-continuous plants depends on the availability of the bulk of feeding material and the hydraulic retention time.

Based on another categorisation of the biogas plants, related to their designs, these are the fixed-dome and the floating-drum types. Nationally, the most popular one of the former type is the *Deenbandhu* and the later type is the KVIC model.

Biogasification of paddy straw

The State of Punjab produces 21MMT paddy straw annually which is mostly burnt in the field itself. Due to this, a lot of smoke and atmospheric pollution causes health issues amongst the contiguous population. The whole of North India has been crippled with such grave incidents particularly in the winter months in the last couple of years. The Punjab Agricultural University (PAU), Ludhiana has developed technologies to utilise paddy straw for various useful purposes. One of such technologies is utilisation for the purpose of energy (biogas) through anaerobic digestion in biogas plants.

The paddy straw can be digested anaerobically to produce biogas as a household fuel, as well as for power generation. The technology adopted is the latest method of anaerobic digestion, *i.e.*, dry fermentation of organic wastes. It is carried out with little labour and it produces a large amount of biogas for a period of 3–4 months. The digested material thus produced is good quality manure for the fields. The disposal of such digested material is easy as it can be lifted from the plant site with the help of semi-automatic system. Thus, there is no issues in generating biogas from paddy straw.

PAU has constructed masonry structure of the digester. This digester is constructed and made completely air tight unlike conventional biogas plant digester in which only the gas holder is air tight. Unlike the digestion taking place in the presence of large amount of water and the gas bubbles moving only in the upward direction in the conventional biogas plants, the quantity of water used is in these digesters very small and thus the gas bubbles freely move in all the directions. In case the digester is not air tight, the anaerobic digestion is affected. Also, this shall affect the safe storage of the biogas.

Dry fermentation biogas plant

The details of the plant are given in the Fig. 1. A brick masonry digester has been designed and developed at the PAU. The opening of the digester is closed and the digester is made air tight with a steel cover. The gas outlet pipe is also fitted with a top cover. The biogas produced in this plant is stored in a steel gas holder connected to the digester through HDPE pipe with the gas inlet pipe. The approximate cost for construction of the plant is Rs.1.20lakh. Fig. 2 provides a photographic view of the plant.

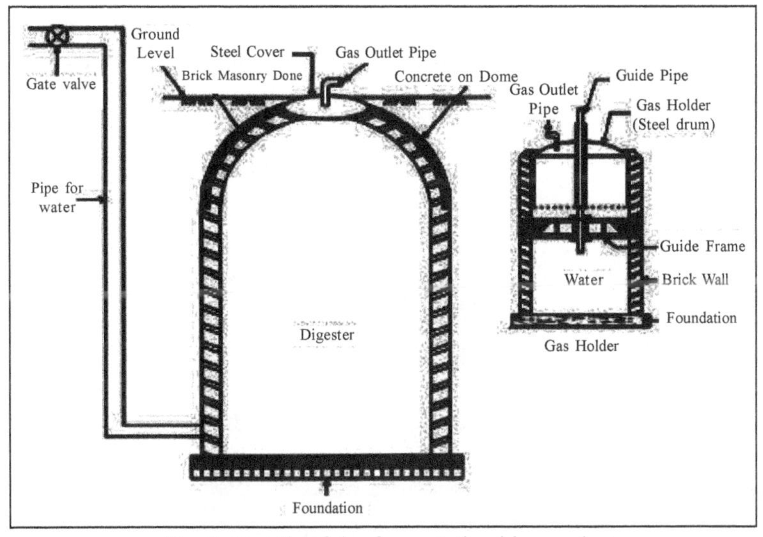

Fig. 1: Details of dry fermentation biogas plant

Fig. 2: View of a dry fermentation biogas plant

Process of commissioning of the plant

The feeding with paddy straw is done in layers of paddy straw and little quantity of cattle dung (Fig. 3) till it fills to the top. The feeding material fed to the digester is, paddy straw 1.6MMT and cattle dung/bio digested slurry 0.4MMT. The opening of the digester is closed air-tight with the help of a GI sheet cover. Water is added in the digester with the help of a pipe connected at the bottom of the digester till the water flows out through the gas outlet pipe fitted on the top. The paddy straw becomes wet after absorbing water and anaerobic digestion starts. Biogas production that starts after about 7–10 days is stored in a steel gas holder connected to the digester through HDPE pipe. Upto 4–5m^3/d for 3–4 months of biogas is produced, which is equivalent to monthly 3–4 LPG cylinders.

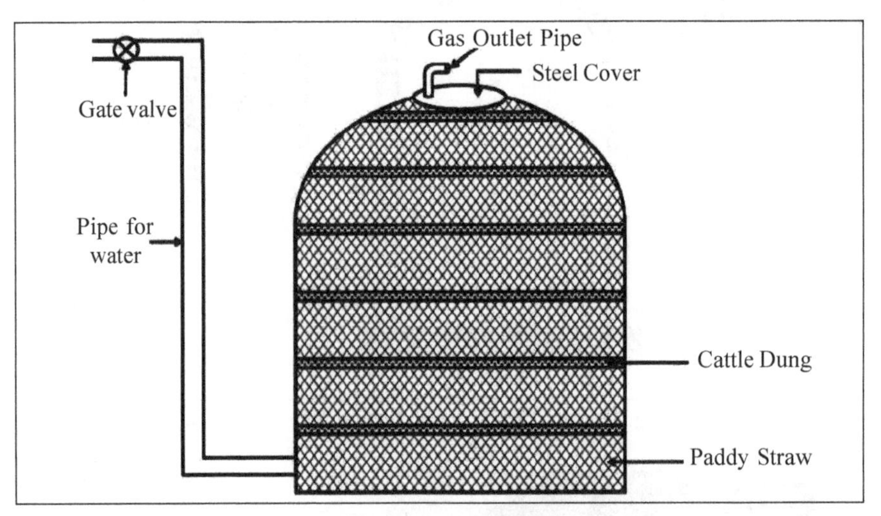

Fig. 3: Feeding in the 'dry fermentation' digester with paddy straw and cattle dung in layers

Daily-fed biogas plants

As of the current scenario, there are basically two categories of the daily-fed biogas plant models popular in India, the KVIC (fixed-drum) plants and the fixed-dome plants.

KVIC plants

This plant was designed by KVIC (Khadi and Village Industries Commission), Mumbai in 1962. It is an underground well-shaped digester with inlet and outlet pipes located on either sides of a partition wall. An inverted drum of mild steel is placed in the digester resting on the wedge shaped support and the guided frame at the level of the partition wall. It floats up and down with the support of a guide pipe complementing with gas accumulation and its use. The different parts of this biogas plant are shown in Fig. 4.

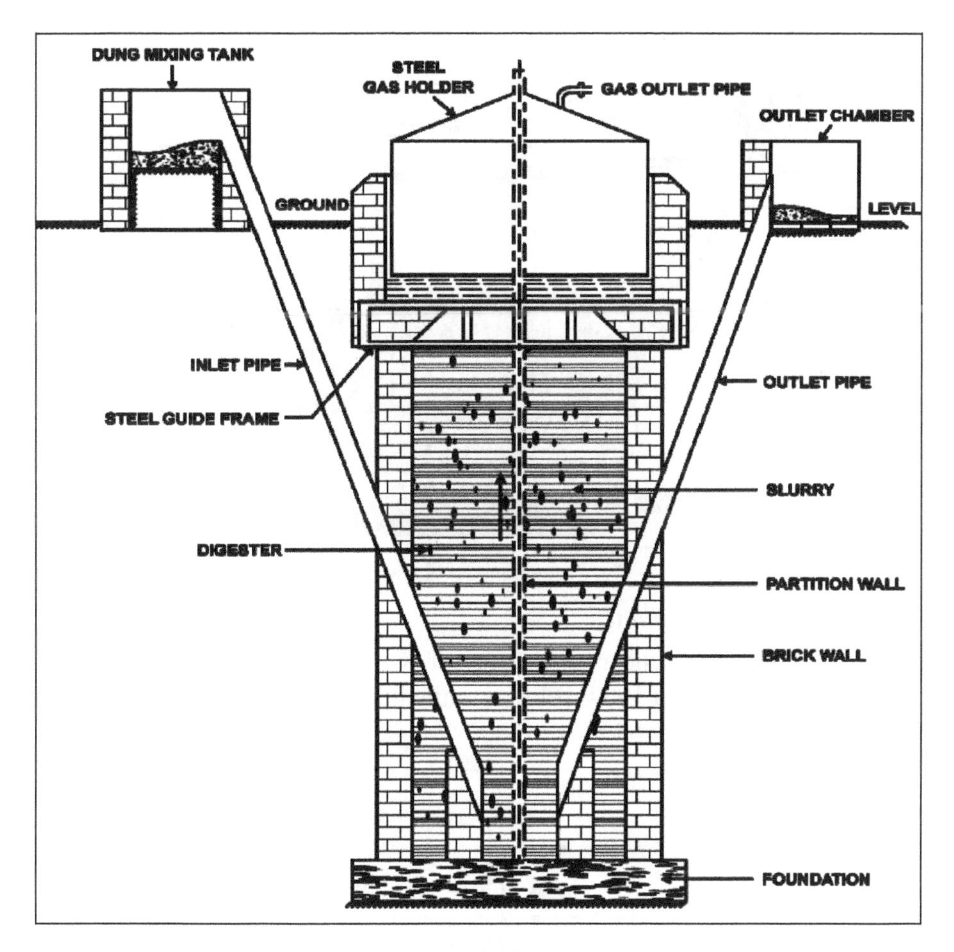

Fig. 4: VIC model biogas plant

Fixed-dome plants

There are two types of fixed-dome biogas plants. They are, *Janta* biogas plants and *Deenbandhu* biogas plants.

Janta *biogas plant*

The first fixed-dome biogas plant was designed by Gobar Gas Research Station, Ajitwal, in 1978. In this plant, the digester and the gas holder are integrated parts of the brick masonry structure. The digester is made of a shallow well with a dome-shaped roof, and the inlet and outlet chambers are connected with the digester through inlet/outlet openings marked in the digester. The gas pipe is fitted on top of the dome. An opening on the outlet wall of the outlet displacement chamber is kept for the discharge of the digested slurry. The capacity of these plants is limited to 15m^3/d. The different parts of this plant are shown in Fig. 5.

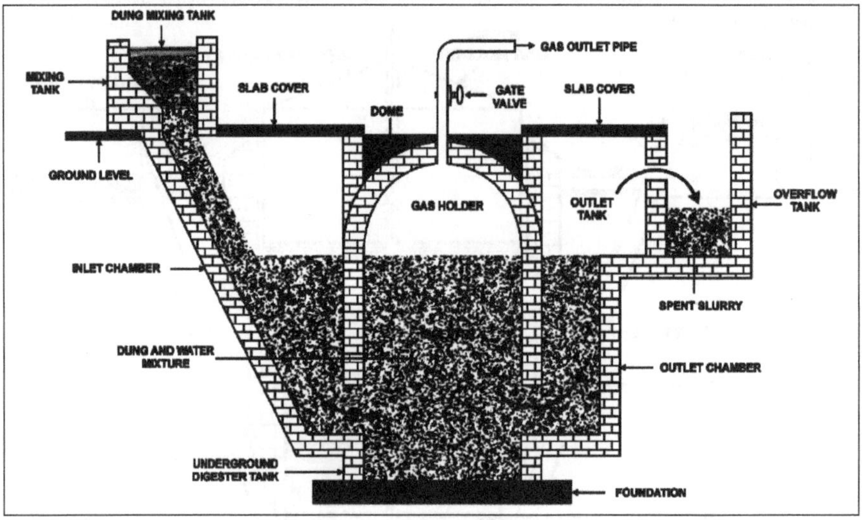

Fig. 5: Janta model biogas plant

Deenbandhu *biogas plant*

Considered as the 'friend of the poor', it was designed and developed by AFPRO, New Delhi, in 1984. It has a limited 6m^3/d capacity, and its various parts are shown in Figs. 6a,b.

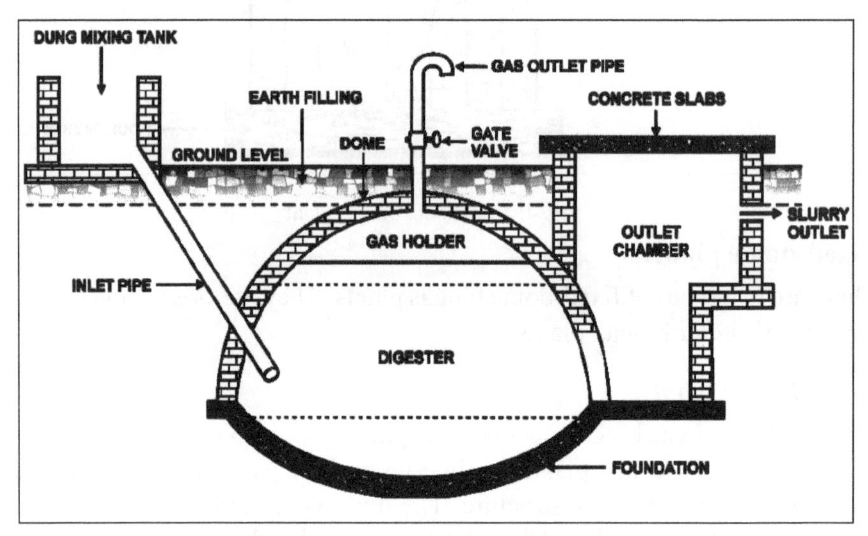

Fig. 6a: *Deenbandhu* model biogas plant

The surface area is reduced without sacrificing the plant's efficiency to minimise installation cost. This plant consists of two hemispheres of varying diameters, joined at their base. This structure acts both as the digester as well as the gas storage chamber. The digester is connected with an inlet pipe, and the outlet is connected to the digester.

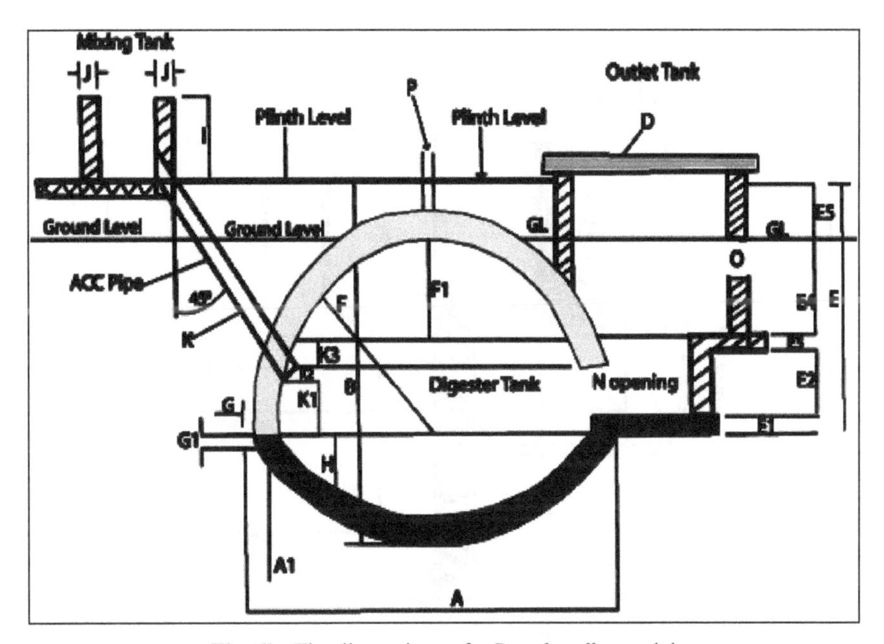

Fig. 6b: The dimensions of a *Deenbandhu* model

Cheaper biogas plants

Rising cost of steel, cement and bricks increases the relative cost of above-discussed biogas plants. This necessitates to dispense with, or to reduce the requirement of costly construction materials, altogether to reduce construction cost. Based on this, PAU developed inexpensive drum-type and dome-type models. Technological and designing interventions brought down the cost by 25–40%. In these models, digester lining is an ordinary dug pit instead of masonry bricks. Their technical details are discussed below.

Drum-type biogas plants

The plant details are shown in Figs. 7a,b. A pit circular is dug as per the size of the steel gas holder keeping the depth at about 5m (0.5m is above ground level to shed away rain water). The part in which the steel gas holder drum is to move up and down is kept 0.3m wide. 15cm space is kept all around for the drum to freely rotate. The lower part is kept smaller by 0.3m so that the drum could rest on it. The inlet is of AC pipe similar to that of the conventional type, and the outlet is made near the top of the pit. The steel guided frame is fixed above the top of the pit (Fig. 7a) or inside the pit (Fig. 7b) for the movement of the steel gas holder.

The inexpensive (*Katcha–Pucca*) model biogas plant has been approved by the Research Evaluation Committee of PAU, Ludhiana. Till now, the Department has covered different districts in the state to install this plant which shows quite an encouraging response to the adoption of this plant.

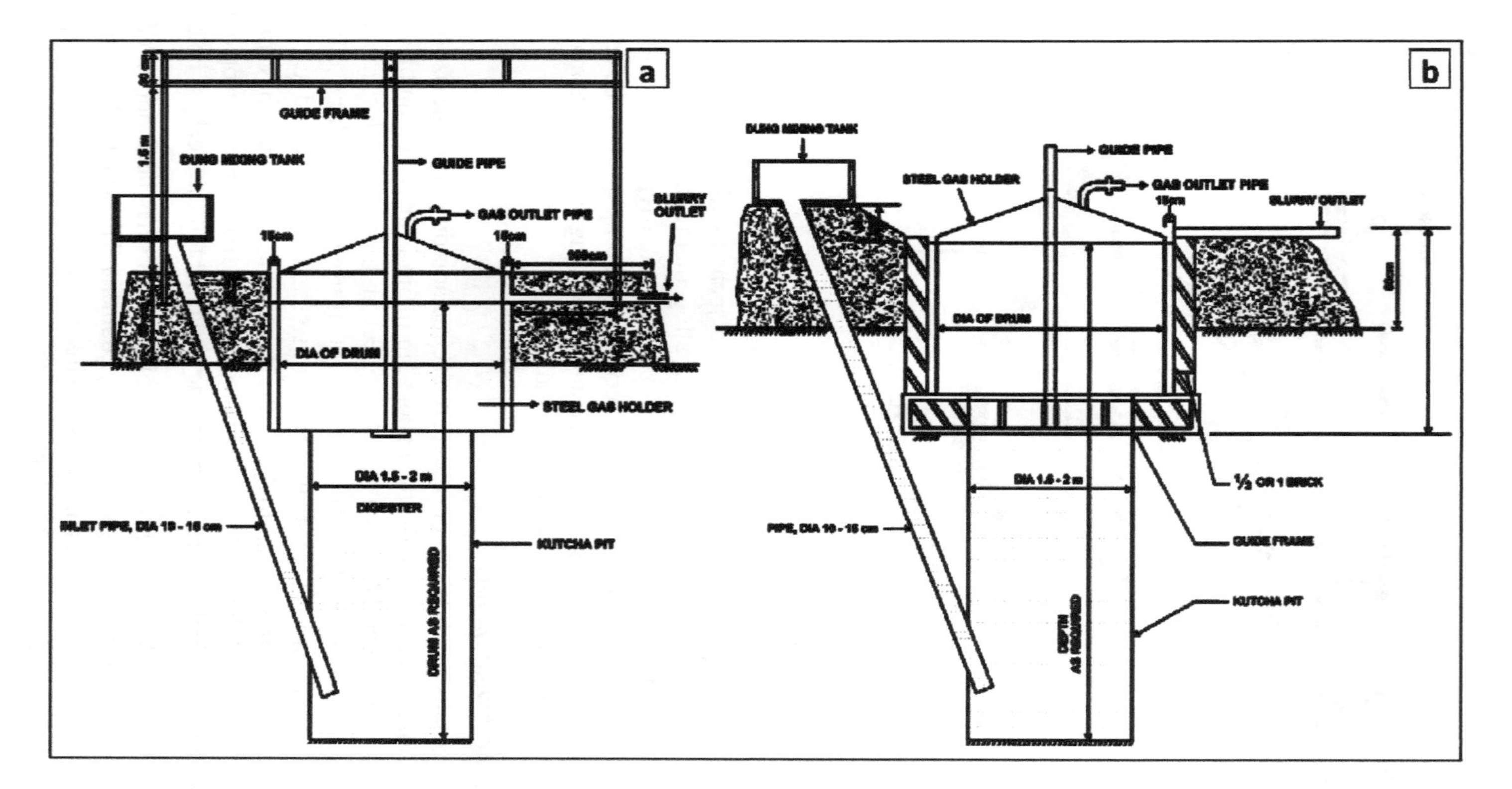

Fig. 7: The *Katcha–Pucca* model. The steel guide frame fixed a: above the pit, or b: inside the pit

Drumless-type biogas plants

A pit of the required diameter and depth similar to the *Janta* biogas plant is dug. The top of the pit is widened by 60cm to a depth of 30cm from the ground level for masonry work. The plant details are shown in Fig. 8. The masonry dome in the widened part is constructed as in the conventional *Janta* plant. The inlet and outlet for the entry and exit of the dung slurry are also similar. The dome is plastered with 1:3 cement mortar from inside to make it gas tight. Precautions are taken to ensure that the dome is always below the ground level particularly the bottom part, and should directly touch the undisturbed soil so as to absorb the impact of the horizontal thrust from the dome to avoid cracks.

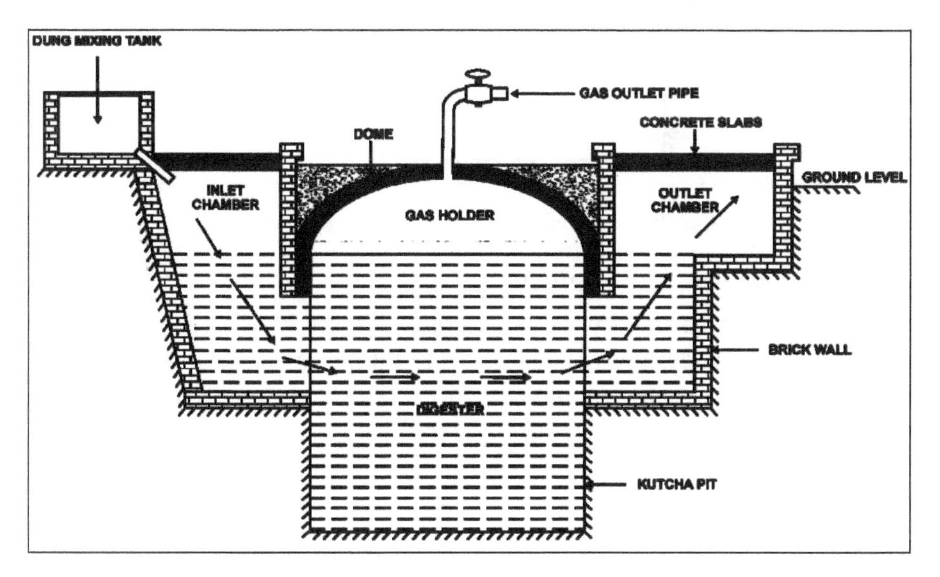

Fig. 8: The PAU *Janta* biogas plant model

It is observed during extension trips that many conventional *Janta* plants fail as the dome developed cracks when constructed above ground level rendering the plant unfunctional. In sandy soil, pit diameter is slightly less than the dome size to avoid its caving in. Modified PAU *Janta* plant can be installed keeping the depth of the plant above the water table in any soil. The model's cost was further reduced by fixing the inlet chamber with AC pipe (Fig. 9).

In case the plant is to be kept above ground up to certain height, 1cm diameter steel ring can be provided at the junction of the of dome and the wall to help the dome take lateral thrust.

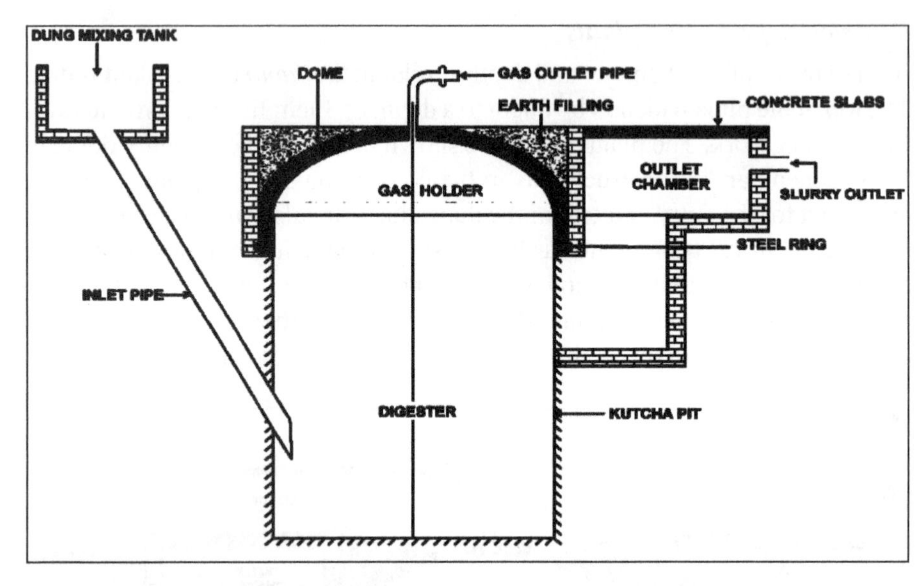

Fig. 9: The modified PAU *Janta* biogas plant model

Flexible dome-type biogas plants

Keeping in view the stability of the *Kutcha* pit in the PAU models with cattle dung slurry, a new design was developed to further lower the plant cost. This design does not need any amount of cement, brick and steel. The construction detail of this plant is shown in Fig. 10.

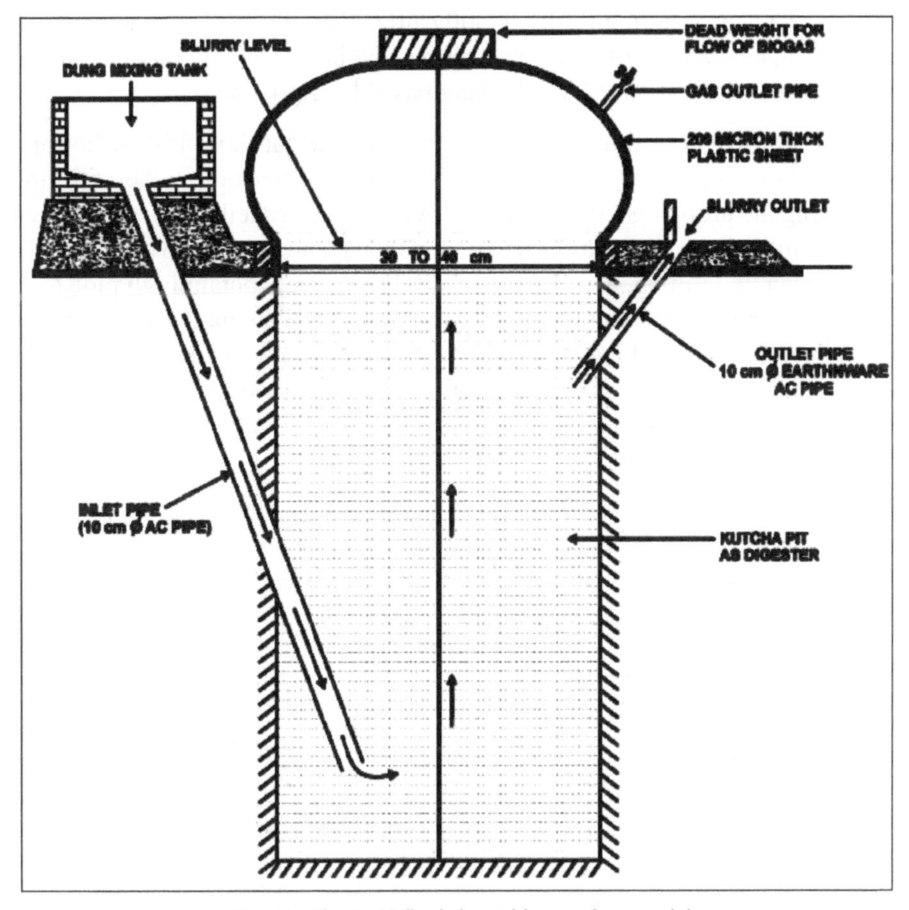

Fig. 10: The PAU flexi-dome biogas plant model

It consists of a *Katcha* pit of the size depending on the desired biogas output. The top of the pit is covered with a plastic impermeable sheet for the collection of the biogas. To generate adequate biogas pressure, weights (sand bags) are placed on the gas holder (plastic sheet). The inlet and outlet pipes are fixed in such a way that the biogas produced does not escape to the atmosphere through them. A tap is provided in the plastic sheet for biogas supply for use. The materials required for the construction of a flexi-dome biogas plant are, 6.5m^2 high density polyethylene sheet, 200 nos. bricks, and 2 nos. (each 60cm long) earthenware pipes (10cm dia).

Large capacity biogas plants

Thousands of family size biogas plants are satisfactorily functioning for the last two decades in Punjab. From 5–10 cattle heads, the concept is shifting towards large herds of cattle as a fulltime job. There are around 4000 dairy farms, each having a capacity ranging from 50–500 cattle. Thus, there is huge quantity of cattle dung available for biogas production, and hence, a large scope to install large capacity (50–500m^3/d) cattle dung based biogas plants.

In India, the hen and broiler populations are estimated at about 150 million and 1700 million respectively. Annually, the wastes generated during their farming (including the droppings, feathers, spill-over feed, egg shells etc.) is estimated to be 3.6 million tonnes whose disposal and management needs to be addressed for sustainable farming. The droppings (cage manure) contain high moisture, 70–80% organic matter, and substantial nitrogen. In Punjab, there are around 15,000 poultry farms (of 2,000–50,000 birds capacity each) for meat, and around 2,000 poultry farms (of 10,000–3,00,000 birds capacity each) for eggs. Thus, there is enough scope to install large capacity (50–500m^3/d) biogas plants based on poultry droppings/litter.

With these in mind, PAU developed a large capacity (modified PAU *Janta* model; Fig. 11) biogas plant that could be installed in dairy farms, poultry farms, cattle farms, institutes, and industries. The plant of any capacity from 20–500m^3/d can be installed.

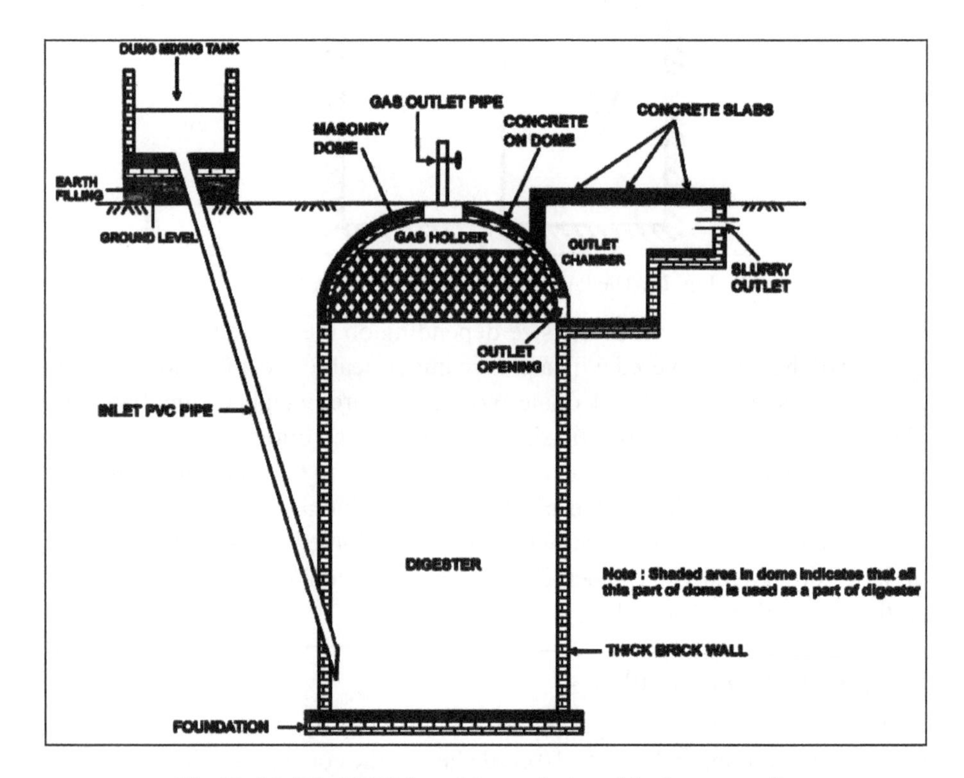

Fig. 11: Modified PAU *Janta* biogas plant model – large capacity

Construction of biogas plants

This discusses the category/model-wise construction expenses of various biogas plants.

A. Batch (periodically-fed) system: The digester and gas holder dimensions are presented in Fig. 12. The estimated cost for the construction is given in Table 1 (2016 market rates). Table 1A details the cost against the digester and gas-holder well construction, 1B the cost estimate of plant parts fabrication, and 1C lists the estimated cost of gas pipeline and supply system.

Table 1A: Estimated cost of construction of digester and gas holder well (2016 market rate)

Sl.No.	Particular	Unit	Quantity (Rs.)	Rate (Rs.)	Amount
1.	Bricks	Nos.	5,000	5.5	27,500/-
2.	Cement	Bags	30	300	9,000/-
3.	Stone ballast/Bajri	ft^3	50	40	2,000/-
4.	Sand	m^3	250	30	7,500/-
5.	Steel required for foundation	Kg	20	50	1,000/-
6.	Pit digging for digester and gas holder	Lump sum	-	5,000	5,000/-
7.	Shuttering for construction work	Lump sum	-	1,500	1,000/-
8.	Labour for plant construction	Lump sum	-	20,000	20,000/-
	Total cost				Rs. 73,000/-

Table 1B: Estimated cost of fabrication of the plant parts (2016 market rate)

Sl.No.	Particular	Unit	Quantity	Rate (Rs.)	Amount (Rs.)
1.	Steel cover for the digester	Lump sum	-	10,000	10,000/-
2.	Gas holder	Lump sum	-	25,000	25,000/-
3.	Guide frame made up of GI Pipe (2" dia 30' long GI pipe + other expense)	Lump sum	-	6,000	6,000/-
4.	GI pipe (½" dia 25' long + 30' gate valve) to supply water to digester etc.	Lump sum	-	1,500	1,500/-
	Total cost				Rs. 42,500/-

Table 1C: Estimated cost of the gas pipe line and supply system (2016 market rate)

Sl.No.	Particular	Unit	Quantity	Rate (Rs.)	Amount (Rs.)
1.	Gas pipe line system including 100' long PVC gas pipe, drip trap, nipples, gate valve, burner and other accessories	Lump sum	-	4,500	4,500/-
	Total cost				Rs. 4,500/-

Thus, the total construction cost of the plant = Cost of (A + B + C) = Rs. 1,20,000/-.

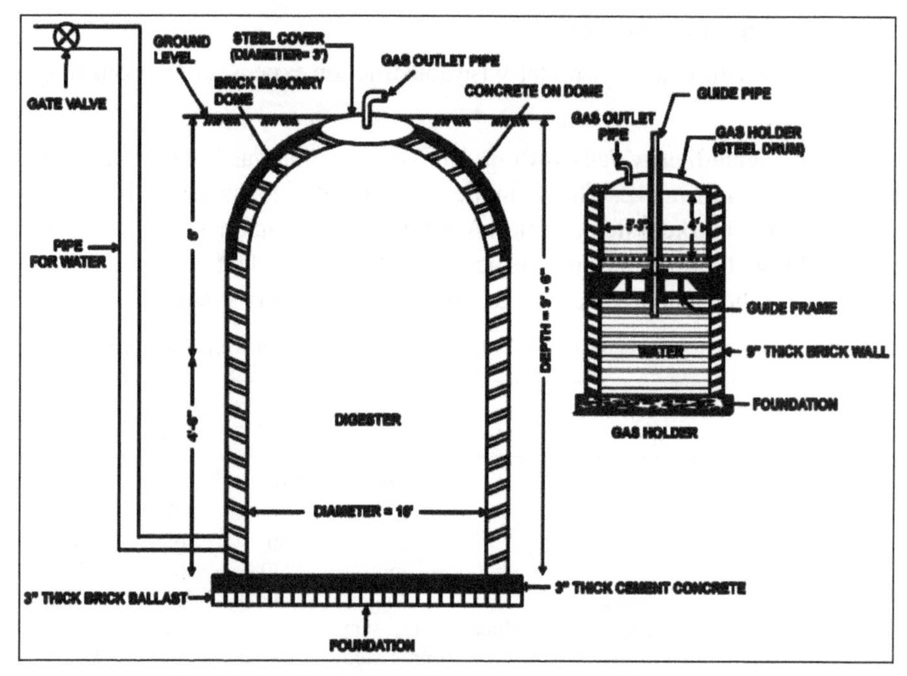

Fig. 12: The dimensions of a dry fermentation biogas plant

B. Daily-fed biogas plants: Capacity-wise, the design, material estimates and construction details of the various (KVIC, *Janta* and *Deenbandhu,* as the case may be) models of daily-fed (continuous or semi-continuous) biogas plants are detailed under the sub-heads family-size plants and large capacity (institutional/community) plants.

For family-type (individual household type) biogas plants

Material requirements for KVIC model: The design of the KVIC model plants is shown in Fig. 13, and the dimensions of the different size biogas plants are given in Table 2. The materials required for the construction of the plant are detailed in Table 3.

Table 2: Dimensions of KVIC model biogas plants (40-day HRT)

Sl.No.	Dimension (cm)	Capacity of biogas plant (m^3)						
		1	2	3	4	6	8	10
Digester								
1.	Excavation width (A)	181	196	221	241	281	301	336
2.	Excavation depth (B)	172	272	292	307	307	347	332
3.	Foundation width (C)	181	196	221	241	281	301	336
4.	Height of digester (D)	180	280	300	315	315	355	340

Contd.

5.	Height of digester below guide frame (D1)	90	150	170	185	185	185	185
6.	Height of digester above guide frame (D2)	60	100	100	100	100	125	125
7.	Digester dia. (inner) (E)	120	135	160	180	220	240	275
8.	Digester dia. (Outer) (F)	166	181	206	226	266	286	321
Gas holder								
9.	Gas holder dia. (G)	105	125	150	165	200	225	260
10.	Gas holder height (H)	60	100	100	100	100	125	125
Inlet tank								
11.	Inlet tank length (L)	38	40	75	75	90	90	90
12.	Inlet tank width (W)	38	40	75	75	90	90	90
13.	Inlet tank height (J)	38	40	45	45	45	52	60
14.	Length of inlet pipe (K)	200	300	350	330	350	400	370
15.	Length of outlet pipe (M)	70	90	110	300	300	340	370
16.	GI pipe for guide pipe of 40 mm dia (O)	130	195	20	205	205	235	240
17.	MS pipe for gas holder (P)	70	115	115	125	125	145	150

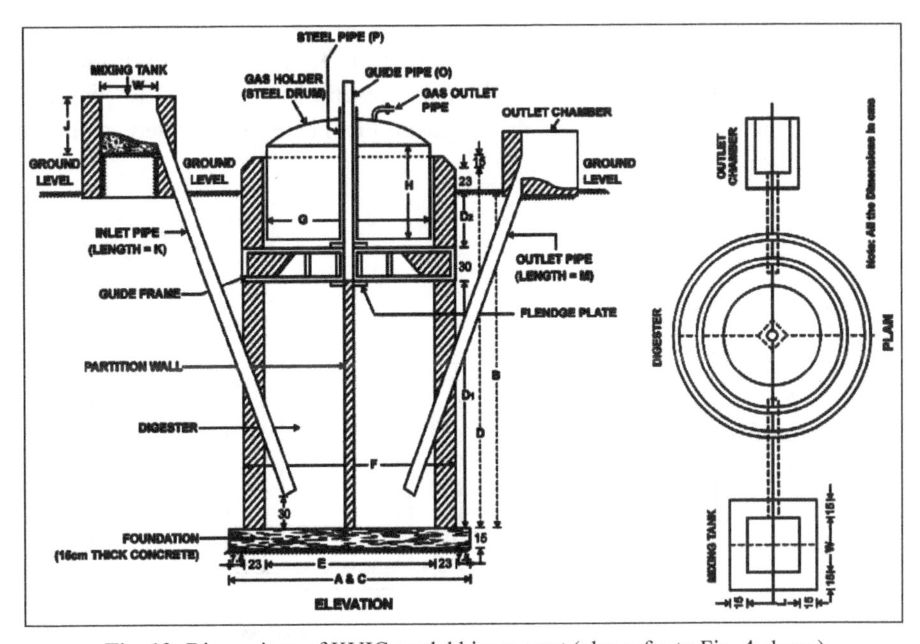

Fig. 13: Dimensions of KVIC model biogas pant (also refer to Fig. 4 above)

Table 3: Details of the materials required for the construction of KVIC model plants (40-day HRT)

Sl.No.	Particular	Capacity of biogas plant (m³)						
		1	2	3	4	6	8	10
1.	Bricks (nos.)	2460	2765	3200	3730	4200	4800	5500
2.	Cement (bags)	13	17	19	23	27	32	40
3.	Sand (m³)	2.00	2.55	2.90	3.40	4.00	4.25	5.00
4.	Brick ballast (m³)	0.60	0.90	0.95	1.25	1.50	1.75	2.00
5.	AC pipe of 10 cm dia.	3.90	4.60	6.30	6.50	7.10	7.70	8.50
6.	Steel drum (gas holder)	1	1	1	1	1	1	1
7.	Steel guide frame alongwith guide pipe	1	1	1	1	1	1	1
8.	Paint (litre)	2.0	2.5	2.5	3.0	3.0	3.5	4.0
9.	Labour for digging the pit (no. of days)	8	8	10	12	15	18	21
10.	Masons (no. of days)	12	15	20	25	30	34	40
11.	Labour for the plant construction (no. of days)	24	30	40	50	60	65	70

A detailed sketch of biogas collection steel drum (gas holder) is presented in Fig. 14, and the dimensions of gas holder are presented in Table 4. The thickness of GI sheet to fabricate the steel drum (gas holder) is recommended as per the plant size as presented in Table 5.

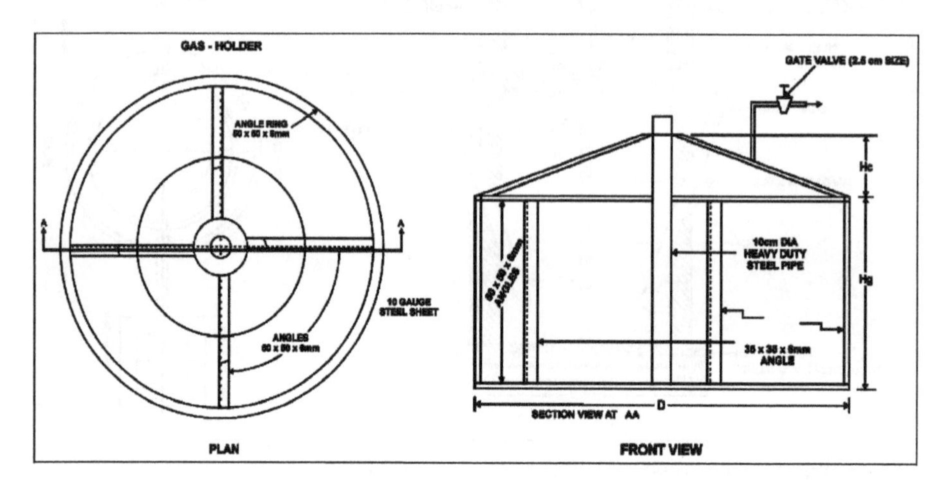

Fig. 14: The details of the steel gas-holder for KVIC model biogas plant

Table 4: Dimensions of steel gasholder for KVIC model biogas plants

Sl.No.	Capacity of the plant (m^3)	All dimensions in metre		
		Diameter (Dg)	Height (Hg)	Rise of gas holder (Hc)
1.	2.84	1.53	0.92	0.80
2.	4.25	1.83	0.92	0.080
3.	7.09	1.99	1.22	0.080
4.	9.92	2.60	1.07	0.080
5.	14.17	3.05	1.07	0.080

Table 5: Thickness of the GI sheet for fabrication of the steel gas-holder

Sl.No.	Size of the biogas plant (m3)	Thickness of GI sheet (SWG)
1.	1-3	14
2.	4-10	12
3.	11-85	10
4.	>85	8

Material requirements for *Janta* model: A detailed sketch of this model is presented in Fig. 15. The dimensions of the different capacity of plants of this model are given in Table 6, and the materials required for the installation of the plant is given in Table 7.

Table 6: Dimensions of the *Janta* model plants (40-day HRT)

Sl.No.	Dimension (cm)	Capacity of biogas plant (m^3)						
		1	2	3	4	6	8	10
1.	A	173	222	250	275	315	346.5	373.5
2.	B	211	260	288	313	376	407.5	434.5
3.	C	37	50	52	60	67	72.5	78
4.	D	24	48	60.6	64.6	67.3	69.8	72.6
5.	E	61	61	61	61	61	100	100
6.	F	14	17	20.4	22.7	25.7	28.2	30.4
7.	G	61	61	61	70	90	100	110
8.	H	99	126	142	157	183	198	213
9.	H_1	15	15	15	15	23	23	23
10.	J	11.5	11.5	11.5	11.5	23	23	23
11.	K	61	61	61	61	61	61	75
12.	L	80	80	80	80	80	80	94
13.	M	94.5	111	113	100	100	100	100
14.	N	61	61	61	61	61	100	100
15.	P	61	75	100	135	182	257.5	250
16.	P_1	84	98	123	158	205	285	273
17.	S	30	42	45	53	60	60	60
18.	T	15	15	15	15	15	23	23

Contd.

19.	U	22.7	22.7	22.7	22.7	22.7	22.7	33.2
20.	X	32.7	56.7	69.3	73.3	84	79	81.8
21.	W	142.7	158.7	164.1	183.1	213.4	238.9	256.6
22.	W_1	75	78	81.4	92.4	115.7	128.2	140.4
23.	Y	67.7	80.7	82.7	90.7	97.7	110.7	116.2
24.	Z	175.4	233.4	233.4	256.4	297.4	317.9	338.4

Table 7: Materials required to construct *Janta* model biogas plants including labour (40-day HRT)

Sl.No.	Dimension (cm)	Capacity of biogas plant (m^3)						
		1	2	3	4	6	8	10
		2	3	4	6	8	10	15
1.	Bricks (nos.)	1500	2500	3500	4000	4500	6000	7500
2.	Cement (bags)	20	25	30	40	45	65	75
3.	Sand (m^3)	2.0	2.5	3.5	4.25	5.0	5.5	6.5
4.	Bricks ballast (m^3)	1.0	1.25	1.5	2.0	2.25	2.75	3.5
5.	Stone ballast (m^3)	1.0	1.25	1.5	2.0	2.25	2.75	3.5
6.	GI pipe with socket having 25mm dia. (cm)	45	45	45	45	45	45	45
7.	Steel for slabs for inlet-outlet chambers (kg)	18	18	20	25	30	35	45
8.	Paint (litre)	1.5	2.0	2.5	3.0	3.0	3.5	4.0
9.	Gate valve	1	1	1	1	1	1	1
10.	Labour for pit-digging (no. of days)	8	8	10	12	15	18	24
11.	Masons (no. of days)	12	15	20	25	31	37	50
12.	Labour for plant construction (no. of days)	24	30	40	50	60	75	100

Material requirements for *Deenbandhu* model: The details of this plant model are given in Fig. 16. The dimensions of the different capacity plants are given in Table 8 and the materials requirements for the installation of the plant is given in Table 9.

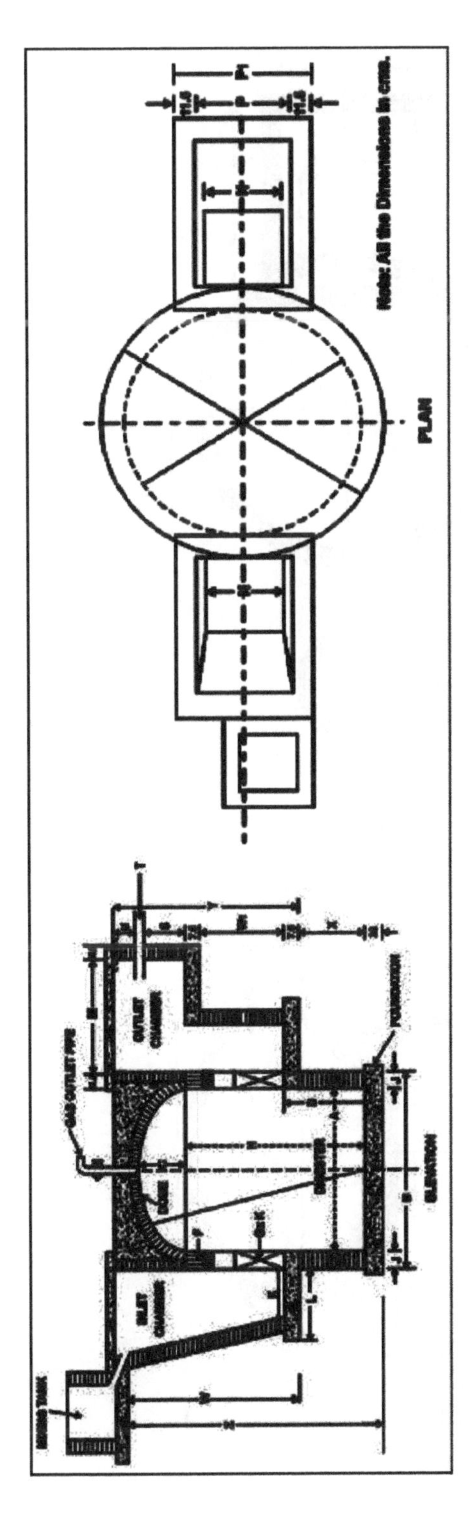

Fig. 15: Dimensions of *Janta* model biogas plant (also refer to Fig. 5 above)

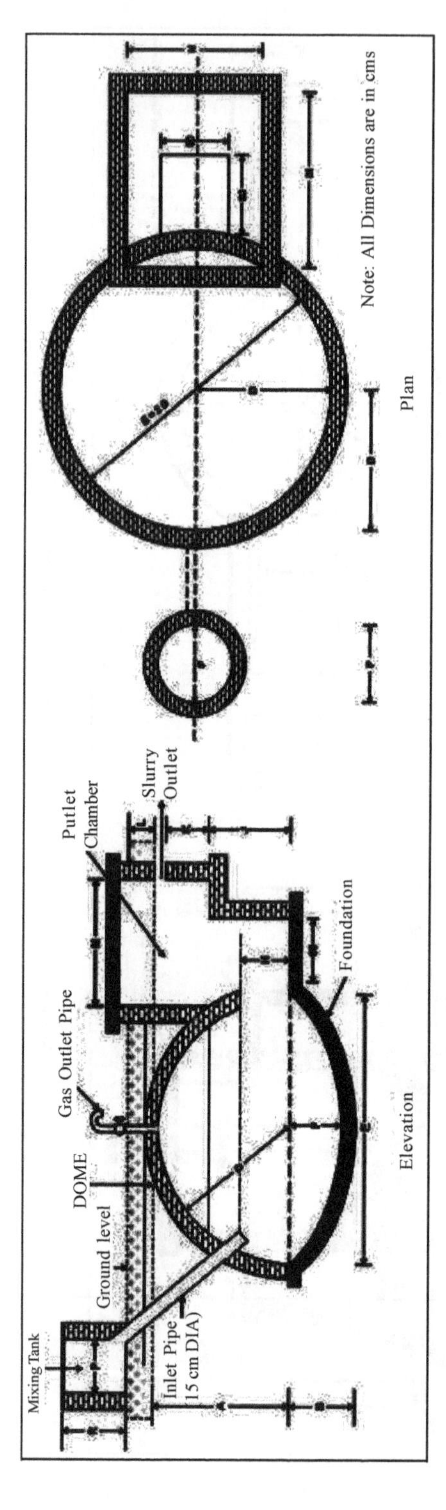

Fig. 16: Dimensions of *Deenbandhu* model biogas plant (also refer to Figs. 6a,b above)

Table 8: Dimensions (in inch) of *Deenbandhu* model plants (figures in parentheses are cm equivalent)

Sl.No.	Description	Capacity of the biogas plant (m^3)			
		70 ft^3 ($2m^3$)	105 ft^3 ($3m^3$)	140 ft^3 ($4m^3$)	210 ft^3 ($6m^3$)
Digester					
1.	Diameter of digester (E)	8'6" (255)	9'6" (285)	10'6" (315)	12'0" (360)
2.	Diameter of pit	10'6" (315)	11'6" (345)	12'6" (375)	14'0" (420)
3.	Depth of digester (A)	4'3" (127.5)	4'9" (144.5)	5'3" (157.5)	6'0" (180)
4.	Inner radius of digester (D)	4'3" (127.5)	4'9" (142.5)	5'3" (157.5)	6'0" (180)
5.	Height of outlet gate (H)	1'6" (45)	2'0" (60)	2'0" (60)	2'0" (60)
6.	Inner depth of foundation (F)	2'3" (67.5)	2'3" (67.5)	2'6" (75)	2'9" (82.5)
7.	Outer depth of foundation (B)	2'6" (75)	2'6" (75)	2'9" (82.5)	3'0" (90)
Outlet chamber					
1.	Height of smaller portion (J)	2'0" (60)	2'3" (67.5)	2'9" (82.5)	3'3" (97.5)
2.	Height of bigger portion (K)	1'6" (45)	1'6" (45)	1'6" (45)	1'9" (52.5)
3.	Width of bigger portion (M)	3'6" (105)	3'6" (105)	4'0" (120)	4'0" (120)
4.	Length of bigger portion (N)	5'6" (165)	6'9" (202.5)	7'9" (232.5)	10'0" (300)
Mixing tank					
1.	Diameter (P)	2'0" (60)	2'0" (60)	2'0" (60)	2'6" (75)
2.	Height (R)	1'6" (45)	1'6" (45)	2'0" (60)	2'0" (60)

Table 9: Materials required to construct *Deenbandhu* model biogas plants

Sl.No.	Particular	Capacity of biogas plant (m^3)				
		2	3	4	5	6
1.	Bricks (no.)	1000	1300	1600	1900	2200
2.	Cement (bags)	8	10	12	14	15
3.	Stone ballast (m^3)	45	0.71	0.86	1.00	1.15
4.	Sand (m^3)	80	2.90	3.50	4.10	5.00
5.	GI pipe 25mm dia. with socket (cm)	30	30	30	30	30
6.	PVC pipe, 15 cm dia. (cm)	180	180	180	210	210
7.	Steel for slab of outlet chamber (kg)	10	10	12	13	15
8.	Paint (litre)	1	1.5	2	2.5	3
9.	Labour for digging the pit (man-day)	3	4	4	5	6
10.	Masons (man-day)	3	4	5	5	6
11.	Labour for plant construction (man-day)	6	8	10	10	12
12.	Gate valve (no.)	1	1	1	1	1
13.	PVC pipe, 2.5 cm (1") dia. (m)	50	50	50	50	50
14.	Drip-trap (no.)	1	1	1	1	1
15.	Burner (no.)	1	1	1	1	1

For large-capacity (community/institutional type) biogas plants

Material requirements for (large capacity) KVIC model: The dimensions of this model are provided in Table 10, and the material requirements for the construction of the same biogas plants of different capacities are provided in Table 11.

Table 10: Dimensions of KVIC (community/institutional) model (40-day HRT)

Sl.No.	Capacity of biogas plant (m^3)	Capacity of digester (m^3)	Digester diameter (m)	Capacity of gas holder (m^3)	Dimensions of gas holder		
					Depth (m)	Diameter (m)	Height (m)
1	15	30	3.37	7.5	3.37	3.00	1.06
2	20	40	3.71	10	3.71	3.25	1.20
3.	25	50	3.99	12.5	3.99	3.50	1.30
4.	30	60	4.25	15	4.25	3.75	1.36
5.	35	70	4.47	17.5	4.47	4.00	1.39
6.	40	80	4.67	20	4.67	4.25	1.40
7.	45	90	4.86	22.5	4.86	4.50	1.41
8.	50	100	5.03	25	5.03	4.75	1.41
9.	55	110	5.19	27.5	5.19	4.75	1.55
10.	60	120	5.34	30	5.34	5.00	1.52
11.	65	130	5.49	32.5	5.49	5.00	1.65
12.	70	140	5.63	35	5.63	5.25	1.62
13.	75	150	5.76	37.5	5.76	5.25	1.73
14.	80	160	5.88	40	5.88	5.50	1.68
15.	85	170	6.00	42.5	6.00	5.50	1.78

Table 11: Materials required to construct large-capacity KVIC model biogas plants

Sl.No.	Details of the Material	Capacity of biogas plant (m^3)	
		85	140
I. General materials for construction			
1.	Bricks for plant (nos.)	38000	870
2.	Bricks for manure pit (nos.)	12,600	12,600
3.	Sand (m^3)	32.20	38.60
4.	Stone chips ¾" (m^3)	17.75	70.75
5.	Cement (bags)	–	246
6.	AC pipe 150mm dia.	14.15 RM	14.00 RM
7.	10mm dia. rod @ 5" C/C		–
8.	10mm dia. rod @ 6" C/C		–
9.	12mm dia. rod @ 5" C/C		–
10.	10mm dia. rod for ledge portion	–	98.00 RM
11.	Stirrups 6mm dia. @ 6" C/C	–	123.00 RM
II. Requirements for central guide			
1.	10mm dia. rod @ 15cm C/C (for footing)	12.00 RM	12.00 RM
2.	20 mm dia. rod for main reinforcement, for column	20.00 RM	20.00 RM
3.	Stirrups 6mm dia. @ 15cm C/C	32.00 RM	–
4.	Stirrups 10mm dia. @ 15cm C/C	–	32.00 RM
5.	10mm dia. rod 6 Nos. (for ledge portion)	98.00 RM	–
6.	Stirrups 6mm dia. @ 15cm C/C	123.00 RM	–
7.	125mm dia. MS pipe	3.25 RM	3.75 RM

Contd.

8.	230mm x 300mm plate 6mm thick (no.)	1	1
9.	12mm dia J. bolt (nos.)	4	4
III.	**Requirements for gas holder**		
1.	Angle Iron (75mm x 75mm x 8mm)	229.00 RM	–
2.	Angle Iron (80mm x 80mm x 8mm)	–	357.00 RM
3.	M.S. Pipe 150mm dia.	2.25 RM	2.50 RM
4.	1350mm dia. and 8mm thick flange plate (no.)	2	–
5.	1500mm dia. and 8mm thick flange plate (no.)	–	2
6.	Gas outlet pipe flange 50mm dia. (no.)	1	1
7.	GI bend 50mm dia. (no.)	1	1
8.	Heavy-duty gate valve 50mm dia. (no.)	1	1
9.	Union joint 50mm dia. (nos.)	2	2
10.	Pipe nipple 50mm dia. 100mm long (no.)	1	1
11.	Pipe nipple 50mm dia. 150mm long (nos.)	2	2
12.	Clip (nos.)	2	2
13.	10 meter polythene or reinforced rubber pipe 50mm dia. (no.)	1	1
14.	3.15mm thick (10 range) mild steel sheet (2.5m x 1.25m) (nos.)	22	33

Source: Khadi and Village Industries Commission (KVIC), Mumbai; RM: running meter; 10% of the total cost covered in parts I, II and III is assumed as the cost of the gas pipe line system.

Material requirements for modified PAU *Janta* model: The major salient feature of this model is, the plant is a brick masonry structure that is suitable for the whole country. Further, the maintenance cost of this plant is very low. The cost of the plant is also upto 50% less compared to the KVIC model. The payback period for this plant is from 3–4 years.

So far, a total of 125 numbers of these plants are installed and functioning well in India. The space requirement for plant-construction (of various sizes; Fig. 17) are given in Table 12. Fig. 18 provides a labelled diagram of the plant including plane and sectional elevation. Important dimensions of 25–500m^3/d rated capacity plant with 40-day HRT are furnished in Table 13.

The materials estimate for construction of plants of 25–500m^3/d capacities are provided in Table 14, and their costs of construction are detailed in Table 15. It may be noted that the cost of construction per m^3 of rated capacity varies between Rs. 12,172 to 5,768, and the rate per m^3 capacity increment gradually reduced as the size of plant increased.

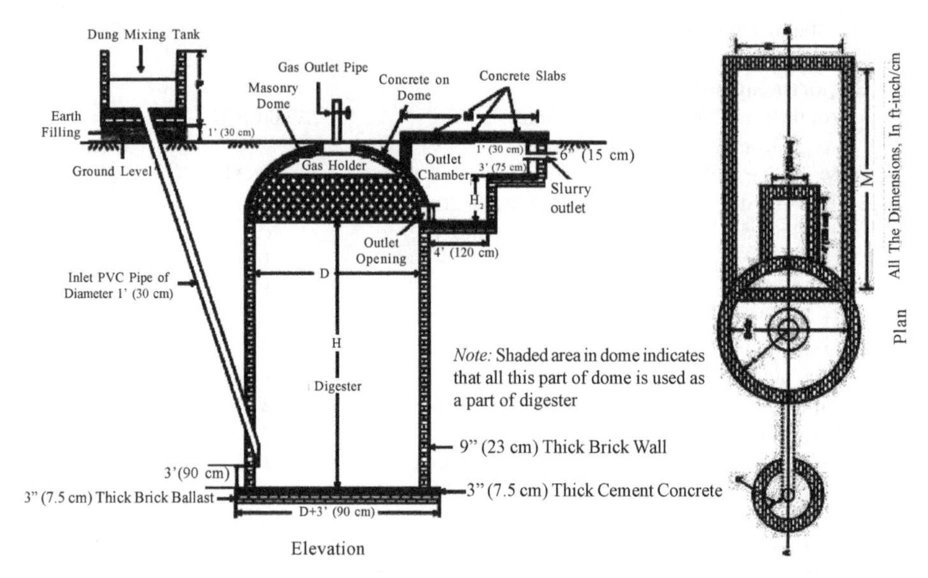

Fig. 17: Measurements of the modified PAU *Janta* model plant (also refer to Fig. 11 above)

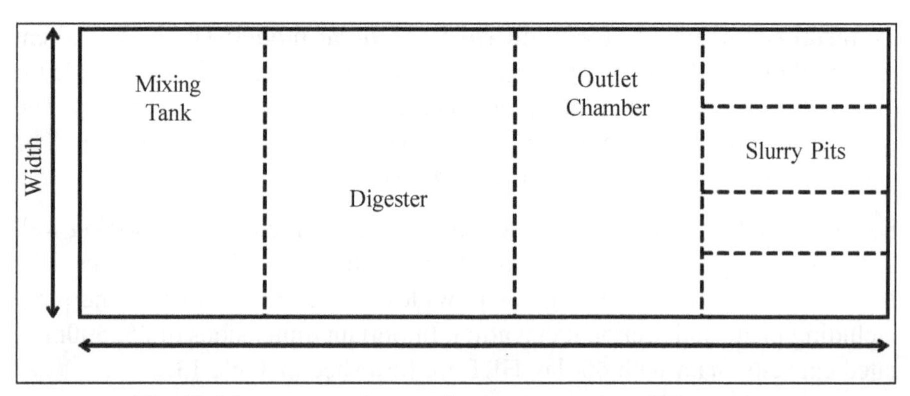

Fig. 18: The space requirement for the construction of biogas plant

Table 12: Space requirements for plant construction (in ft.; figures in parentheses are m equivalent)

Sl.No.	Plant capacity (m^3/d)	Space requirement for the plant			
		Length	Width	Area in Sq. ft. (Sq. m.)	Area in Marla* (Punjabi measurement)
1.	25	65 (19.5)	20 (6.0)	1,300 (117)	6.00
2.	50	70 (21.0)	25 (7.5)	1,750 (157.5)	8.25
3.	75	75 (22.5)	30 (9.0)	2,250 (202.5)	10.50
4.	100	80 (24.0)	35 (10.5)	2,800 (252)	13.00
5.	125	85 (25.5)	37 (11.1)	3,145 (283)	14.50
6.	150	90 (27.0)	40 (12.0)	3,600 (324)	16.75
7.	175	95 (28.5)	42 (12.6)	3,990 (359)	18.50
8.	200	100 (30.0)	45 (13.5)	4,500 (405)	21.00
9.	225	105 (31.5)	47 (14.1)	4,935 (444)	23.00
10.	250	110 (33.0)	50 (15.0)	5,500 (495)	25.50
11.	275	115 (34.5)	52 (15.6)	5,980 (538)	27.75
12.	300	120 (36.0)	55 (16.5)	6,600 (594)	30.75
13.	325	125 (37.5)	57 (17.1)	7,125 (641)	33.00
14.	350	130 (39.0)	60 (18.0)	7,800 (702)	36.25
15.	375	135 (40.5)	62 (18.6)	8,370 (753)	39.00
16.	400	140 (42.0)	65 (19.5)	9,100 (819)	42.25
17.	425	145 (43.5)	67 (20.1)	9,715 (874)	45.25
18.	450	150 (45.0)	70 (21.0)	10,500 (945)	48.75
19.	475	155 (46.5)	72 (21.6)	11,160 (1044)	52.00
20.	500	160 (48.0)	75 (22.5)	12,000 (1080)	55.00

*20 Marlas or 1 Kanal is equivalent to 604.99996 Sq.yds (say, 605 Sq.yds)

Conclusion

The various popular models of nationally adopted (as approved by the MNRE) models, their design and utility aspects as also the designing, construction and the economics of various modified models (as conceptualised, tried and tested on the field, particularly in Punjab) have been detailed. Conscious effort has been made to put the technology as it evolved. Various models and the economics have been discussed in such a way that it provides relevant and equally beneficial information to the researchers, extension workers, beneficiaries and the line department officials.

Table 13. Dimensions (in ft-inches) of modified PAU *Janta* model plants (figures in parentheses are mm equivalent)

Dimension and symbol	Capacity biogas plant (m^3/d)																			
	25	50	75	100	125	150	175	200	225	250	275	300	325	350	375	400	425	450	475	500
Digester diameter (D)	13'6"	15'6"	19'0"	20'6"	23'6"	24'0"	26'0"	28'0"	29'6"	30'6"	33'0"	35'0"	36'0"	36'6"	37'0"	37'6"	38'6"	39'0"	39'6"	40'0"
	(405)	(465)	(570)	(615)	(705)	(720)	(780)	(840)	(885)	(915)	(990)	(1050)	(1080)	(1095)	(1110)	(1125)	(1155)	(1170)	(1185)	(1200)
Digester inner radius (r)	6'9"	7'9"	9'6"	10'3"	11'9"	12'0"	13'0"	14'0"	14'9"	15'3"	16'6"	17'6"	18'0"	18'3"	18'6"	18'9"	19'3"	19'6"	19'9"	20'0"
	(202.5)	(232.5)	(285)	(307.5)	(352.5)	(360)	(390)	(420)	(442.5)	(457.5)	(495)	(525)	(540)	(547.5)	(555)	(562.5)	(577.5)	(585)	(592.5)	(600)
Digester depth (H)	11'6"	14'6"	14'6"	14'6"	14'0"	16'3"	15'0"	14'0"	14'6"	14'6"	12'6"	11'0"	11'0"	11'0"	12'0"	13'6"	12'0"	14'0"	15'0"	15'0"
	(345)	(435)	(435)	(435)	(420)	(487.5)	(450)	(420)	(435)	(435)	(375)	(330)	(330)	(330)	(360)	(405))	(360)	(420)	(450)	(450)
Height of outlet opening (H_1)	3'0"	4'0"	5'0"	5'0"	5'0"	5'0"	5'0"	5'0"	5'0"	5'0"	5'0"	5'0"	5'0"	5'0"	5'0"	5'0"	5'0"	5'0"	5'0"	5'0"
	(90)	(120)	(150)	(150)	(150)	(150)	(150)	(150)	(150)	(150)	(150)	(150)	(150)	(150)	(150)	(150)	(150)	(150)	(150)	(150)
Height of smaller portion of	3'3"	4'3"	6'0"	6'9"	8'3"	8'6"	10'0"	10'6"	11'3"	11'9"	13'0"	14'0"	14'6"	14'9"	15'0"	15'3"	15'9"	16'0"	16'3"	16'6"
outlet chamber (H_2)	(97.5)	(127.5)	(180)	(202.5)	(247.5)	(255)	(300)	(315)	(337.5)	(352.5)	(390)	(420)	(435)	(442.5)	(450)	(457.5)	(472.5)	(480)	(487.5)	(495)
Length of bigger portion of	14'0"	18'0"	22'0"	25'0"	31'0"	31'0"	32'0"	34'0"	36'0"	37'0"	38'0"	40'0"	43'0"	45'0"	48'0"	50'0"	53'0"	55'0"	58'0"	60'0"
outlet chamber (M)	(420)	(540)	(660)	(750)	(930)	(930)	(960)	(1020)	(1080)	(1110)	(1140)	(1200)	(1290)	(1350)	(1440)	(1500)	(1590)	(1650)	(1740)	(1800)
Width of bigger portion of	10'0"	13'0"	17'0"	19'0"	20'0"	23'0"	25'0"	28'0"	30'6"	32'0"	35'0"	35'6"	36'6"	37'0"	37'6"	38'0"	38'6"	38'6"	39'0"	39'6"
outlet chamber (N)	(300)	(390)	(510)	(570)	(600)	(690)	(750)	(840)	(915)	(960)	(1050)	(1065)	(1095)	(1110)	(1125)	(1140)	(1155)	(1155)	(1170)	(1185)
Mixing tank diameter (R)	5'0"	8'0"	8'0"	8'0"	8'0"	8'0"	8'0"	8'0"	8'0"	8'0"	8'0"	8'0"	8'0"	8'0"	8'0"	8'0"	8'0"	8'0"	8'0"	8'0"
	(150)	(240)	(240)	(240)	(240)	(240)	(240)	(240)	(240)	(240)	(240)	(240)	(240)	(240)	(240)	(240)	(240)	(240)	(240)	(240)
Mixing tank height (P)	2'0"	2'0"	3'0"	3'0"	3'0"	3'0"	3'0"	3'0"	3'0"	3'0"	3'0"	3'0"	3'0"	3'0"	3'0"	3'0"	3'0"	3'0"	3'0"	3'0"
	(60)	(60)	(90)	(90)	(90)	(90)	(90)	(90)	(90)	(90)	(90)	(90)	(90)	(90)	(90)	(90)	(90)	(90)	(90)	(90)

Table 14: Estimate of material requirements for the construction of different large-capacity biogas plants

Sl. No.	Material	Unit	Capacity of biogas plant (m^3/d)																			
			25	50	75	100	125	150	175	200	225	250	275	300	325	350	375	400	425	450	475	500
1.	Brick	Nos.	10000	13500	21000	27000	32000	36000	42000	46000	52000	56000	62000	66000	72000	76000	82000	86000	92000	96000	110000	120000
2.	Cement	Bags	110	150	190	220	275	325	400	450	525	575	650	700	775	825	900	950	1025	1075	1150	1200
3.	Brick ballast	m^3	2.5	3.25	4.5	5.25	6.5	8	11	13	16	18	21	23	26	28	31	33	36	38	41	45
4.	Stone ballast	m^3	8	9.5	11.5	13	16	18	21	23	26	28	31	33	36	38	41	43	46	48	51	55
5.	Sand	m^3	11.5	16	27	32	40	45	60	70	85	95	110	120	150	160	175	185	200	210	230	240
6.	Inlet pipe (PVC; 300mm dia.)	m	6	6	6	6.25	6.5	6.5	8	8	9	9.5	10	10.5	11.5	12	12.5	13	14	14	15	15
7.	Steel to cover outlet chamber slab	Kg	200	300	450	550	700	800	950	1050	1200	1300	1450	1550	1700	1800	1950	2050	2200	2300	2450	2500
8.	Feeding system (lump sum)	No.	1	1	1	1	1	1	1	1	1	1	1	1	1	1	1	1	1	1	1	1
9.	Gas pipeline system		As per requirements																			
10.	Digging and pit filling		As per requirement of the structure given in Table 15																			
11.	Shuttering for construction work		As per requirements																			

Table 15: Estimate of cost of construction (Rs. in Lakh) of various large-capacity biogas plants

Sl. No.	Material	Rate (Rs.)	Capacity of biogas plant (m^3/d)																			
			25	50	75	100	125	150	175	200	225	250	275	300	325	350	375	400	425	450	475	500
1.	Bricks	5.5/unit	0.55	0.742	1.155	1.485	1.76	1.98	2.31	2.53	2.86	3.08	3.41	3.63	3.96	4.18	4.51	4.73	5.06	5.28	6.05	6.60
2.	Cement	300/Bag	0.33	0.45	0.57	0.66	0.825	0.975	1.20	1.35	1.575	1.72	1.95	2.10	2.325	2.475	2.70	2.85	3.075	3.225	3.45	3.60
3.	Brick ballast	800/m^3	0.02	0.026	0.036	0.042	0.052	0.064	0.088	0.104	0.128	0.14	0.168	0.184	0.208	0.224	0.248	0.264	0.288	0.304	0.328	0.36
4.	Stone ballast	1100/m^3	0.088	0.104	0.126	0.143	0.176	0.198	0.21	0.253	0.286	0.30	0.341	0.363	0.396	0.418	0.451	0.473	0.506	0.528	0.561	0.605
5.	Sand	1000/m^3	0.115	0.16	0.27	0.32	0.40	0.45	0.60	0.70	0.85	0.95	1.10	1.20	1.50	1.60	1.75	1.85	2.00	2.10	2.30	2.40
6.	Inlet pipe (PVC; 300mm dia.)	1000/m	0.06	0.06	0.06	0.062	0.065	0.065	0.08	0.008	0.09	0.09	0.10	0.105	0.115	0.12	0.125	0.13	0.14	0.14	0.15	0.15
7.	Steel for outlet chamber cover slabs	45/Kg	0.09	0.135	0.202	0.247	0.315	0.36	0.427	0.472	0.54	0.58	0.652	0.697	0.765	0.81	0.877	0.922	0.99	1.035	1.102	1.125
8.	Mixer for feeding material (lump sum)		0.30	0.50	0.50	0.75	0.75	0.75	0.75	0.75	0.90	1.00	1.00	1.00	1.25	1.50	1.50	1.75	1.75	2.00	2.00	2.00
9.	Gas pipeline system	Lump sum (Rs. in Lakh)	0.30	0.50	0.50	0.75	0.75	0.75	0.75	0.75	0.90	1.00	1.00	1.00	1.25	1.50	1.50	1.75	1.75	2.00	2.00	2.00
10.	Digging & pit filling		0.75	1.00	1.00	1.25	1.50	1.50	1.50	1.75	1.75	1.50	1.75	1.75	1.75	2.00	2.00	2.25	2.25	2.50	2.75	3.00
11.	Shuttering for construction work		0.30	0.40	0.50	0.50	0.75	0.75	0.75	0.85	0.95	1.50	1.50	1.50	1.50	1.50	1.50	1.50	1.50	1.50	1.75	2.00
12.	Labour cost		0.50	0.60	0.75	1.00	1.25	1.50	1.75	1.75	2.00	2.00	2.00	2.00	2.25	2.50	3.00	3.25	4.00	4.25	4.50	5.00
	Total cost (Rs. in Lakh)		14.971	15.529	3.403	4.677	5.669	14.971	15.529	9.342	10.415	11.267	14.971	15.529	17.269	18.827	20.161	14.971	15.529	21.719	23.309	24.862

Further readings

Jha AK, Li J, Nies L, Zhang L, 2011. Research advances in dry anaerobic digestion process of solid organic wastes. *African Journal of Biotechnology*, **10(64)**: 14242–53.

Anon., 2014. Progress Report of AICRP on 'Renewable Sources of Energy for Agriculture and Agrobased Industries (Demonstration/Pilot Level Introduction of Renewable Energy Systems)', Indian Council of Agricultural Research. VIII Workshop at GBPUAT, Pantnagar, 29th Oct. to 1st Nov., 2014: 14–23.

Fischer T, Krieg A, 2001. About dry fermentation in Agriculture. *Biogas Journal*, **1**: 12–16.

Grewal NS, Sooch SS, Ahluwalia S, Brar GS, 2000. Hand Book Biogas Tech. PAU, Ludhiana: 106 pp.

Jewell WJ, Kabrick RM, Dell'orto S, Fanfoni KJ, 1980. Low cost approach to methane generation, storage and utilization from crop and animal residues. *In*: American Institute of Chemical Engineers Symposium 'Biotechnology in Energy Production', Chicago, Illinois, Nov. 17, 1980: 1–22.

Mittal KM, 1996. Biogas Systems: Principles and Applications. New Age International (P) Ltd, New Delhi: 412 pp.

Singh B, Sooch SS, Grewal NS, 2006. Study for Travel of Pollution from PAU model biogas plant to the Sub Soil. *Journal of the IPHE India*, **1**: 49–55.

Singh KJ, Sooch SS, 2004. Comparative study of economics of different models of family size biogas plants for state of Punjab, India. *Energy Conversion and Management*, **45(9-10)**: 1329–41.

Singh SK, Sooch SS, Ahluwalia S, 2000. Biogas plants development programme in India. Journal of the Institution of Public Health Engineers India, **1**: 36–43.

Sooch SS, 2010. Biogas plants – an essential part of modern dairy farming. *Renewable Energy (Akshay Urja)*, Newsletter of the MNRE, **3(2)**: 38–39.

Sooch SS, 2012. Ready-recknor of biogas technology. *School of Energy Studies for Agriculture*, Punjab Agricultural University, Ludhiana: 64 pp.

Sooch SS, Gautam A, 2013. PAU kucha–pucca model of biogas plants in Punjab – a case study. *Agricultural Engineering Today*, **37(3)**: 6–11.

Sooch SS, Gupta D, Singh I, 2005. Economic appraisal of biogas for cooking and electricity generation – A case study. *Indian Journal of Environment and Ecoplanning*, **10(2)**: 459–64.

Sooch SS, Gupta D, Singh S, 2005. Comparative study of community biogas plant versus family size biogas plant – a case study. *Indian Journal of Environment and Ecoplanning* **10(2)**: 383–86.

Sooch SS, Jain R, Gupta U, 2009. Large capacity low cost biogas plants for digestion of cattle dung. Coodination Cell, AICRP on 'Renewable Sources of Energy for Agriculture & Agrobased Industries', *Central Institute of Agricultural Engineering*, Bhopal: 1–17.

Sooch SS, Jain R, Khullar NK, 2008. Role of regional biogas development and training centre in uplift of biogas plants in punjab – a case study. *Indian Journal of Environment and Ecoplanning*, **15(3)**: 695–700.

Sooch SS, Jain R, Khullar NK, 2009. Installation of large capacity fixed dome type biogas plants – a case study. *Proceeding International Conference on Changing Environmental Trends and Sustainable Development* (CETAS-2009), GJ University of Sciences and Technology, Hisar, 9-11 February, 2009: 46–8.

Sooch SS, Jain R, Khullar NK, Gupta U, 2009. Design aspects of new modified PAU Janta large capacity fixed dome type biogas plants – a case study. *Proceedings International Congress on Renewable Energy*, Solar Energy Society of India (SESI), New Delhi, 6-7 October, 2009: 255–60.

Sooch SS, Soni R, Singh S, Saimbhi VS, Gautam A, 2013. Renewable energy in Punjab: status and future prospects, *School of Energy Studies for Agriculture*, PAU, Ludhiana: 98 pp.

Vijay VK, 2007. Biogas refining for production of bio-methane and its bottling for automotive application and holistic development. In: *Proceeding International symposium on Eco Topia Science*. Nagoya University, Nagoya, Japan: 23–25.

Vijay VK, Chandra R, Subbarao PM, Kapdi SS, 2006. Biogas purification and bottling into CNG cylinders: Producing Bio-CNG from biomass for rural automotive applications. *In*: *The 2nd Joint International Conference on 'Sustainable Energy and Environment'* (IRCSEEME), Walter de Gruyter GmbH: 1–6.

15

Socioeconomic Analyses of Biogas Technology Towards the Upliftment of Rural India

***Biswajit Paikaray and Snehasish Mishra*[1]**

BDTC, Bioenergy Lab, School of Biotechnology, KIIT, Bhubaneswar, Odisha, India
[1]Corresponding Author: snehasish.mishra@gmail.com

ABSTRACT

It is impossible to envision a world without energy provisions and has become inevitable and an integral part of a nation's prosperity. It has unparallel contributions towards the socioeconomic empowerment, industrial growth, and technological advancements, as also in mitigating greenhouse gas (GHG) emissions. Judicious utilisation of energy sources without infringing upon the sustainable energy demand is the need of the hour. Taking India's huge bovine strength into account, biogas technology has the potential to transform the standard of living of rural millions. Effectively implementing the Govt. of India's National Biogas and Organic Manure Programme (NBOMP) could save precocious demand on electricity, fertiliser and firewood while promoting self-sustainable organic farming. In this chapter, the cost and benefit analyses of biogas technology (involving biogas plants of different capacities) are discussed along with the long-term approach at the grassroots level to promote digested slurry for organic farming. The commercialisation and carbon crediting of biodigester technology is also narrated. It is necessary to chalk out a revenue model to commercialise slurry at the grassroots level to economically empower the beneficiaries, from family-type to industrial-scale biogas plants. Small-scale biofertiliser hubs at district level may be organised to collect dry slurry from the beneficiaries and process them further to make it market-ready through entities operating at the national level developing manure for agricultural practices, as also feeding materials for animal husbandry practices.

Keywords: Family-type biogas plants, Organic manure, Green house gas emission, Carbon credit, Digested slurry, Commercialising biogas byproduct

Abbreviations used

CNG : Compressed natural gas
GHG : Greenhouse gas
FYM : Farmyard manure
MNRE : Ministry of new and renewable energy
BDTC : Biogas development and training centre
CHP : Combined heat power
BPGP : Biogas-based power generation system
UNIDO : United Nation Industrial Development Organisation
KVIC : Khadi and village industry commission
LPG : Liquified petroleum gas

Introduction

With increasing research on biogas, the concept/scope of biogas generation from conventional sources like cattle dung has been widened, backed by technological advances. Recently, research objectives on other aspects of biogas technology including making it more user-cum-environment friendly, contributing towards national energy grid thereby ensuring energy security, and also making it economically accessible across all sections of the society have taken the centre stage. This has become more imminent and imperative with the recent flagship 'Swacchh Bharat' and 'GOBAR(Galvanising Organic Bio-Agro Resources)dhan' initiatives of the Govt. of India.

The evolving biogas sector in terms of application of latest technology has made it simple to adopt, broader in application, and accurate in lab testing. Apart from the conventional cattle dung which is rich sources of methanogens (that is responsible for CH_4 production), various other potential anaerobically biogasifiable substrates that exhibit tremendous potential to harbour active methanogens exist. These are essentially organic wastes/refuse, such as, agro-residues, household wastes, municipal solid wastes (excluding non-degradable ones), and the kitchen refuse. The scope of application of these diverse substrates in anaerobic digestors, prompted researchers across the world to take this technology even to the vulnerable/weaker sections of the society for a much-needed inclusive growth while ensuring energy self-sufficiency at the national level.

Biogas is an energy product that can be obtained anaerobically (in the absence of oxygen) from decomposable organic materials like cattle dung, food and vegetable waste, municipal waste and agro-residues that are abundantly available. The utilisable component in the biogas is CH_4 which can be used directly for cooking (thermal applications), lighting at the household level, and for off-grid applications such as generating electricity and Bio-CNG.

Biogas is a mixture of gases (such as, methane, carbon-di-oxide, and traces of hydrogen sulphide (H_2S), carbon monoxide (CO), ammonia (NH_3), nitrogen oxides (NO_x) etc. and moisture. Reports suggest that the percent composition of biogas is CH_4: 50–75; CO_2: 25–50; H_2S: 0–3; nitrogen, nitrous oxide (N_2O) andhydrogen: 0–1; and water vapour. It is generally agreed that the composition is 55–75% methane and 44–24% carbon-di-oxide, and 1% or less of the mixture of various other undesirable gases (out of which the H_2S is the most obnoxious). The calorific value of biogas is 21.48mJ/m^3, and density is 1.21Kg/m^3. When mixed with oxygen biogas could become explosive, the lower limit being 5% methane and the upper limit being 15% methane. Methane gas is largely responsible for combustion whereas interestingly the other major component carbon-di-oxide (CO_2) is a fire extinguisher. Further, H_2S is a pungent smelling problematic gas with corrosive nature to the devices that runs on biogas. The likely traces of NO_2 and NH_3 are relatively obnoxious unless present in appreciably large amount.

Anaerobic digestion (the soul of biogasification) of organic material primarily involves a series of four microbial reaction stages: hydrolysis, acidogenesis' and methane formation. Initially, the complex organics degrade to simple monomers through hydrolysis. Resultant sugar, alcohols and amino acids are converted to organic acids by the acid-producing microbes. Degradation of organic material involves a range of microorganisms that specialise in reducing the intermediate products. In the next sequence of reaction, these products are microbially transformed to CH_4. The biodigestion depends on the efficiency of the biochemical reactions in these four stages. Efficient digestion with a shorter retention time ensures better biogasification.

Global bioenergy market

The global biogas market is segregated into organic waste from landfills, municipal waste water, commercial organic waste, agricultural waste and energy crops. Municipal wastewater in global biogas market is set to grow appreciably pertaining to intensive population growth and developing food industry on a global scale (Ojha *et al.*, 2015). The demand for domestic and industrial waste management would positively drive the sewage gas technology across the world.

Initiatives by the governments towards waste management coupled with the demand for cost-effective clean energy shall positively drive the global renewable resource industry by 2024. Further, the depleting conventional resource is poised to steer the global biogas market due to the rising energy demands. The inconsistency in composition and complex facility designs to handle wastes is likely to hamper global biogas market, though. The technology pertaining to waste handling and improved treatment to optimise the generation rates have become more imperative as urbanisation and economic growth diversify.

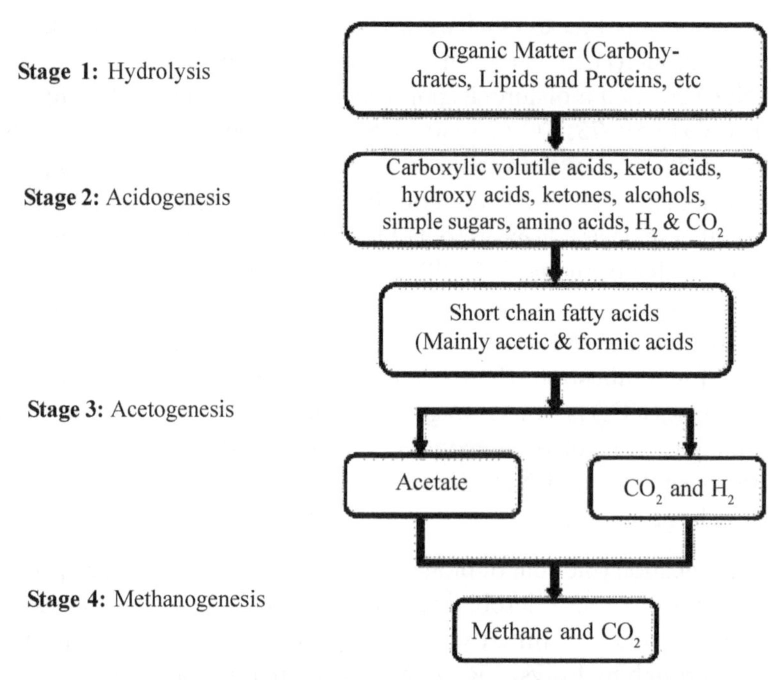

Fig. 1: Stages of bioprocessing in biogasification (*Source*: Vaish *et al*., 2017; Ojha *et al*., 2012)

With a major objective to curb greenhouse emissions, government-led stringent regulation is expected to provide substantial growth to global biogas market. Commercial biogasification of organic wastes is predicted to grow notably owing to rapid industrialisation and growing infrastructure across the globe.

The vertical and horizontal growth of biogas production is dependent on the plant capacity and availability of various local feeding materials. Biogas production by individual digestor also varies greatly owing to the various operational factors. As a result, there is no specific dependable database or standardised parameter to measure the growth rate for biogas production in India. However, there are sporadic growth rate estimates and projections of family-type biogas plants installed in the country. Further details on the biogas potential and its recent advances in India are discussed later. This social justice and equality programme ('biogas programme') is being steered by the Ministry of New and Renewable Energy through the central flagship scheme the New National Biogas and Organic Manure Programme (NNBOMP), formerly known as the National Biogas and Manure Management Programme (NBMMP). About 49.6 lakh household biogas plants against the estimated potential of 2.1 crore have been installed since the inception of the biogas programme in the country till 31st March 2017. The figures of the last six years are detailed in Table 1.

Table 1: Installation of family-type biogas plants in the last six years under NBMMP (MNRE)

Year	Target allotted (units upto 6m^3)	Achievement (units ranging up to 6m^3)	% Installation (against the target)
2012-13	1,35,000	1,15,377	85
2013-14	1,06,000	83,540	78
2014-15	1,10,000	84,882	77
2015-16	1,11,000	47,490, as by 31/01/2016	*42
2016-17	1,00,000	57504	57
2017-18	65,180	20,125, as by 31/12/2017	*30

*% installation may slightly go up due to increase in the final figure beyond 31st March of the financial year

For the European Union, the German biogas market provides moderate scope with growing climatic concern in compliance to security of supply and government initiatives. In 2016, German Biogas Association (GBA) upgraded the capping for the technology installations from 150MW in 2017–2019 to 220 MW per year from 2020–2022. As the regulatory protocols and demand to effectively implement waste management escalates, it is anticipated that use of landfill gas would substantially grow. The US-based Mas Energy LLC collaborated with the Republic Services Inc. in 2016 to set up three landfill gas plants adhering to the operational dynamics of the Republics Hickory Ridge Landfill, Georgia.

The regulated reforms and cogeneration technology for electricity generation coupled with the greenhouse emission legislations are providing thrust to this technology in Chinese biogas market too. China has announced incorporating 3150 large-scale methane and 172 new waste-to-energy projects in its 13th five-year plan (2016–2020), at an estimated investment of about US$7.3 billion. The Asian Development Bank (ADB) funded waste-to-energy (WTE) projects of Dynagreen Environmental Protection (China) in 2017. The Australian government has also launched a US$2 million programme in Victoria to support waste-to-energy technologies in the same year. The Singapore government initiated directives pertaining to carbon emission reduction by 36% below 2005 level by 2030, as recently as in 2016.

Biogas market in Africa is poised to witness substantial growth due to the rapid growth in cogeneration technology and security of supply reforms across off-grid remote areas. In 2017, the United Nations Industrial Development Organisation (UNIDO) launched an industrial waste generated biogas project in South Africa in collaboration with the Global Environment Facility, Washington. In this year, the Africa-based farm Biojoule installed an on-grid biogas powered CHP power plant that supplies electricity and heat to the gorge farm in Kenya.

Few key global biogas market players are, A2A Energia (Italy), AEB Amsterdam (the Netherlands), Babcock & Wilcox, North Carolina (the US), Veolia Environnement S.A. (France), Shenzhen Energy, Guangdong (China), Aterro Recycling Pvt. Ltd. (India). Wide technological offerings, effective turnkey project implementations as also mergers and acquisitions are the key market penetration strategies for many of these companies.

Indian biogas scenario

In India, biogas market offers a wider scope owing to the highly supportive government initiatives and intensive demand in this huge country. The technology which was once restricted to the rural India has now spread in the urban and industrial belts. However, the Government support and thrust is still restricted to the economically backward areas and population. For instance, the Tambaram Municipality launched a sewage gas based biomethanation plant in 2013 to generate methane from public toilet waste in Tamil Nadu. Additionally, there are numerous research and development organisations in India that are either working solely on alternate energy, or other research priorities combined. In 2017, the Government has reduced the basic customs and countervailing duty levied on the balance of systems operational in biogas, hydrogen byproduct and biomethane.

Indian villages exhibit tremendous bioenergy potential owing to the available huge quantity of cattle dung and agro-wastes. Harnessing such energy and delivering it for household and industrial use is the need of the hour. KIIT-BDTC (as also other seven BDTCs in India and the state renewable energy agencies and the Khadi and Village Industry Commission, KVIC, working closely with the stakeholders) through its training courses, encourages beneficiaries to adopt biogas technology for multiple benefits, and is striving to create awareness about the damaging consequence of environmental emissions and the unhealthy disposal practices of agro and other biodegradable wastes. With minimal maintenance, a biogas plant can perform up to 20 years with one time investment. The expenses incurred in the purchase of LPG cylinders and payment of electricity bills for households could well be compensated to a great extent through a family-type biogas plant. A family-type biogas plant at the household level is enough to cater to the cooking needs of 4-5 members. Also, if treated and administered properly in agricultural fields, the slurry of the plant becomes useful as biofertiliser/manure. The Ministry of New and Renewable Energy (MNRE), Govt. of India through its popular public welfare scheme National Biogas and Organic Manure Programme (NBOMP) extends Central Financial Assistance (subsidies) of upto Rs.12,000.00 (for a 2m^3 biogas plants) for installation.

Biogas potential and its recent advances in India

As per MNRE database, 3.22 lakh plants were installed during the 12th Plan period (upto 31.12.2015). The estimated average daily biogas generation capacity of these cumulatively is about 6.46 lakh m^3. This means that they give an estimated average annual savings of about 70.90 lakh LPG cylinders equivalent while simultaneously producing annually about 88.4 lakh tonnes of organically enriched biomanure (about 31,100 tonnes urea equivalent). Alternately, during the last three years, the daily saving by these installed plants could be about 4.00 lakh litres kerosene equivalent. This also means that, there would be a direct annual saving of about 8.20 lakh tonnes firewood, thereby helping in preventing emission of about 1,615,000 tonnes of CO_2 annually.

Upto December 2015, the cumulative figure of the installed family-type biogas plants stood at 48.7 lakh, about 39.58% achievement against the estimated potential of 2.2. Installation of 1,11,000 family-type biogas plants more could result in an estimated annual saving of about 2,81,498 tonnes of firewood equivalent and production of about 10,700 tonnes of urea equivalent. The digested biogas slurry would give 30,38,625 tonne organic manure annually. It is estimated that the related activities including construction and commissioning would generate about 33.0 lakh person-days of employment for skilled and unskilled workers in rural and semi-urban localities.

The figures (Figs. 3–4) below are compiled from the information collected from various reliable sources, like, National Biogas and Manure Management Programme, Comptroller and Auditor General (CAG) Report on NBMMP renewable energy sector in India, 2015 (https://cag.gov.in/sites/default/files/audit_report_files/Union_ Civil_ Performance_ Renewable_ Energy_Report _34_2015_ chap_8.pdf) Chapter-viii; https://www.indiastat.com/power-data/26/non-conventional-energy/184/biomass-biogas-1985 2019/452705/stats.aspx owned by Datanet India of the Ministry of state, Department of statistics, planning and public grievances (https://www.indiastat.com/aboutus/ourwebsites.aspx); and the Ministry of New and Renewable Energy, India (http://pib.nic.in/newsite/pmreleases.aspx?mincode=28). It appears that still there are numerous gaps and inconsistencies in the existing data on the manner the biogas technology extension activities are building up and unfolding year after year. Although the trends are clear and are very encouraging, however the activities need to be expedited through various policy, administrative and coordinated-approach interventions.

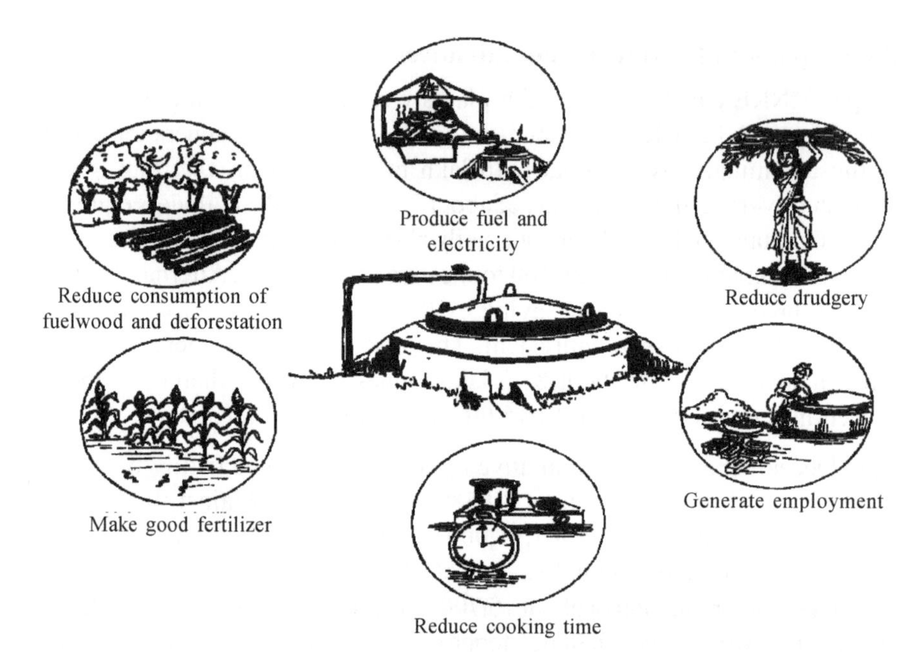

Fig. 2: Multiple benefits from a biogas plant

It is observed that, about 49.6 lakh household plants have been installed till 31st March 2017 against the estimated potential of 2.1 crore since the inception of Biogas Programme in India.

As observed in the figure, there is a decline in the trend particularly in the *fy* 2015-16. For this as well as to expedite the biogas extension activities in India the possible cited reasons which may call for attention are the following:

1) The central National Biogas and Manure Management Programme scheme was revised as the New National Biogas and Organic Manure Programme during 2017-18. The process was initiated during 2015-16 which could have consumed the productive time of various stakeholders.

2) Regulatory measure taken for direct benefit transfer (DBT) to the beneficiary during this period (2015-16) might have also consumed the working months. This could have had spilled over effect during 2016-17 and 2017-18 as well. However, it is slowly but steadily getting stabilised each subsequent year.

3) Inadvertent administrative delay in the disbursement of CFA from the MNRE to State Nodal Agencies (SNAs), and other stakeholders.
4) Delay in CFA disbursement from the State Nodal Agencies to the beneficiaries.

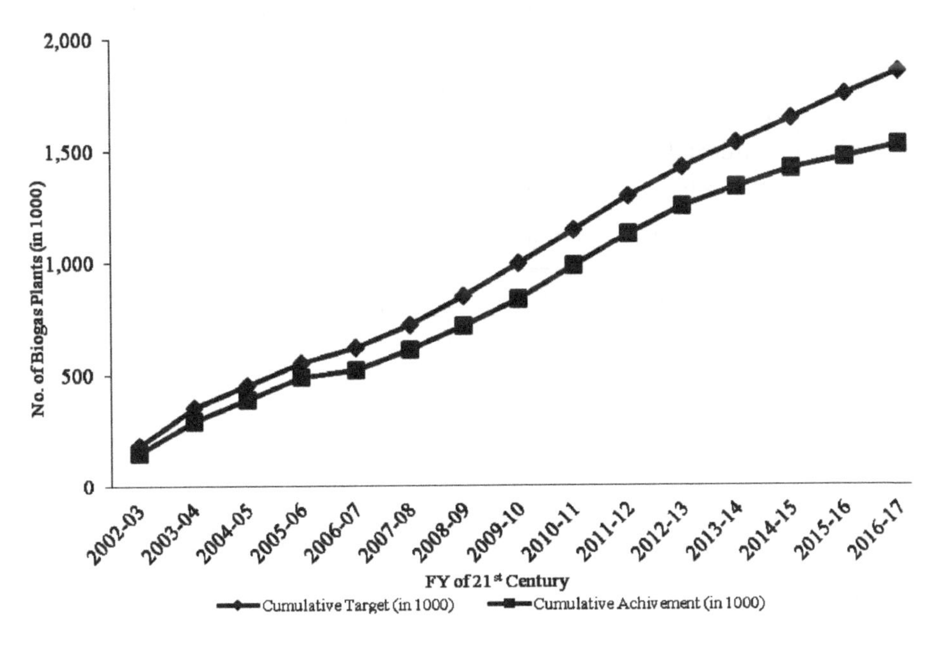

Fig. 3: Graph representing the trend of the cumulative target vis-à-vis installation of the family-type biogas plants in (rural) India in last 15 years through MNRE initiatives

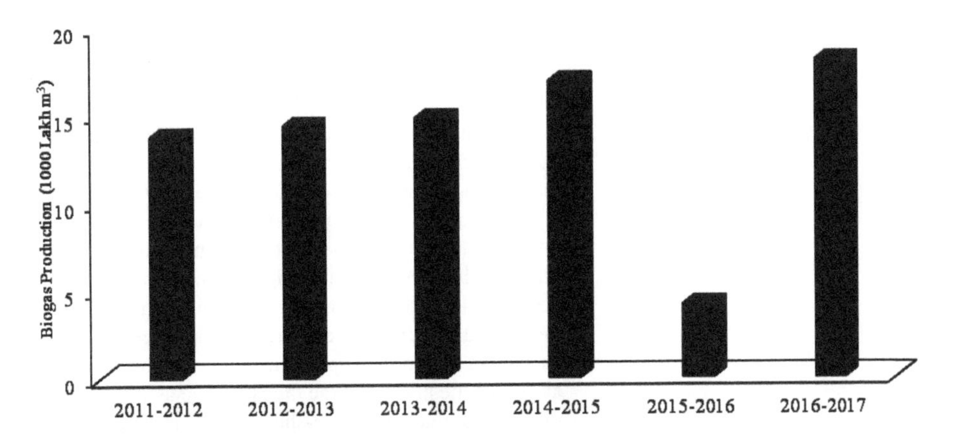

Fig. 4: Graph showing cumulative biogas production trend (1000 lakh m^3) in India in the last six years

5) Delayed payment to the turnkey workers by the State Nodal Agency.
6) Non-uniform geographic and diverse climatic factors.

Economics of biogas technology - family-type biogas plants

Depending on the capacities, the family-type biogas plant models (both fixed-dome and floating-drum digesters) approved in India range from 1–6m^3. In addition to the conventional ferro-cement and brick-masonry biogas plants, pre-fabricated models made of High Density Polyethylene (HDPE) and Fibre-glass Reinforced Plastic (FRP) and Reinforced Cement Concrete (RCC) material are also considered. Some hugely popular biogas models in India are Deenabandhu, KVIC, Pragati, Janata, Konark and Nisarguna models.

Table 3: Some approved models of biodigesters in India (MNRE) in operation

Sl.No.	Name of Plant (1–6m^3)	Specification
1.	Fixed-dome Deenabandhu	Deenabandhu fixed-dome model with brick masonry constructionDeenabandhu ferro-cement model with *in situ* techniquePrefabricated RCC fixed-dome model
2.	Floating-drum KVIC plant	KVIC floating-drum steel metal dome with brick masonry digesterKVIC floating-drum type plant with Ferro-Cement digester and FRP gas holder
3.	Pragati model	Floating-drum type
4.	Prefabricated plant	Prefabricated reinforced cement concrete (RCC) digester with KVIC floating drum
5.	The Nisargruna model	It is a promising technology for municipal corporations, dairy farms, public canteens, etc.Basically, it is for industrial-level operations to handle large quantity of food (and other such organic) wastes, municipal waste and other biodegradable refuses.Setting up a 2m^3 Nisargruna plant may not be viable for small /medium farmers in rural areas due to the limited cattle dung and kitchen waste available with themThe details of the model are discussed elsewhere.
6.	The Janta Model	It is a family-type biogas plant constructed without the use of steel, with no moving part, and therefore, the maintenance cost is low compared to the KVIC model.The principle of gas production is same as that of the KVIC and Deenabandhu Models.The plant consists of an underground hemispherical digester made of bricks and cement with a dome-shaped roof which remains below the ground level.There are two rectangular openings facing each other and coming up to a little above the ground level at almost the middle of the digester, one as inlet and the other as outlet.The dome-shaped roof is fitted with a pipe at its top which is the gas outlet of the plant.The

Contd.

	pressure of the collected biogas in the space of the fixed dome is much higher (around 90 cm water).
7. The Konark Model	This model is a spherical digester having enough empty space to accommodate more gas in strained conditions.The spherical structure remains firm with no residual effect on the structure as it is loaded from the convex portion (the internal load due to the self-weight of the slurry and the gas pressure are counterbalanced).Based on a study indicating usual gas consumption pattern of rural households, high gas storage capacity was considered.The storage capacity is kept at 50% to match the demand of the user and zero-consumption period.Construction material is site-specific (can be constructed with all types of building materials) although it is usually similar to the construction of fixed-dome biogas plants. The ferro-cement constructed one is encouraging on economy and cost effectiveness.

Estimate of the volume of a plant to construct as per available inputs

The volume of the biogas plant is expressed in cubic meter (m^3). A $2m^3$ biogas plant means that the plant generates $2m^3$ gas in 24h. The three major factors necessary to consider while estimating the biogas plant's capacity are: 1. Daily availability of cattle dung to feed a plant, 2. Gas requirement for the beneficiary, and 3. Location aspects including the agro-climatic conditions where the plant is to be installed. Table 3 lists some of the commonly accessible feedstock at the household level for the daily operation of a family-type biogas plant.

Table 4: The mean average household organic inputs available for feeding a family-type biogas plant

Organic input	Available per head (in kg)	Gas yield/kg (in m^3)
Cattle dung (Hybrid/local)	12/8	0.037
Pig	0.9 (900g)	0.1
Sheep and lamb	0.8 (H≈1.0kg)	0.01
Human night soil	200g	0.05
Kitchen waste	1.0	0.02 (H≈0.04)

Generally for every $1m^3$ plant, the heat generation is estimated at 4700–6000kcal, the density of the biogas is 0.94, and the temperature of the (burning) blue flame is 870°C. $1m^3$ of biogas is equivalent to 0.62l kerosene, 3.47kg firewood, 12kg dry cakes, 1.6kg coal, 0.8l petrol, and 0.43kg LPG. Based on the beneficiary's requirement and the available feed resource, a survey work for the size/capacity of the biogas plant to be constructed is carried out on specific queries such as: 1. Number of cattle heads with the beneficiary, 2. The purpose of gas (*e.g.*, for lighting, cooking, etc.), 3. The type of biogas plant suitable to construct (considering the quality and quantity of the available feed and the soil characteristics of the location), 4. Estimated expenditure, 5. The amount of slurry (digestate) likely to be generated, and its likely use, and 5. The scope for plant operation and maintenance.

Tables 5 and 6 detail the basic minimum requirements of organic feedstock (with cattle dung as a case example) for uninterrupted running of biogas plants of various capacities (1–6m³).

Table 5: Recommended plant feeding for initial recharge (for commissioning)

Plant capacity (m³)	Hybrid cattle	Local cattle	cooking	Lighting	Compost (tonne/annum)
1	2–3	3–4	4	1	2.25
2	4–5	7–8	7	2	4
3	6–8	11–12	10	4	7
4	9–11	16–18	14	5	10
6	17–18	30–34	30	10	20

A typical 2m³ household plant would need a total feeding of 50kg X 365 = 18,350kg or annually 18.35t (with a daily feeding of 50kg solid + 50l water). After the decomposition and biogasification is completed, the digestate generated would be H≈4.95t (18.35 X 0.27) manure, considering that 0.27kg compost would be generated per every tonne of organic input. Table 7 provides a list of possible benefits that could be derived from a biogas plant (Deenbandhu model as a case) at the household level.

Recent R&D works on biogas technology in India

As mentioned earlier, dedicated research and development bioenergy institutes are active in India. Their activities include, but not necessarily restricted to, 1. development, testing and standardisation of alternate plant construction materials to reduce the unit cost of setting-up of a plant, 2. optimisation and standardisation of plant design and process for alternative and mixed feeds for gas production, 3. qualitative approach to treat, handle, and commercialise the slurry as a readymade bioorganic fertiliser at national level, 4. approaches for advanced landfill management, and 5. development of engines that works on BioCNG. However, the entrepreneurial translational research in this sector is still in its infancy. On the economy front, based on following commercial aspects of biogas technology, a potential market for the trading of biogas and its value-added products may be developed:

Table 6: An approximate calculation to set-up a biogas plant at the household level and the estimated slurry generation (from a Deenbandhu biogas plant)

No. of person (assuming 2 children equivalent to 1 adult)	Estimated consumption (m^3; 0.27m^3 per capita consumption)	No. of cattle heads	Dung generated (in kg)	Estimated gas produced (m^3 gas produced per kg, as per above table)	Suggestion(s)	Remark	Compost generation[2]
1	2	3	4	5	6	7	8
(4P + 3C) or 5.5A	5.5 X 0.27 = 1.48	7 (3HC + 1LC + 3CL)	3 X 12 = 36 1 X 8 = 8 3 X 3.5 = 10.5 Sum = 56	54.5 X 0.037 = 2.01 1.5 X 0.02 = 0.03 Sum = 2.04 (KW=1.5)	1. 2m^3 is suggested 2. (1.48/2.04) X 100 = 73%	Remaining 27% of of the gas may be used for lighting a bulb for 1 hr in the evening (10–15% annual saving on electricity bill)	Daily feeding: 56kg + 60l water Monthly feeding: 56kg X 30 = 1680kg Annual feeding: 1680kg X 12 = 20160kg Compost yield: 20160 X 27/100 = 5443kg (5.4 tonne)
(4P + 5C) or 5.5A	5.5 X 0.27 = 1.48	7 (4 HC + 3 CL)	4 X 12 = 48 3 X 3.5 = 10.5 Sum = 60	58.5 X 0.037 = 2.16 1.5 X 0.02 = 0.03 Sum = 2.19 (KW = 1.5)	1. 2m^3 is suggested 2. (1.48/2.19) X 100 = 68%	Remaining 32% of the gas may be used for lighting a bulb for 1.5 hr in the evening (15–18% annual saving on electricity bill)	Daily feeding: 60kg + 60l water Monthly feeding: 60kg X 30 = 1800kg Annual feeding: 1800kg X 12 = 21600kg Compost yield: 21600 X 27/100 = 5832kg (5.8 tonne)
(7P+ 3C) or 8.5A	8.5 X 0.27 = 2.29	10 (6 HC + 4 CL)	6 X 12 = 72 4 X 3.5 = 14 Sum=87.5	86 X 0.037 = 3.18 2 X 0.02 = 0.04 Sum = 3.22 (KW= 2)	1. 3m^3 is suggested 2. (2.29/3.22) X 100 = 71%	-	Daily feeding: 87.5kg + 90l water Monthly feeding: 87.5kg X 30 = 2625kg Annual feeding: 2625kg X 12 = 31500kg Compost yield: 31500 X 27/100 = 8505kg (8.5 tonne)

Contd.

No. of person (assuming 2 children equivalent to 1 adult)	Estimated consumption (m^3; $0.27m^3$ per capita consumption)	No. of cattle heads	Dung generated (in kg)	Estimated gas produced (m^3 gas produced per kg, as per above table)	Suggestion(s)	Remark	Compost generation[2]
(11P+5C) or 13.5A	13.5 X 0.27 = 3.64	20 (11C + 4 CL + 3 SP + 2 Pig)	11 X 8 = 88 4 X 3.5 = 14 3 X 1 = 3 2 X 0.9 = 1.8 Sum= 110.3	102 X 0.037 = 3.77 3 X 0.01 = 0.03 1.8 X 0.1 = 0.18 3.5 X 0.02 = 0.07 Sum = 4.05 (KW= 3.5)	1. 4 m^3 is suggested 2. (3.64/4.05) X 100 = 90%	-	Daily feeding: 110.3kg + 110l water Monthly feeding: 110.3 X 30 = 3309kg Annual feeding: 3309kg X 12 = 39708kg Compost yield: 39708 X 27/100 = 10721kg (10.7 tonne)

Figures are approximate values; HC-Hybrid Cow, LC-Local Cow, CL-Calf, P-Person, C-Children, A-Adult, SP-Sheep.

Table 7: Benefits from household-level Deenbandhu biogas plant[3]

Plant capacity	Substrate requirement	Avg. gasyield/hr	No. of dependants	Savings	Additional profits
$2m^3$	Cattle dung+Kitchen waste (1:1)	0.463mg oil equivalent	04	1. Saving on 16.4kg LPG per month and firewood needs 2. Lighting a lamp for 1½ hr in the evening	1. Digestate for use in own paddy fields 2. Locally sell the digestate at Rs.2–3/kg 3. Ensure hygienic environment 4. Respite from heart and eye diseases among the rural folks

1. Increasing the share of renewable energy from the biogas sector to the total energy mix in the country, at the national level.
2. Encouraging the use of biogas at household and industrial level to reduce consumption of subsidised LPG, electricity and petrol/diesel; on successful implementation, it would help the nation reduce the subsidy budget in 5–10 years.
3. Promoting biogas off-grid technology in rural/peri-urban areas to reduce the pressure on conventional power plants.
4. Legislating, formulating and designing India as a global biogas-biofertiliser hub. The huge amount of slurry from family-type to industrial plant may be processed by the fertiliser companies and made available similar to urea, DAP fertiliser, etc. It would set the stage for biofertiliser-based market at national level where both the producer (beneficiary) and the processor (companies) benefit.
5. Forming vermicompost market at local level; it could encourage organic farming practices. Vermicomposting is value-added commercial slurry with treatments at or below 20°C in a manure yard of at least $2X5ft^2$ dimension adjacent to the outlet chamber of a family-type biogas (such as the Deenabandhu model plant).
6. Scoping the use of treated biogas digestate as a feeding material for aquaculture, etc.
7. Mechanism to distribute bottled biogas units under the midday meal scheme in schools; it could reduce the cooking cost.

Percent NPK of the digestate

The benefits derived from a biogas plant are not only in terms of the useful bio-energy but also the manure generated in the process. It is noteworthy that the biogas digestate is rich in NPK by more than four times compared to the ordinary cattle dung. The average percent N, P and K contents of the farmyard manure (FYM) is about 0.5, 0.2 and 0.5 respectively. The nutritional composition of a typical biogas digestate is presented in Table 8.

Table 8: Nutritional status of the biogas digestate

Nutrient	Percent content
Nitrogen (N)	1.50–2.00
Phosphorus P (P_2O_5)	1.00–1.50
Potash K (K_2O)	1.00

Thus, biogas digestate could replace chemical fertilisers to a great extent, especially in view of the fact that the country spends valuable foreign exchange to import chemical fertilisers, especially P and K. Other established benefits in addition to the nutritional benefits are that, biogas digestate facilitates N_2-fixation, improves soil porosity, structure and texture.

Livelihood supported biodigestor (biogas plant) of 25–85m^3 capacity

When developing nations like India are on the path of escalating GDP growth, expanding urbanisation and growing consumerism, bioenergy can occupy the driver's seat. It exhibits tremendous potential towards socioeconomic advancements for future. This energy and its allied sector account for 35% in terms of contribution towards revenue generation to the state exchequer, and economic empowerment of people (based on field surveys and the inputs from various stakeholders). Thus, it is designated as a pivotal component in the socioeconomic infrastructure of the country serving as an important input to small, medium and large-scale industries. Table 9 below provides a bird's eye view of the financial benefits the technology offers at the micro- (household) level.

Livelihood-supported economically feasible biodigestor (LSE-B), a case study

A 25m^3 capacity LSE-B can be installed (in about 1200–1500ft^2 area) in a rural set-up having adequate bovine population, water points, and requisite space. For the operation and maintenance of the digester, the involvement of 12 stakeholders (basically members from SHGs and small farmers' society) is proposed. The stakeholders are imparted maintenance training after successful commissioning of the LSE-B unit. The following are the point-wise prequalifying conditions:

1. Stakeholder's eligibility: 12–15 members of self help group (SHG) and small farmers having overall cattle strength of 70–80 heads.
2. Each member should have the ability to read and write having exposure to rural banking.
3. Feedstock availability: About 650kg cattle-dung and 50–60kg of other biodegradable household wastes is required on a daily basis.
4. Land: Suitable govt. land of about 1200–1500ft^2 area (with prior approval from Sarpanch/ other authority) adjacent to the households not exceeding 60ft from the place of installation, constituting a cluster.
5. Instruments requirements: a) A Generator set (5HP dual fuel/2HP electrical) unit having power generation capacity of 3.5kW, b) a dual rotor blade to

Table 9: A comparison of the household expenses on different energy sources vis-à-vis biogas

Energy source	Calculation	Compared economic benefits
Biogas	Cost of construction and commissioning of a $2m^3$ fixed-dome plant: Rs.22,000 Cost for 5 years (including operational cost): Rs.22,000 + Rs.2,000 = Rs.24,000	Daily feeding: 50kg + 50litre water Monthly feeding: 50kg X 30 = 1,500kg Annual feeding: 1,500kg X 12 = 18,000kg Annual compost/organic manure yield: 18,000kg X 27/100 = 4,860kg (4.8 tonne) Revenue generated on selling compost @2/kg is: 4,860kg X 2 = Rs.9,720.00 (US$147) annually or Rs.810.00 (US$12) monthly Thus, in 5 years: Rs.9,720.00 X 5 = Rs.48,600 (US$736) Payback: The cost of installation is recovered from the first 2½ years from the date of commissioning. *Note*: Compost may sell at a premium if value-added through vermicomposting or as granular dried material for feeding to poultry, aquaculture and horticulture. In such a scenario, above calculations might change accordingly.
Firewood	Cost of firewood per quintal: Rs.250 Monthly expenditure @ 7kg/day = 2.1 quintals = Rs.525 Annual expenditure @ 2.1 quintals/month = 25.2 quintal = Rs.6,300 (US$100) Thus, 5 years expenses: Rs.31,500 (US$477)	Yearly savings: Rs.6,300 (US$100) In 5 years: Rs.31,500 (US$477) along with the sale of organic manure in local/regional market.
LPG	Monthly refill cost: Rs.480 (at current subsidised rate) Annual cost: Rs.480 X 12 = Rs.4,800 (US$77) Thus, 5-year expenses: Rs.4,800 X 5 = Rs.24,000.00 (US$363) So, with biogas plant one saves 80% annually on at least two refills	If beneficiary installs a $2m^3$ biogas plant,there would be saving of Rs Rs.4,800 (USD77) yearly and Rs.24,000.00 (USD363) in 5 years on account of cooking expenses along with sale of organic manure in regional market

Contd.

Energy source	Calculation	Compared economic benefits
Electricity	Cost of one unit at rural household level: Rs.3.00	A $2m^3$ biogas plant lights 2 bulbs for 2hrs (assuming that every 8–10hrs consumes a unit). Thus, monthly saving on electricity would be 6 units = 6 X Rs.3 = Rs.18 **....B**
	Avg. monthly expense: Rs.3 X 60 units/month = 180 + others = Rs.250 (US$4)	A water-heater consumes a unit every 30 min. So, consumption in warming water is 10 units a month = 10 X Rs.3 = Rs.30 **....C**
	Annual expense: Rs.250 X 12 = Rs.3,000 (US$45) **...A**	Thus, the annual saving on electricity would be: (18 + 30) X 12 = Rs.576 (US$9) **....D** So, it is an annual saving of around 20% on electricity bill.

chop vegetable wastes and other easily accessible biodegradable waste available free-of-cost, c) A scrubber unit, d) a lamp post, and e) a 1.2HP water (or slurry) pump.

Socioeconomic benefits of the LSE-B

The set-up could ensure a hygienic environment especially in terms of removing fermentable organic residues and reduction of odour, not only at the households but also the nearby areas since the various household wastes are used in the plant on mass-scale apart from the daily cattle-dung feed requirement. Utilisation of the digested slurry in agricultural practices reduces dependence on fertilisers (NPK, Di-ammonium phosphates, etc.) and, consequently such initiative leads to consolidating the soil nutrients/fertility for years to come. Further, as the dependence on firewood decreases, reforestation at remote locations may be promoted through this. The cleanliness with the successful operation of the LSE-B ensures prevention of contagious and vector-borne diseases among inhabitants/rural folk, reduces the vulnerability to retina (cataract), cardiovascular, dysentery, diarrhoea and respiratory complications thereby reducing the incidence of occurrence of such diseases drastically.

Industrial interactions & cooperation

Large- (commercial) scale biogas plant could be set in urban/semi urban/ metropolitan areas wherever a large quantity of organic biomass is stashed in municipal areas like waste dumps, and public utilities. Industrial interventions could be realised by outsourcing such huge amount of biomass for recycling and biogas generation through Public Private Partnership (PPP) mode. The source of institutional credit to install portable- to industrial-type biogas plants may also be augmented not only from the banking sector but also from biofertiliser companies, bioconsultants, the intermediaries and various other stakeholders including start-up funds. The commercialisation and industrialisation to this effect will lead to the production of green energy while minimising the environmental menace.

In order to make it commercially viable and technically acceptable, the Technology business incubation units (TBIs) may be promoted in urban areas inviting young researchers, experts and consultants to develop the state-of-art technology as well as formulate marketing strategies of the biogas plant outputs. The livelihood factor may be realised after the commissioning of the setup but the commercial aspect of the plant will have to be balanced with the initial nominal investments from beneficiaries in terms of providing space for common purpose, and human capital.

Table 10: Benefits of biogas technology at the community level

Capacity of plant	Substrate required	Ag. hourly gas yield	No. of dependants	Savings	Economic benefits
>$25 m^3$	Cattle dung+Kitchen waste+agro-residue+Water (@1:1)	0.463mg oil equivalent	20–30 families	1. Savings on LPG/ fuel-wood consumption 2. Partial savings on electricity; lights a lamp for 2hrs 3. Savings on fertilisers	1. Slurry can be commercialised. 2. Hygienic environment. 3. Avoiding cardio-vascular diseases and cataract 4. Eco-friendly

Bank loan for installation of biogas plants

Scheduled commercial (public sector) banks, NABARD (through its central and district cooperative banks, primary agricultural societies, self-help groups), Regional Rural Banks (RRB's) need to be proactively involved in the implementing NBOMP by pursuing with them to disburse short-term credit at concessional rates to install family-type plants. A beneficiary of a $2m^3$ plant is entitled to receive a loan of approx. Rs.10000.00 against the total installation cost of approx. Rs.19000.00 over and above the CFA/subsidy (Rs.9000.00) awarded to the beneficiary, subject to fulfilling certain criteria/parameters (bovine strength, credit worthiness of the beneficiary, etc.) as set by the SNAs/BDTCs/ KVIC operating in the region. The guidelines from the stage of identification of beneficiaries to the commissioning of plants must be framed out in close liaison among officials of the SNAs, BDTCs, KVIC and bank officials. A detailed scheme to avail financial support from banking institutions is provided separately as an annexure (towards the end of the volume).

Environmental implications with other socioeconomic concerns

Capturing CH_4 from substrates which would inevitably emit it to the atmosphere, and utilising it for beneficial purposes would obviously reduce the GHG impact. Biogas energy is considered carbon neutral since carbon emitted by its combustion comes from the organic matter that fixed the carbon from the atmospheric CO_2 in recent times. Additionally, biogas utilisation would reduce the use of fossil fuels, a principal contributor of green house gases and thereby global warming.

Biogas doesn't have any reported adverse impact on the user unlike the firewood which increases the risk of vision loss, respiratory and cardiovascular health issues. Further, reduced CO_2 level at the household due to the use of biogas would ensure a healthy and balanced environment. Also, biogas option paves the way for restoration of greenery as the dependency of the poor households around on the firewood would reduce. From the viewpoint of affordability of biogas, installing a biogassifier is assumed as the choice for rural India by default in terms of its material cost and its adoptability compared to other energy sources. For instance, as per the prevalent market prices, a $2m^3$ biogas plant saves 26kg LPG or 37l kerosene or 88kg coal or 210kgs of firewood every month. The digested slurry obtained from the digester, it is nutrient-rich, odourless and nontoxic, and a safe and useful agri-input.

Being one of the largest importer of petroleum and natural gas in the world, universalising the use of biogas in India at national level would drastically/ significantly reduce our dependence on petroleum and allied products, and also

Table 11: Central Financial Assistance (CFA) of the Ministry of New and Renewable Energy (MNRE), Govt. of India to promote family-type biogas (1–6m^3) plants

Particulars of the CFA	Biogas plants under NBOMP				
CFA/Subsidy applicable per plant (in Indian Rupee)	1m^3	2–6m^3	8–10m^3	15m^3	25m^3
North-Eastern states including Sikkim and SC/ST category of Sikkim	17000	22000	24000	25000	35000
Jammu and Kashmir, Himachal Pradesh, Andaman and Nicobar Islands and SCs and STs in all states	10000	13000	18000	21000	28000
General category for all states	7500	12000	16000	20000	25000
Additional subsidy if biogas plant linked with toilet	1600	1600	1600	Nil	Nil

would save revenue to the state exchequer in the form of forex reserve. Popularising biogas would create local/regional employment opportunities for semi-skilled workers and masons in construction and maintenance of biogas plants.

Table 12: Biogas potential of various organic residues in rural India and their biogas potential

Agro residue	Gas yield (l/kg VS)	Methane (%)
Wheat straw	200-300	50-60
Rye straw	200-300	59
Barley straw	250-300	59
Oats straw	290-310	59
Corn straw	380-460	59
Elephant grass	430-560	60
Sunflower leaves	300	63
Algae	420-500	47
Sewage sludge	310-470	-
Animal dung	Gas yield (l/kg VS)	Methane (%)
Pig	340-550	65-70
Cow	90-320	65
Poultry	310-620	60
Horse	200-300	42
Sheep	90-310	48
Barnyard	175-280	51

Larger size biogas plants

Industrial biogas plants having a capacity more than 85m^3 can generate more than 8–10kW power that is crucial to address the power requirement of cottage/ micro industry, dairy firm (goshala), rice mill, or public canteen. The MNRE extends central financial assistance subsidy of Rs.40,000/kW to the beneficiary of a biogas-based power generation plant (BPGP), subject to the declaration made by the beneficiary of using it 8–10hrs a day.

Power generation through integrated solid waste management

The economic development of a country heavily depends on its industrial growth and sustainable utilisation of natural resources. Industrial advancement without depleting the natural resource treasure has certainly become a challenge and biogas technology seems to be an effective solution. Considering the volume of municipal waste generated in urban areas, Indian cities offer tremendous potential of power generation from bio-waste. Thus, it reduces the energy requirements by at least 30% from the grid. The intensity of greenhouse gas emissions also can significantly be reduced by adopting bio-CNG in automobile and bottled-CNG in cooking and elsewhere.

Increasing milk demand forced Indian dairy to intensify operations for enhanced production and India is the currently globally the largest producer of milk. To stay in the business, dairy operators are required to herd large cattle population which has led directly to an increase in dung production. In the process, the once manageable volume of manure (when the herd size was smaller and farms less densely populated) has become a social and environmental issue. As a fallout, inadequate collection, storage, and treatment strategies degrade the air, soil and water quality. The pollutants thus generated by the mismanaged livestock manure increase the biochemical oxygen demand (BOD), pathogens, nutrient loading, methane, and ammonia in the environment. This affects the health of not only the dairy operators and their families, but also the local economy.

Dairies consume significant energy in chiller systems to cool milk, air compressors to operate milking equipment, heaters/boilers to provide hot processed water, and various pumps. Like other business owners, dairy operators face increasing and uncertain energy costs. Dairies could reduce/eliminate the energy bills through on-farm energy production using biogas system. Electricity thus produced from biogas in an engine-generator can be used on-farm, or sold to local grid. By directly burning the biogas in a boiler/furnace or from a heat recovery system connected to engine-generator, thermal energy can be used for water/building heating.

Economic viability

The economics of the biogas system based facility in the dairy farm (as example) is discussed below in Tables 13–16. The viability of a plant would depend on whether the gas and slurry output could substitute for fuels and fertilisers/feeds that were earlier purchased. The savings that result could be used to repay for the capital and maintenance cost.

Table 14: Scenario 2: 12kW (15kVA) biogas for electricity generation

Total investment on the project	Rs.15,00,000/-
Daily biogas production	85m^3/day
Daily requirement of poultry litter (kg) to mix with equal quantity of water	1Tonne
Capacity of biogas Genset	12kW
Daily replacement of electricity 12kWh X 8–10hrs	96–120 units
Daily saving of electricity @ Rs.10/- X 96 units	Rs.960/-
Manure production per day	330 kg.
Annual saving of electricity Rs.960/- X 365 days	Rs.3,50,000/-
Annual maintenance expenses on the project	Rs.30,000/-

Table 13: Scenario 1: biogas utilised as cooking fuel

Capacity of biogas plant	$45m^3$/day	$85m^3$/day	$200m^3$/day	$300m^3$/day
Total investment (approx.)	Rs.8,00,000/-	Rs.12,00,000/-	Rs.18,00,000/-	Rs.26,00,000/-
Daily feed requirement for biogas plant	540kg	1000kg	2400kg	3600kg
Daily approximate LPG replacement	18kg	30kg	80kg	120kg
Daily cooking for number of persons	135	255	600	900
Annual saving on LPG (12 months) @Rs.70/- 360 days	Rs.4,59,900/-	Rs.8,68,700/-	Rs.20,44,000/-	Rs.30,66,000/-
Daily manure production	178kg	330kg	800kg	1200kg
Annual maintenance (daily 2-3h material feeding to digester)	Rs.15,000/-	Rs.20,000/-	Rs.25,000/-	Rs.25,000/-

Table 15: Scenario 3: 24 kW (30kVA) biogas for electricity generation

Total investment on the project	Rs.28,00,000/-
Daily biogas production	200m³/day
Daily requirement of poultry litter (kg) to mix with equal quantity of water	2Tonne
Capacity of biogas Genset	24kW
Daily replacement of electricity 24kWh X 6-8hrs	192 units
Daily saving on electricity @ Rs.10/- X192 units	Rs.1,920/-
Manure production per day	600kg
Annual saving of electricity @ Rs.1,920/- X 365 days	Rs.7,00,800/-
Annual maintenance of project	Rs.30,000/-

Table 16: Scenario 4: 40kW (50kVA) biogas used for electricity generation

Total investment of the project (approximately)	Rs. 36,00,000/-
Daily biogas production	300 m³/Day
Daily requirement of poultry litter (kg) to mix with equal quantity of water	3.6 tons
Capacity of biogas Genset	48 kw
Daily replacement of electricity 48kW X 6–8hrs	384 units
Daily saving of electricity @ Rs.10/- X 384 units	Rs.3,840/-
Manure production per day (dry weight basis of daily feed stock)	1200kg
Annual saving of electricity Rs.3,840/- X 365 days	Rs.14,01,600/-
Annual maintenance of project	Rs.40,000/-

Note: As seen from above, the investment on gas plant can be recovered in 1-2 year period in Scenario 1 and in 2-3 year period in Scenario 3-4; Manure production @33% dry weight basis of daily feeding. Above data are computed based on the consumption of biogas 1HP consumption $0.425m^3/15ft^3/HP/hour$ ($1m^3 = 35ft^3$).

Plant capacity needs to be increased by installing additional unit (subject to enough feed stock being available) to run a higher capacity genset or for increased duration of electricity supply. Accordingly, the manpower also should compliment; ideally, such plants need little maintenance, only to maintain general cleanliness and overall inspection.

Carbon credit through biogas plant

A carbon credit is a type of a tradable greenhouse gas emission reduction unit issued to projects under the Kyoto Protocol. One carbon credit is equivalent to one tonne of carbon dioxide (CO_2) mitigated. Setting up of Clean Development Mechanism (CDM) projects to control global warming has served the purpose.

Certified Emission Reduction (CER) is different for CO_2 emission, choloflourocarbon (CFC) and N_2O. It is traded in Euro and US $, and fluctuates. Rural as well as urban biogas plants in India exhibit enormous potential to earn carbon credits.

Table 17: An estimated carbon credit potential of a $3m^3$ Deenbandhu biogas plant

If a $3m^3$ domestic plant is commissioned	It would replace:
	11-12kg firewood/day
	1.86l kerosene/day
	38kg dry cattle cake/day
	2.5l petrol/day
	1.29kg LPG/day
	10.5kg daily slurry produced contains enough NPK to replace urea, DAP, and chemical pesticide

The main components in biogas from digester are CH_4 and CO_2. Released to the environment uncombusted it certainly enhances the global warming potential as CH_4 is 20 times more potent greenhouse gas than CO_2. Considering only firewood, the global warning mitigation potential (GWMP) or the annual carbon credit from a $3m^3$ biogas plant may be calculated as (the equation is indicative in nature):

GWMP (CO_2 equivalent) = GWP_{FW}(CO_2 + CH_4 emission) + GWP_F(CO_2 emission + CH_4 & N_2O emission) – GWP_L(CH_4 Leakage) where, FW: Firewood, L: Leakage, GWP: Global warning potential.

The GWMP is calculated based on the following:

1) The amount of CO_2, CO, and N_2O emitted/kg or litre of the otherwise used material as the energy source is calculated. If a $3m^3$ plant is not installed it means one is consuming these commodities (listed in an earlier-mentioned table).
2) For example, CO_2 emission from firewood stands at 1.83kg/kg.
3) Firewood also emits CH_4 @ 3gm carbon/kg.
4) CO_2 emission from Kerosene burning stands at 2.41kg/l.
5) CH_4 leakage from biogas plant (in %) stands at 10.
6) CO_2 emission from N-fertiliser production stands at 1.3kg/kg.
7) CO_2 emission from P-fertiliser production stands at 0.2kg/kg.
8) CO_2 emission from K-fertiliser production stands at 0.2kg/kg.
9) N_2O emission from N-fertiliser production stands at 0.007kg/kg.

(Source: *Proc. of the International Conference on Science and Engineering*, ICSE 2011)

Digested slurry: Towards sustainable organic farming and economic empowerment

The digested biogas slurry contains 91% water and 8% dry matter on an average of which 6.5% is organic and 2.5% is inorganic matter. The % NPK (Nitrogen, Phosphorus and Potassium) constituents is 0.25, 0.13 and 0.12 on wet weight basis, and 3.6, 1.8 and 3.6 on a dry weight basis, respectively. Apart from these biofertilising agents, trace amount of manganese, copper, zinc and iron are also found. This odourless digested slurry could be used as a liquid fertiliser with high fertilising ability than fresh dung or the farmyard manure (FYM). For some crops it is superior to chemical fertilisers because the nitrogen in it is available in a form that is immediately absorbed. Upon composting of the bioslurry, nutrient value in it gets enriched. The digested slurry is an excellent material for accelerated rate of composting compared to refuse, crop waste and garbage etc. Table 18 shows the N, P, K values in various organic fertilisers.

Table 18. The N, P, K values in different types of organic fertilisers/manures

Nutrient	Compost manure		Farmyard manure		Digested bioslurry	
	Value range (%)	Average value (%)	Value range(%)	Average value (%)	Value range (%)	Average value (%)
Nitrogen	0.50-1.50	1.00	0.50-1.00	0.80	1.40-1.80	1.60
Phosphorus	0.40-0.80	0.60	0.50-0.80	0.70	1.10-2.00	1.55
Potassium	0.50-1.90	1.20	0.50-0.80	0.70	0.80-1.20	1.0

It is seen from the table that the digestate bioslurry has higher nutrient values compared to its counterparts. Although the effect of compost on crop production would depend on the soil type and condition, the seed quality, the climate and other factors, its application would bring positive changes in the soil, such as, improving the physical structure of soil, increased soil fertility, increased soil water-holding capacity, and enhanced activity of the soil microflora and an overall better soil health.

Studies have revealed that the digestate improves soil fertility and increases crop production by 10-30% compared to FYM, if stored and applied properly. Application of digestate is reportedly very successful in cereal crops like paddy, wheat and maize, and vegetables like cabbage, tomato, etc. The most responsive crops, however, are root vegetables (such as, carrots and radish), potatoes, and fruiting trees.

Application of slurry-based organic fertiliser is crucial in the cyclic utilisation of biomass resources, and has relatively high economic interest, combined with ecological and social benefits. A beneficiary of biogas plant may sell biogas slurry as a marketable commodity. However, due to issues, such as, lack of awareness and market for organic manure as well as the absence of a definitive commercial and legal framework, such organic farming practices is still not realised to full-scale. Hence, in order to make a profitable trade at grassroots level, realistic measures may be initiated not by the government only but also by the public and private set-ups like the biofertiliser companies and agricultural universities.

A cooperative society (or dung bank) dealing with the biogas and manure commercialisation (considering 25–30 beneficiaries) could be proposed to create for sale of the commodities. If feasible thereafter, bottled biogas may be utilised in mid-day meal scheme under the *Sarva Sikhya Abhiyaan* to reduce the cost of expenditure on cooking.

Toilet-cum-garbage digester facilities may be set up at many institutions, and a collection point for biofertiliser for the purpose is suggestible. A part of the cost of such installation may be sourced from the *Swacch Bharat Mission*.

Conclusion

Taking stock of the current energy scenario and the need to conserve the natural resources for future, a focus on bioenergy sources and harnessing the potential associated with is needed. A long-term sustainable energy policy needs to be adopted to restabilise the environmental setup by bringing down the CO, CO_2 and other depleting gases generated from coal and thermal power plants, automobile exhausts, firewood and infrastructure sector back to the '90s levels. Building on its goal of installing 175gigaWatt (GW) of renewable power capacity by 2022, India's INDC is setting a new target to increase the share of non-fossil-based power capacity with international support from the existing 30% to 40% by 2030. India has also committed to reduce the emissions intensity by 33–35% per unit GDP to below 2005 levels by 2030. Additionally, it envisages creating an additional carbon sink of 2.5–3 billion tonnes of CO_2 through additional green cover. It also prioritises the efforts to build resilience to climate change impacts, and gives a broad indication of the amount of financing necessary to attain its goals. Decades of collective experience gained through various guided research, development, extension and training programmes related to the biogas (and bioenergy) sector are incorporated lucidly for the benefit of the reader.

Acknowledgements

The chapter is prepared with valuable contributions in terms of field information from field officials of the BDTC supported by the Ministry of New and Renewable (MNRE), Govt. of India, along with Non-government Organisations (NGOs) working with BDTC. Inputs from www.oredaodisha.com, www.mnre.gov.in, https://energypedia.info, research articles on biogas technology are duly acknowledged herewith. The support and encouragement received from various quarters at individual and institutional levels are also acknowledged.

Further readings

Abbas T, Ali G, Adil, SA, Bashir MK, Kamran MA, 2017. Economic analysis of biogas adoption technology by rural farmers: The case of Faisalabad district in Pakistan. *Renewable Energy*, **107(2017)**: 431–9.

Achinas S, Achinas V, Euverink GJW, 2017. A technological overview of biogas production from biowaste. *Engineering*, **3(3)**: 299–307.

Ayodele TR, Ogunjuyigbe ASO, Alao MA, 2018. Economic and environmental assessment of electricity generation using biogas from organic fraction of municipal solid waste for the city of Ibadan, Nigeria. *Journal of Cleaner Production*, **203**: 718–35.

Bera SR, Jain P, 2016. Contribution of carbon credits for financial feasibility of a biogas project. *International Journal of Engineering Research in Mechanical and Civil Engineering*, **1(1)**: 51–8.

Cucui G, Ionescu C, Goldbach I, Coman M, Marin E, 2018. Quantifying the economic effects of biogas installations for organic waste from agro-industrial sector. *Sustainability*, **10(7)**: 2582.

Dale BE, Sibilla F, Fabbri C, Pezzaglia M, Pecornia B, Veggia E, Baronchelli A, Gattoni P, Bozzeto S, 2016. Biogasdoneright™: An innovative new system is commercialized in Italy. *Biofuel, Bioproducts and Biorefining*, **10(4)**: 341–5.

Fuess LT, Zaiat M, 2018. Economics of anaerobic digestion for processing sugarcane vinasse: Applying sensitivity analysis to increase process profitability in diversified biogas applications. *Process Safety and Environmental Protection*, **115**: 27–37.

Ganthia BP, Sasmita S, Rout K, Pradhan A, Nayak J, 2018. An economic rural electrification study using combined hybrid solar and biomass-biogas system. *Materials Today: Proceedings*, **5(1)**: 220–5.

Ghosh SB, Bhattacharya K, Nayak S, Mukherjee P, Salaskar D, Kale SP, 2015. Identification of different species of *Bacillus* isolated from Nisargruna Biogas Plant by FTIR, UV–Vis and NIR spectroscopy. *Spectrochimica Acta Part A: Molecular and Biomolecular Spectroscopy*, **148**: 420–6.

Giarola S, Forte O, Lanzini A, Gandiglio M, Santarelli M, Hawkes A, 2018. Techno-economic assessment of biogas-fed solid oxide fuel cell combined heat and power system at industrial scale. *Applied Energy*, **211(2018)**: 689–704.

Haberl H, Erb KH, Krausmann F, Running S, Searchinger TD, Smith WK, 2013. Bioenergy: how much can we expect for 2050? *Environmental Research Letters*, **8(3)**: 031004.

Hamzehkolaei FT, Amjady N, 2018. A techno-economic assessment for replacement of conventional fossil fuel based technologies in animal farms with biogas fuelled CHP units. *Renewable Energy*, **118(2018)**: 602–14.

Haque A, 2013. Bio slurry ultimate choice of biofertilizer. *Bangladesh Agric. Univ*, **2(4)**: 738.

Herbes C, Halbherr V, Braun L, 2018. Factors influencing prices for heat from biogas plants. *Applied Energy*, **221(2018)**: 308–318.

Joshua OS, Ejura GJ, Bako IC, Gbaja IS, Yusuf YI, Fundamental principles of biogas product. *International Journal of Scientific Engineering Research*, **2(8)**: 47–50.

Kabyanga M, Balana BB, Mugisha J, Walekhwa PN, Smith J, Glenk K, 2018. Economic potential of flexible balloon biogas digester among smallholder farmers: A case study from Uganda. *Renewable Energy*, **120(2018)**: 392–400.

Kale SP, 2004. Nisarghruna plant for urban and rural waste management, energy conservation, better environment and restoration of soil fertility. *Bioenergy News*, **8(1)**: 1–5.

Koido K, Takeuchi H, Hasegawa T, 2018. Life cycle environmental and economic analysis of regional-scale food-waste biogas production with digestate nutrient management for fig fertilisation. *Journal of Cleaner Production*, **190**: 552–62.

Kumar S, Malav LC, Malav MK, Khan AS, 2015. Biogas slurry: source of nutrients for eco-friendly agriculture. *Journal of Extensive Research*, **2**: 42–6.

Lee DH, 2017. Evaluation the financial feasibility of biogas upgrading to biomethane, heat, CHP and AwR. *International Journal of Hydrogen Energy*, **42(45)**: 27718–31.

Ligus M, 2017. Evaluation of economic, social and environmental effects of low-emission energy technologies in Poland–multi-criteria analysis. *Energy Procedia*, **136(2017)**: 163–8.

Lorenzi G, Lanzini A, Santarelli M, Martin A, 2017. Exergo-economic analysis of a direct biogas upgrading process to synthetic natural gas via integrated high-temperature electrolysis and methanation. *Energy*, **141(2017)**: 1524-1537.

Ma C, Liu C, Lu X, Ji X, 2018. Techno-economic analysis and performance comparison of aqueous deep eutectic solvent and other physical absorbents for biogas upgrading. *Applied Energy*, **225(2018)**: 437–47.

Mittal S, Ahlgren EO, Shukla PR, 2018. Barriers to biogas dissemination in India: A review. *Energy Policy*, **112**: 361-370.

Mohanty PK, Mishra S, Suar M, Ojha S, VK Malesu, Paikaray B, 2013. Konark model-the spherical biogas plant – a field based approach, Biogas Forum, New Delhi, India.

Offermann R, Seidenberger T, Thrän D, Kaltschmitt M, Zinoviev S, Miertus S, 2011. Assessment of global bioenergy potentials. *Mitigation and adaptation strategies for global change*, **16(1)**: 103–15.

Ojha SK, Mishra S, Kumar S, Mohanty SS, Sarkar B, Singh M, Chaudhury GR, 2015. Performance evaluation of vinasse treatment plant integrated with physico-chemical methods. *Journal of Environmental Biology*, **36(6)**: 1269–75.

Ojha SK, Mishra S, Nayak S, Suar M, 2012. Molecular biology and biochemistry for enhanced biomethanation. *Dynamic Biochemistry Process Biotechnology and Molecular Biology* (Spl issue 1): 48–56.

Paakkonen A, Tolvanen H, Rintala J, 2018. Techno-economic analysis of a power to biogas system operated based on fluctuating electricity price. *Renewable Energy*, **117(2018)**: 166–74.

Rahman MM, Hasan MF, Saat A, Wahid MA, 2017. Economics of biogas plants and solar home systems: For household energy applications. *Journal of Advanced Research in Fluid Mechanics and Thermal Sciences*, **33(1)**: 14–26.

Rotunno P, Lanzini A, Leone P, 2017. Energy and economic analysis of a water scrubbing based biogas upgrading process for biomethane injection into the gas grid or use as transportation fuel. *Renewable Energy*, **102(2017)**: 417–32.

Ruiz D, San Miguel G, Corona B, Gaitero A, Domínguez A, 2018. Environmental and economic analysis of power generation in a thermophilic biogas plant. *Science of the Total Environment*, **633(2018)**: 1418–28.

Searle S, Malins C, 2015. A reassessment of global bioenergy potential in 2050. *Global Change Biology Bioenergy*, **7(2)**: 328–36.

Smeets EM, Faaij AP, 2007. Bioenergy potentials from forestry in 2050. *Climatic Change*, **81(3-4)**: 353–90.

Ubwa ST, Asemava K, Oshido, B, Idoko A, 2013. Preparation of biogas from plants and animal waste. *International Journal of Science and Technology*, **2(6)**: 480–5.

Vaish B, Singh P, Singh PK, Singh RP, 2017. Biomethanation potential of algal biomass. *In*: Gupta S, Malik A, Bux F (Eds) Algal Biofuels. Springer, Cham: 331–46.

Vo TT, Wall DM, Ring D, Rajendran K, Murphy JD, 2018. Techno-economic analysis of biogas upgrading via amine scrubber, carbon capture and ex-situ methanation. *Applied Energy*, **212(2018)**: 1191–202.

Voß A, 1988. Commercialisation of renewable energies issues and actions. *Proceedings of the Euroforum – New Energies Congress*, University of Stuttgart, **2**: 1–12.

Yasar A, Nazir S, Rasheed R, Tabinda AB, Nazar M, 2017. Economic review of different designs of biogas plants at household level in Pakistan. *Renewable and Sustainable Energy Reviews*, **74(2017)**: 221–9.

Zhang CA, Guo H, Yang Z, Xin S, 2017. Biogas slurry pricing method based on nutrient content. *In*: *IOP Conference Series: Earth and Environmental Science*, **94(1)**: 1–7.

Other internet sources:

https://cag.gov.in/sites/default/files/audit_report_files/Union_Civil_ Performance_Renewable_ Energy_Report_34_2015_chap_8.pdf, REPORT: Comptroller and Auditor General (CAG) Report on NBMMP Renewable Energy Sector in India, 2015. Chapter-viii: National Biogas and Manure Management Programme: 141–9.

https://www.indiastat.com/power-data/26/non-conventional-energy/184/biomass-biogas-1985 2019/452705/stats.aspx: Launched on 14th November 2000 owned by datanet India.

http://pib.nic.in/newsite/pmreleases.aspx?mincode=28: The Ministry of New and Renewable Energy, Govt. of India.

Annexures

Annexure-I: A Model Bankable Scheme for Biogas Commercialisation Venture

1. Energy crisis

Energy is a necessary concomitant of human existence. Although many energy sources exist in nature, it is coal, electricity and fossil fuel which are commercially exploited for many useful purposes. This century has witnessed phenomenal growth of various industries based on these energy sources. They have applications in agricultural farms and have domestic use in one form or the other. Fossil fuel in particular has played the most significant role in the growth of industry and agriculture, which would be recorded in golden words in the history of progress of human race. Whether it is flying in the air or speeding automobiles on the roads or heating and prime moving in the industry or petrochemicals and fertilisers for farms or synthetics for daily use or cooking at home, all have been made possible by a single source - fossil fuel. It has penetrated so deep into the human living that it's uneasy to accept that this useful energy source is not going to last long. Once available easily and at lower prices irrespective of its origin of supply, fossil fuel has now been scarce and costly. The immediate effect of this is that the world is in a grip of inflation and rising prices. Energy crisis has mainly emerged from the fear that the boons of fossil oil may turn into a bane as fossil fuel would disappear and compel the habits and practices of the society to change. This is a crisis and it is a compulsion for search of alternate energy source.

2. Biogas as an alternate renewable energy source

It is evident that no single energy source having diverse applications is capable of replacing fossil fuel completely. On the other hand, dependence on fossil fuel would have to be reduced at a faster pace so as to stretch its use for long and in critical sectors till some appropriate alternate, preferably, energy sources are made available. Presently, the country is spending a fortune in importing fossil fuel that is hardly affordable for long on the face of developmental needs. Popularly known as biogas, CH_4 is one such alternate energy source that is identified as a useful hydrocarbon with combustible qualities as that of other hydrocarbons. Though its calorific value is not high as some products of fossil oil and other energy sources, it can meet some needs of household and farms. Following table provides an idea of comparative heat values and thermal efficiency of commonly used fuels in the household and farms.

Table 1: Calorific values and thermal efficiencies of some commonly used domestic fuels

Commonly used fuel	Calorific value (Kilo calories)	Thermal efficiency
Biogas	$4713/M^3$	60%
Dung cake	2093/Kg	11%
Firewood	4978/Kg	17.3%
Diesel (HSD)	10550/Kg	66%
Kerosene	10850/Kg	50%
Petrol	11100/Kg	-

These calorific (or heat) values indicate that biogas can perform works similar to fossil fuel in domestic cooking, lighting, etc., with better efficiency depending upon the CH_4 content in it. Biogas also has the potential for use in internal combustion engines for pumping water, etc. for which research and development works are in progress. Thus, biogas has a bright future as an alternate renewable energy source for domestic and farm uses.

3. Biogas, its production process and composition

It would be useful to know the properties of biogas:

i) **Biogas**: It mainly comprises of hydrocarbon which is combustible like any hydrocarbons and can produce heat and energy when burnt. Chemically, this hydrocarbon is CH_4, where C stands for carbon and H for hydrogen, and chemically the gas is termed as methane gas. The chemical formulae of some other commonly used hydrocarbons derived from fossil fuel are, *viz*., petrol (C_6H_{14}), kerosene (C_9H_{20}), diesel ($C_{16}H_{34}$). Unlike these hydrocarbons derived through fractional distillation, biogas is produced through a microbial process wherein some bacteria convert biological wastes into useful biogas comprising methane through chemical interaction. Methane gas is renewable through continuous feeding of biological wastes that is available aplenty in rural areas in the country. Since this useful gas originates from biological process, it has been termed as biogas in which methane gas is the main constituent.

ii) **Production process**: The process of biogas production is anaerobic and takes place in two stages, acid formation and methane formation. In acid formation stage, the biodegradable complex organic (solids and cellulosic) compounds present in waste materials are acted upon by acid-forming bacterial group present in the dung and reduce them to organic acids, CO_2, H_2, NH_4 and H_2S. Since organic acids are the main products in this stage, it is known as acid forming stage, and this serves as the substrate for CH_4 production by methanogenic bacteria.

In the second stage, a group of methanogenic bacteria act on the organic acids to produce CH_4 gas, and also reduce CO_2 in the presence of H_2 to form methane (CH_4). There are four types of methanogenic bacteria, *Methanobacterium*, *Methanospirillium*, *Methanococcus* and *Methanosarcina*. They are oxygen- and photo-sensitive and don't perform as effectively in presence of light and oxygen. As the process ends, the amount of oxygen demanding materials in the waste product reduces to within the safe level for handling by humans.

The gas thus produced by the above process in a biogas plant does not contain pure CH_4, and has several impurities. A typical composition of such gas is as follows:

Table 2: Mean average composition of biogas

Component	Percentage
Methane	60.0
Carbon-di-oxide	38.0
Nitrogen	0.8
Hydrogen	0.7
Carbon-monoxide	0.2
Oxygen	0.1
Hydrogen sulphide	0.2

The calorific value of methane is 8400 kcal/m^3 and that of the above mixture is about 4713 Kcal/m^3. However, the biogas gives a useful heat of 3000 kcal/m^3. If similar heat values are to be obtained from other sources of fuel, the equivalent quantities of those fuels have to be substantial as may be seen from Appendix III. It is not the quantity which is so important, but while biogas is renewable others are not.

4. Scope of biogas plants

The basic feed material for biogas plants in India has been considered to be cattle dung which is available aplenty. The estimated cattle population of 238 million in the country has the potential to produce about 1000 million tonnes of dung every year. According to an estimate (1977) of Khadi and Village Industries

Commission (KVIC), biogas plants of average family size may provide energy equivalent to 5432 million liter of kerosene, which in terms of current prices may annually cost well over Rs. 1000 crore. Although cattle dung is recognised as the chief raw material for biogas plants, others like night soil, poultry litter and agricultural waste are also used based on their social acceptability. In addition to combustible gas, biogas plants would also be a source for conversion of NPK-rich organic manure. It is estimated that biogas plants are very useful as the recoverable dung from 236 million cattle adds about 3.5 mmt nitrogen to the soil each year and for ensuring its conservation. Thus, the scope for biogas plants in India is substantial if the accruable benefits from these are exploited well.

5. Major benefits of installing biogas plants

It is estimated by the Ministry of Energy, Govt. of India, that alternative energy sources like biogas plants, wind mills etc. may reduce the dependence on conventional energy sources by about 20% by the turn of the century, provided promotional efforts are continued. Presently, the cooking media in rural areas consist of burning dung cake, firewood and to some extent kerosene where it is available easily. The installation of biogas plants would directly replace the use of above three, and following gains would be made while saving them:

i) Nearly 30% of available dung that is burnt and wasted could be recovered as biogas plants conserve the dung while producing biogas. Again, the dung preserves more NPK in the dung solids and cellulose after digestion in plant, which otherwise gets lost if heaped in the open.

Table 3: Percent NPK in various (organic) biocomposts

Biocompost	N	P_2O_5	K_2O
Biogas slurry	1.4	1.0	0.8
Farm yard manure (FYM)	0.5	0.2	0.5
Town compost	1.5	1.0	1.5

The benefits derived from biogas plants in terms of manure and useful energy are illustrated at Appendices I & II. The average NPK content of farmyard manure (FYM) is about 0.5, 0.2 and 0.5% respectively. It may be observed that when converted into FYM, biogas slurry is rich in NPK by more than four times than ordinary dung. When the country faces shortage of fertilisers and has to spend enormous amounts for its import, application of biogas slurry can replace chemical fertilisers to a large extent. Biogas slurry or FYM not only adds NPK but it improves the soil porosity and texture. These are the established benefits.

ii) Second major benefit is that rural people would gradually stop felling trees. Tree felling is identified as a major cause of soil erosion and worsening flood situations. Government has started massive afforestation programmes to tackle this. Continued deforestation causes ecological and environmental imbalances. Biogas plants would be helpful in correcting this situation.

iii) Kerosene is used for lighting lantern and cooking in a limited way in rural India wherever its supply has been made possible. Whatever quantity is used can be replaced by biogas as it can be used for lighting and cooking. This would directly reduce dependence on fossil fuel and in saving foreign exchange.

iv) Lastly, the most important social benefit is that there would be no open heap of dung to attract flies, insects and infections, the dung being digested in the digester. The slurry from digester can be transported to the farm for soil application, thus keeping the environment clean for inhabitation. Gas cooking also removes health-related (like respiratory and eye diseases) hazards from dung-cake or fire-wood cooking prevalent in the villages.

6. History of technological development, past achievements, future programmes and the role of institutional finance

Biogas technology is not new to India. Its testing started in 1940 when Dr. SV Desai took up an experimental gas plant at Indian Agricultural Research Institute (IARI) after visiting Dadar sewage puri-fication station at Bombay. Cattle dung fermentation fol-lowed next which was patented in 1951 by Shri Jasbhai J. Patel. The model underwent several modifications, and the plant was named Gramlaxmi III in 1954. The same model is being propagated by KVIC in a nationwide programme since 1962. Although many institutions and indi-viduals kept experimenting for better models and introduced several mod-els this KVIC model has stood the test of time. Although the cost difference of the new Janata model in the late 70s was about 20%, even this model has not affected the popularity of KVIC model.

Past achievements and future programmes

All along since 1962, KVIC was the sole agency to promote biogas plants independent of government programme. To reduce the dependence on fossil fuel due to the embargo threat during the Arab-Israel war in 1973, the Government included biogas plants as alternate energy sources in its 5th five year plan, with a target of one lakh gas plants for the plan period. However, inspite of the fact that 25–50% subsidy was provided to the users, only 80,000 gas plants were reportedly installed. It was more vigorously pursued in the 6th five year plan with a target of 4 lakh plants, and not more than 45% of this was achieved. The target was 1.60 lakh plants in FY 2003-2004. The programme that was till now (2018) covered under the National Biogas and Manure Management Programme (NBMMP) of Govt. of India, Ministry of New and Renewable Energy (MNRE; the then Ministry of Nonconventional Energy Sources – MNES), is set to continue further under the revamped programme of New National Biogas and Organic Manure Programme (NNBOMP).

It is noted from the above that, during the last 21 years, the achievement is not appreciable. The government is convinced that the biogas plant technology isn't a failure. However, social environment has to be more favourable for an expeditious progress. As a case, China has taken a rapid stride in the same biogas technology field due to favourable social environment.

The relevance of Chinese experience

A comparative study of India and China carried out by the Centre for Application of Science and Technology to Rural Areas (ASTRA), Indi-an Institute of Science (IISc), Bangalore gave striking revelation of what China achieved in biogas technology. India started experimenting much earlier and actual programme started in 1967 whereas China took to this technology in 1970. Even after a late start, China overtook India to reach an unbelievable target in a short span of seven years. However, with persistent efforts thereafter, the Indian programme got a fillip though a huge gap still remains to tap the potential. It's observed from the following:

Table 4: Year-wise number of biogas plants in India and China in the last part of 20th Century

Year	India	China
1973	8,000	5,000
1980	80,000	72,00,000
1998	27,50,000	69,00,000

China's achievement deserves all appreciation as they have managed to install so many numbers without much of cattle dung but with organic resources within easy reach. The daily feed in Chinese gas plants consists of 20 kg waste from four pigs, 4 kg waste from 5 hu-mans, 6 kg of straw and poultry litters. All these are available with majority of Chinese families and so the technology became more successful and popular as feed was not a constraint. In India, the rural population would perhaps not adopt such raw materials as their Chinese counterparts do. In rural India, cattle dung is readily handled but hesitantly for other dungs including night soil as raw materials, and only 20% of the household own the minimum required 4–5 cattle. Unless the remaining rural India adopts raw materials like the Chinese have done, success is limited. China has followed the programme with all seriousness in spite of the fact that it is fossil fuel exporter and energy self-sufficient. When Chinese could take advance measures to counter future energy crises, we should be more vigilant in taking the programme seriously. However, a word of caution is added here that, since large number of plants is failing in the field, enough care may be taken to select the plant type and in sound design and construction.

Role of institutional finance

The responsibility to extend technical guidance, installa-tion, supervision and also the subsidy is majorly entrusted on KVIC and the states who have acquired adequate experience due to their long association with the biogas programme by the Government. However, it's observed that most of the plant users fail to raise their own contribution due to lack of resources. This made the financial institutions difficult to mediate in meeting the credit requirements. During 1974-75, commercial banks entered the fray. The Govern-ment has now attached top priority to the biogas programme in which banks have a special role to play in making the programme successful. In this context that the bank officials acquaint themselves with the technology, operations, field problems, post installation maintenance and economic and social aspects of biogas programme so that formulation, appraising and processing of schemes are easier.

7. Main features of a biogas plant

Based on the gas holder, biogas plants are classified into fixed-dome and floating-drum types based on the gas holder. Both the types have the following common functional components:

i) **Digester**: This is the fermentation tank built partially or fully underground. It is generally cylindrical or spherical, and made up of bricks and cement mortars. It holds the slurry within it for the specified digestion period for which it is designed.

ii) **Gas holder**: This is meant to hold the gas after leaving the digester. It is a floating-drum or a fixed-dome, the basis of plant classification. Gas connection is taken from the top of this holder to the gas burners or for other purposes by suitable pipelines. The floating gas holder is made up of mild steel sheets and angle iron that exerts a pressure of 10 cms of water in the gas dome masonry and a pressure upto 1m of water column on the gas.

iii) **Slurry-mixing tank**: This is a tank in which the dung is mixed with water and fed to the digester through an inlet pipe.

iv) **Outlet tank and slurry pit**: An outlet tank is usually provided in a fixed-dome plant from where the slurry in directly taken to the field or to a slurry pit. In case of floating-drum plant, the slurry is taken to a pit where it can be dried or taken to the field for direct applications.

8. Broad basis of plant design

To design a biogas plant, there is a need to match the gas requirement to the available feed material for uninterrupted gas production and supply. For this, the average gas requirement for different uses, daily dung produced and the average gas production potential of various feed is useful to know. Some basic information on these is furnished in Appendices I to VI. The design of a plant may be determined in the following manner:-

a) **Biogas requirement**: Use Appendix IV to decide the total daily gas requirement. In case gas requirement is $3m^3$, follow the design consideration as detailed in paragraphs that follow.

b) **Raw material requirement**: To find out the quantity of dung equiva-lent to produce $3m^3$ gas, refer Appendix V. By dividing $3m^3$ by $0.04m^3$ (that is equivalent to a kg dung) it reveals the total quantity of dung required for the same. In this case it is 3/0.04 = 75kg. Thereafter, refer Appendix VI to know the number of animals necessary to produce this (75kg dung).

c) **Digester design**: Slurry of specific gravity 1.089 is prepared by mixing dung with an equal quantity of water. So, the daily slurry volume fed is $(75+75)/(1000X1.089) = 0.138m^3$. Thus, the digester volume is 0.138X50 or $6.9m^3$, say $7m^3$ equivalent for a 50-day retention plant.

The KVIC recommends a digester volume of 2.75 times the required daily gas volume. The Commission's recommendation for the depth of the plant is between 4 and 6m as per the size. But, for building materials economy reason, a 1.0:1.3 depth to diameter ratio is considered ideal for all plant types. In floating-drum plant, a continuous ledge is built into the digester at a depth 10cm shorter than the height of the gas drum to prevent the gas holder from going down when no gas is left in it thereby preventing the gas inlet being choked. It also guides the gas bubbles rising from the side of the plants into the gas bolder. Slurry is fed at the bottom and removed at the top in some plants. When the digester diameter exceeds 1.6m, a parti-tion wall is provided in it to prevent short circuiting of slurry flow and increasing its retention period. Some standard dimensions of such floating drum plants are given in Appendix VII. In fixed-dome plants, the digester volume comes to between 1.5 and 2.75 times of the daily gas produced. The higher the plant capacity, the ratio of digester volume to daily gas produced becomes lesser.

d) **Gas-holder design**: The gas holder designing is influenced by the digester dia-meter and daily gas distribution. The gas holder capacity is kept at 60% of a day's gas production in domestic plants, and 70% in laboratories. In a floating-drum plant, the gas holder diameter is 15cm less than the digester diameter, and accordingly other dimensions are decided. Gas holder can be rotated around its guide to break the scum formed at the top. In a fixed-dome plant, the dome angle is between 17° and 21° (giving a pressure equivalent to 100cm water). Due to higher pressure, the gas pipeline diameter can be re-duced and the gas can be taken to greater distance. Care is taken to provide earth pressure equivalent to 100cm of water column from the top of the dome. For better stability, '*A*'class brick is preferred for the domes.

e) **Inlet tank**: Before the dung is fed into the plant, it is mixed with water in a tank to make slurry. This tank also helps to remove grass-like floating and non-degradable materials for optimal biogasification and also to prevent excess scum formation. This tank connects to the digester by an asbestos cement pipe. The mixing tank floor is sloppy opposite to the direction of inlet pipe to help heavy nondegradable solid particles to settle and separate from the slurry.

9. Basic prerequisites of a biogas system

i) **Land and site**: Following aspects are considered while selecting a site for a biogas plant:

   a) The land should be leveled and elevated than the surroundings to avoid water runoff.

   b) Soil should not be too loose and should have a bearing strength of $2kg/cm^2$.

   c) It should be nearer to the intended place of gas use.

   d) It should also be nearer to the cattleshed/stable for ease of handling of raw materials.

   e) The water table should not be very high.

   f) Adequate supply of water should be there at the plant site.

   g) The plant should get clear sunshine during the most part of the day.

   h) Plant site should be well-ventilated as methane is very explosive when mixed with O_2.

   i) A minimum distance of 1.5m may be kept between the plant and any wall or foundation.

j) It should be away from any tree to make it free from failure due to any root interference.

k) It should be at least 15m away from any well meant for drinking water purpose.

l) There should be adequate space to construct slurry pits.

ii) **Feed for gas plants**: The plant feed mainly comprises of cattledung in India. Although the quantity of dung per cattle depends on health, age, type and many other factors, generally an average daily yield of about 10kg dung is estimated. On this assumption, the number of cattle required for various plant sizes as recommended by KVIC is provided in the following table. However, it is advisable that the assessment of dung and animal requirement may be made as per approach discussed in Sl. No. 8.

Table 5: Requirement of cattle for various sizes of biogas plants

Plant Size (m^3)	Minimum number of cattle required
2	3
3	4
4	6
6	10
8	15
25	45

iii) **Temperature**: Temperature plays the most important role in the biogas production. The amount of gas production from a fixed weight of organic waste is best when temperature is within the mesophillic (25–37°C) and thermophillic (45–55°C) ranges. Gas yield maximises in the thermophillic region and the digestion period also reduces. It takes about 55 days in mesophillic range whereas it is about 7 days in thermophillic region for digestion.

iv) **Hydrogen ion concentration**: The pH indicates the acidity and alkalinity of feed mixture, and any excess acidity or alkalinity would affect gas production. The pH of slurry in digester may be maintained between 6.8 and 7.2 for optimal gas production which is accomplished by maintaining proper feeding rate. There are a number of ways to correct the pH, in case.

v) **Agitation**: Mechanical agitation of the scum layer and slight stirring of slurry improves gas production although violent stirring retards it.

vi) **Solid content**: For optimum gas production, the solid content in the slurry is maintained between 7.5 and 10%.

vii) **Carbon-Nitrogen ratio**: A carbon–nitrogen ratio of 20:1 to 30:1 is optimum for biogas production. So as to keep C:N ratio of the feed mixture at desired level, the carbon–nitrogen ratio of various materials is given in Table 6.

Table 6: Carbon to nitrogen ratios of various materials

Material	Nitrogen Content (%)	Carbon:Nitrogen
Urine	15.18	8:1
Cow dung	1.7	25:1
Poultry manure	6.3	*
Night soil	5.5-6.5	8:1
Grass	4.0	12:1
Sheep waste	3.75	*
Mustard straw	1.5	20:1
Potato tops	1.5	25:1
Wheat straw	0.3	128:1

*: data not available

10. Biogas application and appliances

Cooking and Lighting: The main use of biogas in rural areas is cooking. Firewood, crop residues or dung cakes are not available on regular basis and so are other conventional sources of energy like kerosene, electricity or coal etc. The other use is gas lamp which glows like any bright lamp. Generally, 15 running meter length (RML) of pipes/tubing are allowed in biogas plant schemes along with burners and lamps for schemes financed by banks.

i) **Dual-fuel engines**: This is a recent appliance where certain modification of air intake system helps carburation of biogas to run diesel engines. It is well-known that diesel engine has wide application in rural areas from irrigation to any stationary operations and these engines can be converted to dual-fuel ones. This dual-fuel engine makes use of about 70% biogas and 30% diesel. The economy of running dual-fuel engine by biogas is undisputable but it has certain operational difficulties. The main problem is that while biogas plants are located near the house, the running of the diesel pumpsets is required in the fields. As such it is impractical and technically unsound to provide long gas pipes to connect the gas plant with the diesel engines in the field. Another aspect that prohibits the use of biogas in diesel engine is that it requires larger gas plants at least 8-10m^3 to enable a low 3 HP diesel engine to run for 4–5h a day. Often, users prefer the sizes of 2-4m^3 gas plants as cattle number is not many. However, considering large biogas plants as key to economical operation of dual-fuel engine, one may see its wide application in future as many such gas plants come into existence.

ii) **Refrigeration, incubation and water boiling**: There are other applications of biogas such as in refrigerators, incubators and water boiler. Trials in this regard are going on in R&D centres whose results are awaited before they become common application items. Average requirement of biogas for these applications is shown in Appendix IV to give an idea of the quantity of gas consumed by these appliances.

iii) **R&D in biogas application**: Unless ways and means are found for diverse application of biogas, it may not receive social appreciation it deserves. Community biogas plant has been mooted to provide cooking gas, street light and drinking water through pumping. Government of India had proposed to install 500 community biogas plants on pilot basis during 1983-84 to study their feasibility to run on a cooperative basis. The results may help in spreading similar plants to other areas. Besides, many institutions/individuals in the country have attempted to provide alternative gas plant designs and some 30 designs have been tried so far. Yet the KVIC design and Janata Model are the two which are recognised by the financing institutions as standard designs. This status would be maintained unless better designs replace. In certain parts of the country, the use of night soil and connecting latrine to biogas plants have already found its social acceptability. However, in most part of the country it is still not socially accepted. When it finds its social acceptability, it would ensure regular supply of feed besides improving sanitation in rural areas.

11. Implementation of schemes

It is stated earlier that the biogas schemes have been recognised as priority schemes under the new 20 point programme of the government. In order to expedite schemes formulation and their imple-mentation, the government has strengthened the state and district nodal agencies with adequate staff. Besides these agencies, the Khadi and Village Industries Commission (KVIC) is also helping banks in formulating the schemes and also in installing and operating of gas plants upto the running stage. They are technically equipped to undertake field work. They are also responsible to release the subsidy (25–75% of the total amount depending on the beneficiary type; scheduled castes/tribes are entitled for higher subsidy) as declared by the government from time to time. Central government provides subsidy to repair defective plants also. The banks process individual cases and provide the loan amount directly to the material supplier/contractor. Approved contractors supply the main components and material under orders from implementing agency and loaning bank. Procedure has been systematised under the National Biogas and Manure Management Programme (NBMMP).

The unit costs of biogas plants vary from district to district due to local taxes and transport costs, and also for variation in site conditions including soil, sub-surface water etc. Though KVIC announces the unit costs of biogas plants of various sizes of their model from time to time for major regions taken as the benchmark of unit costs, but the actual unit costs may sometimes differ from KVIC rates and in such cases sponsors of the schemes should provide detailed cost estimates of material and labour. Recently, biogas plants are covered under banking schemes to give a further boost to the programme.

To simplify the scrutiny of unit costs, National Banks have prepared a proforma (Appendix VIII) to furnish by sponsoring agencies through the help of local KVIC representatives and local bodies associated with the implementation of the schemes. However, in view of the size of the programme, National Bank now fixes unit costs for all plant models taking into account the approved schedule of rates. Besides, National Bank has also circulated a checklist (Appendix IX) which all the banks should complete in order to make sure that all the requisite information is furnished in the scheme, and that National Bank clears the scheme without seeking any further clarification.

12. Field problem in implementation and operation

As observed so far that in spite of biogas plant having good potential to serve as an alternate energy source in the midst of energy crisis, the progress made in installing some 1,65,000, biogas plants during last 21 years is not appreciable enough when we see China's progress of installing 12,00,000 plants within 7 years period. This strengthens the view that biogas has a better chance of success in our country than many other countries in the world where all the basic pre-requisites are present. Especially, when a fossil fuel sufficient country like China feels the urgency, we should feel greater need for it. Unfortu-nately, presently we don't realise the consequences of delaying the adoption of biogas plants. The government's determination to give fillip to the development of renewable energy sources throws a challenge to all the concerned agencies that have to ensure that field bottlenecks are overcome and biogas project is effec-tively implemented. Bank should take special note that field problems and operational difficulties have been mostly identified and are known to implementing agencies.

i) The field problems are mainly technical, socio-economic and organisational in nature. To overcome the technical difficulties and to run the plant effectively, KVIC introduced some 'Dos' and 'Donts' (Appendix X), which takes care of most technical problems of existing biogas plant models if followed. However, there is a dire need to improve the models for easy handling, installation, operation and gas use, and hence the scope of abundant R&D here.

ii) The socio-economic issues are somewhat sensitive and needs to be dealt with much care. In the first place, there are very few farmers who consider biogas as a priority in the face of several other requirements like food, cloth, medical attention, education, social obliga-tions etc. Besides, the minimum cost of a sizable plant even after subsidy is considered high from the viewpoint of rural India. There is a need to develop low-priced biogas plants. The models should ideally be repairing and maintenance free. Then, there are issues of dung collection. Often, the dung is not available in adequate quantities when cattles are free-grazing type. To overcome such diffi-culties the collection of night soils, poultry litters etc. can be very useful. In China, there are no such stigmas attached to night soil handling which is a reason of their success. In India, aggressive extension work needs to be carried out to educate the people and remove social stigma on handling of night soil. Amongst others, the non-availability of adequate backyard space is an important issue as most of the villages have clustered houses.

iii) There are lapses on organisational side also. Extension agencies often fail to communicate the true benefits to the users. Now, since each district nodal agency is provided with technical and other support staff, promotional work should get a boost. Discussion with several state im-plementing agencies revealed that the target-oriented approach has proved detrimental to the cause of biogas programme. Instead, only those potential areas where the programme can be successful should be adopted for implementation. It is also necessary to assess the need for fuel and manure of the individual cases in a scheme area before a programme is envisaged. One dispute which always crops up in the organisational side is the release of subsidy by the state Govt./KVIC and the time taken by banks to sanction the loan amount. It would be useful to come to some under-standing with agencies responsible for releasing subsidies for speedy clear-ance of schemes so that the implementation does not get delayed. Most of the schemes and their formulation should involve the funding agencies from the very beginning so that no further need for clarification may arise. It may be pointed out that govt. desires that banks may take more initia-tive to coordinate the various activities and, if necessary, an inter-agency coordination committee may be set up to ensure speedy disbursements. Banks have also been instructed that mortgaging of land or asking for securities should not be insisted upon for any biogas programme. This means banks may break-away from conventional banking procedure and thus it is a new challenge to banking sector which may have to operate in circumstances where securities etc. may have to be overlooked.

13. Techno-economic feasibility

Technoeconomic feasibility is an important aspect for assessment time to time. Economics of operation, cash-flow and other financial exercises are not the objective of the discussion here. Still, it's necessary to understand the mechanism of techno-economic exercise, particularly the technical assumptions that comprise techno-economic feasibility. A sample calculation of the economics of a $4m^3$ capacity floating-drum plant for guidance is given in Appendix XI.

i) **Gas plant size**: Assessing the exact daily gas requirement is important and, therefore, efforts should be made to ensure that there is no wastage of gas in whatever size is chosen. Otherwise, installing a plant would be uneconomical. It may be noted that gas production is a continuous process and fresh quantity of gas must be available for daily consumption. If one fails to make use of available gas, then assumption on savings on account of gas consumption would lead to misleading Cost-Benefit Ratio (BCR).

ii) **Capital cost of plants and operation cost**: Although the programme is subsidised, the cost of the plants is considered high by the beneficiaries. A $4m^3$ gas plant costing Rs. 4000/- may mean monthly expenditure of Rs. 40–50/ assuming a 10% flat rate of interest on capital without the subsidy (Rs. 20–25/- with 50% subsidy). Rarely, people in rural areas spend such amount in cash for domestic cooking. However, the plants show certain additional social benefits considering the manurial value. Still there is a need to make a comparative study of the existing practice of cooking and lighting in rural India and actual cost incurred on them, and if the biogas use can lead to direct savings. Unless the benefits are demonstrated, it would be difficult to promote biogas programmes.

iii) **Manurial value of the digested dung**: Techno-economic feasibility is incomplete unless the importance of manurial value is understood properly. A notional value of the dung is generally assumed in terms of its heat value vis-a-vis the quantity and cost of kerosene replaced at its prevailing prices and in terms of NPK obtained after digestion. The main consideration for this is that the value of the dung increases several times due to proper preservation of available NPK in the digested solids. Working out the economics, a definite value of such dung is assumed in terms of its NPK content. However, the dung containing NPK is not saleable like chemical fertilisers due to lack of awareness among the beneficiaries and its volume. The difference between the two is that, while chemical fertilisers can be readily applied and instantly assimilated in the plants, the digested dung requires time for assimilation in the soil. So far, manure is not looked upon as a cash return, particularly after the preservation of Nitrogen due to digestion. But the fact remains that this manure is of cash value which a borrower can realise by applying it in his own farm and by reduced application of chemical fertiliser. This would add to his savings. This aspect requires to be communicated to the borrowers so that they may find economics of operation of biogas plant rational.

14. Summary

'Energy Crisis' has engulfed almost all the nations including the comfortably placed energy sufficient countries, the main cause being the depleting fossil fuel reserves, deepening with each advancing year. It is not far when the automobiles and aeroplanes era may come to halt unless suitable alternative fuel is in place. However, where alternatives are available for like domestic cooking, lighting etc., should be replaced without further delay to save the precious fossil fuel for critical uses. Biogas is such renewable alternative source to reduce dependence on fossil fuel to a considerable extent. Manure, the biogas production byproduct, is rich in NPK that demonstrably im-proves the soil fertility for food production and thereby reduces the pressure on the import of chemical fertilisers. Biogas production technology is simple and is successful in India where tropical environment and large cattle population provide ample of dung that favour the installation of a biogas plant. Experiences of the last four decades whereby some 1.65 lakh plants have been installed are not encouraging com-pared to what has been achieved by fossil fuel sufficient countries like China where about 8 million plants were installed in a decade. The Indian government provides subsidy, organisational support and has included it in the 20-point programme to encourage technology adoption in rural India. Banks also support the programme and extend borrower loan. An aspect of particular interest to banks is that biogas plant has its own pay-back capacity even without government sub-sidy as other fuel sources become costly. Its time for the sector provide better banking opportunity due to its self-dynamic nature of development. For this, banks need to update on the evolving technology time to time to assess the returns and prepare to extend needed financial support.

Appendix I: Benefits from a tonne of fresh cattle dung through a biogas plant

1. Manure:

1.1. Fresh dung contains 80% moisture; one tonne fresh dung gives 0.2 tonne dry dung.

1.2. One tonne of fresh dung gives 240kg manure.

1.3. Manure contains 1.6% Nitrogen; one tonne of fresh dung gives 3.84kg Nitrogen.

1.4. Manure contains 1.55% Phosphorous; a tonne of fresh dung gives 3.72kg Phosphorous.

1.5. Manure contains 1.0% Potash; one tonne of fresh dung gives 2.4kg potash.

1.6. Fresh dung has 16% organic solids; a tonne fresh dung gives 96kg humus after digestion.

2. Useful energy: Digestion in a biogas plant converts 40% organic solids to gas; may vary depending on various factors as discussed.

Appendix II: Potential of biogas plants in the country and in village from cattle dung

1. Energy potential in the country:

1.1. 22 X 106 animals X 3.18kg dry dung X collection efficiency = 5.38 X 108kg dung daily

1.2. 5.39 X 108kg dry dung X 365 days = 1.967 X 1011kg dung yearly = 196 X 106 t dung/yr

1.3. 5.39 X 108kg dry dung/day X 0.19m^3 gas/kg dung = 1.02 X 108m^3 gas daily

1.4. 1.02 X 108m^3 gas/day X 4.698 KWh/m^3/day = 4.791 X 108kWh daily

1.5. 4.791 X 108kWh/day X 365 day/yr = 175 million MWh/yr

1.6. 1.02 X 108m^3 gas X 365 days/yr X 62 X 102 kerosene/m^3 gas = 23.08 billion l kerosene

2. Manure potential in the country per year:

2.1. 196 X 106 t dry dung/yr X 1.2 t manure/tonne dry dung = 235 X 106 tonne manure/yr

2.2. 236 X 106 tonnes manure/yr X 0.016 tonne N/tonne manure = 3.76 x 106 tonnes N/yr

2.3. 3.76 X 106 tonnes N/yr X 2.1 tonnee N/tonne naphtha = 1.76 X 106 tonnes naptha/yr

3. Energy potential in a village of 500 persons (considering collection of 100kg dung daily):

3.1. 100 kg dry dung/day X 0.19 m^3 gas/kg dry dung X 4.698kWh/m^3 = 89.26kWh/day

4. Manure potential in the village:

4.1. 109kg dry dung/day X 1.2kg N/kg manure/kg dry dung = 120kg manure/day

4.2. 120kg manure/day X 0.016kg N/kg manure = 1.92kg N/day or 700kg N/yr

4.3. 700kg N/yr X 5kg food grains/kg N = 3.5 tonnes food grains/yr

Appendix III: Equivalent quantity of fuel for 1 m^3 biogas

Name of the fuel	Equivalent quantities to 1 m^3 biogas
Kerosene	0.6201
Firewood	3.474kg
Cowdung cake	12.296kg
Charcoal	1.458kg
Soft coke	1.605kg
Butane	0.433kg
Furnace oil	0.4171
Coal gas	1.177m^3
Electricity	4.698kWh

Appendix IV: Biogas requirements

Utility	Quantity requirement
Cooking	336–430 1/day/person
Gas stove	330 1/hr/5 cm burner
	470 1/hr/10 cm burner
	640 1/hr/15 cm burner
Burner gas lamp	126 1/lamp of lighting equivalent to 100 watt filament lamp
	70 1/hr/1 mantle lamp
	140 1/hr/2 mantle lamp
	1691/1/3 mantle lamp
Dual fuel engine	425 1/hp/hr
Refrigerator	100 1/100 litre capacity
Incubator	80 1/100 litre capacity
Water boiler	110 1/1 litre water

Appendix V: Average maximum biogas production potential from different feedstock

Sl.No.	Feedstock	Litre/kg dry matter	% Methane content
1.	Dung	350*	60
2.	Night soil	400	65
3.	Poultry manure	440	65
4.	Dry leaf	450	44
5.	Sugarcane trash	750	45
6.	Maize straw	800	46
7.	Straw powder	930	46

*: Average gas production from dung may be taken as 40lit/kg fresh dung without any temperature control provided in the plant. One m^3 gas is equivalent to 1000 litre

Appendix VI: Average dung (night soil) production by various living beings

Living being	Quantity of dung/night-soil produced (kg/day)
Cow, heifer	10.0
Bullock	14.0
Buffalo	15.0
Young bovine	5.0
Horse	14.0
Horse, young	6.0
Pigs, over 8 score	2.5
Pigs, under 8 score	1.0
Ewes, rams and goats	1.0
Geese and turkeys	0.2
Lambs	0.5
Duck	0.1
10 hens	0.4
Human beings	0.4

Note: For free grazing animals the availability of dung may be taken as 50% of the quantity mentioned

Appendix VIIA: Dimensions of fixed-dome type biogas plants (dimensions are in mm)

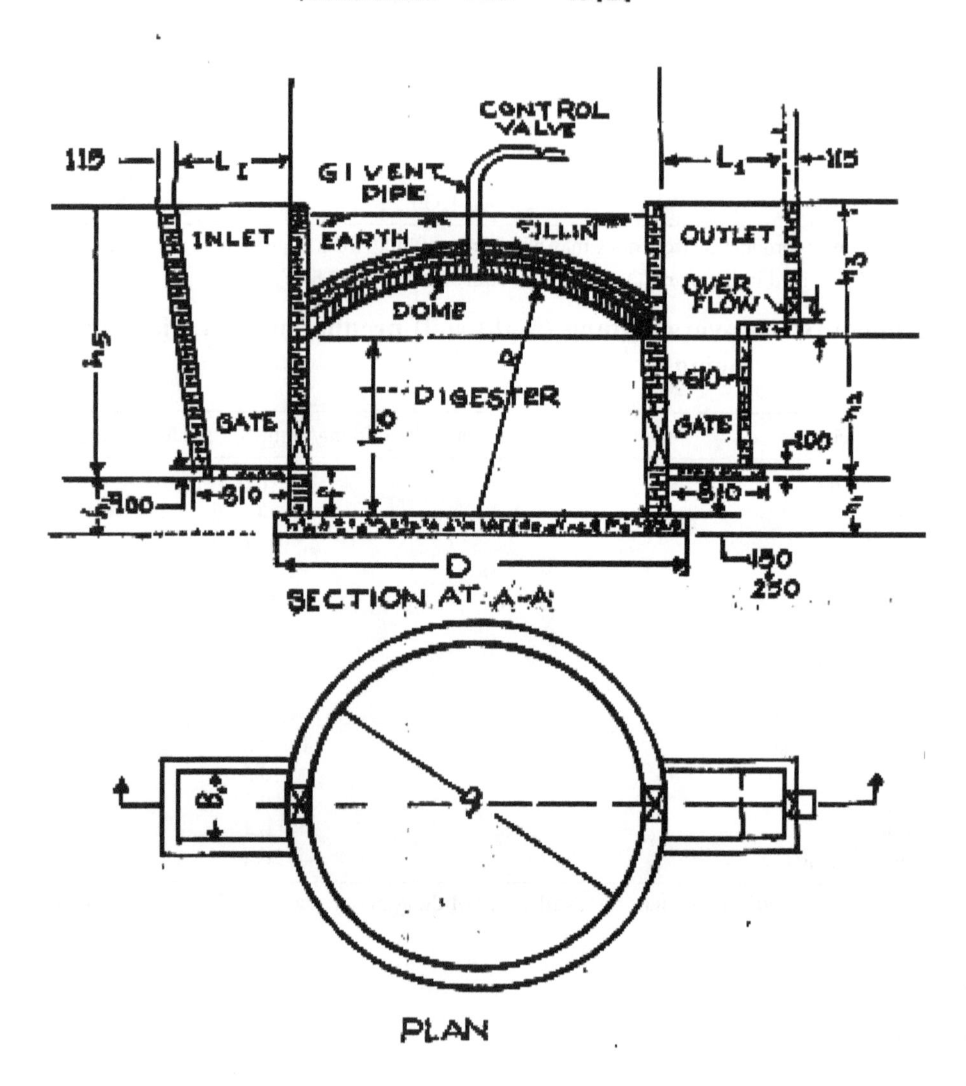

Plant Size	D	D1	D2	L1	B1	F1	h1	h2	h3	h4	h5	h6	h7	R	GATE
2	2750	2500	2370	850	610	580	680	765	810	680	1575	1345	520	1610	610x610
3	3100	2950	2720	1216	900	746	846	828	880	845	1710	1575	590	1865	610x610
4	3610	3400	3000	1372	950	825	1005	990	854	1005	1845	1715	640	2080	610x760
6	4110	3890	3420	1620	1380	660	840	1290	1025	840	2315	1950	735	2360	610x1000

Appendix VIIB: Dimensions of floating-drum type biogas plants (dimensions in cm)

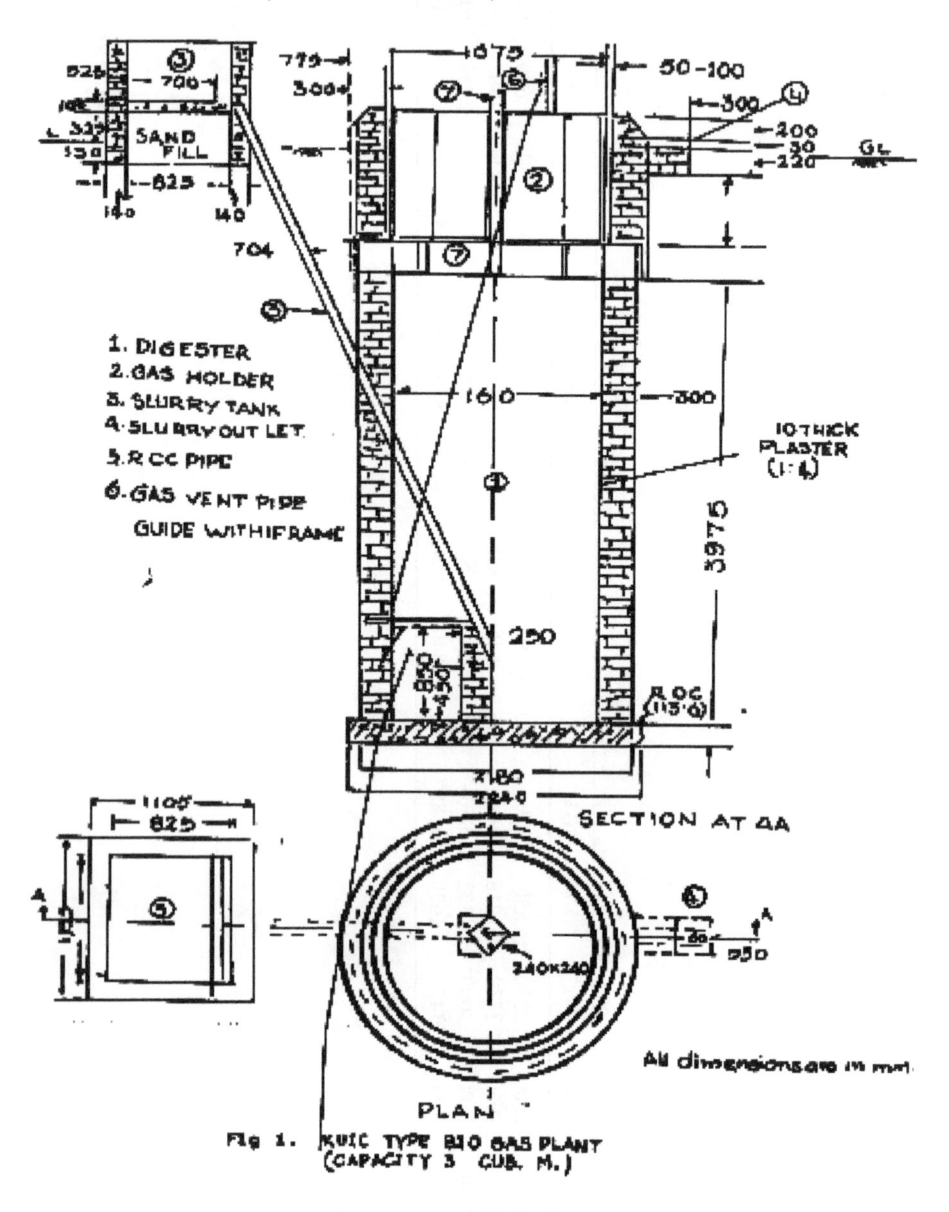

Fig 1. KVIC TYPE BIO GAS PLANT (CAPACITY 3 CUB. M.)

Dimension	Plant capacity of m^3																				
	for 30-d retention period							for 40-d retention period							for 55-d retention period						
	1	2	3	4	6	8	10	1	2	3	4	6	8	10	1	2	3	4	6	8	10
A	120	135	160	181	220	240	275	120	135	160	180	220	240	275	120	135	160	180	220	240	275
B	157	187	202	212	212	242	232	177	257	277	292	292	332	317	227	327	377	427	427	477	477
C	170	95	110	135	135	125	115	170	165	185	200	200	200	200	170	220	270	320	320	345	345
D	112	70	70	70	70	70	70	112	110	100	90	90	90	120	112	175	180	180	210	210	205

Appendix VIII: Cost estimate for biogas plants

Model: **Capacity:**

Sl. No.	Item	Quantity	Rate/Unit quantity	Cost
1.	Earth Work			
2.	Bricks			
3.	Cement			
4.	Sand			
5.	Morrum/stones			
6.	Skilled labour days for construction of plant			
7.	Unskilled labour days for plant construction			
8.	AC pipes (when required)			
9.	Gas holder			
10.	Pipes and fittings with sizes			
11.	Gas burner/chullah			
12.	Gas lamp (when required)			
13.	Any other item, if required with specific details			
14.	Transportation charges			

Note: The rates should be as per the State Government schedule of rates or approved district schedule of rates

Appendix IX: Checklist

Scheme to install biogas plant (completed by Sr executive/officer forwarding the Scheme)
Note: Ticks in boxes signify that the details of relevant information, as per NABARD guidelines, are furnished in the scheme on the following aspects:

1. Objectives
2. Whether the scheme area has been specified?
3. Whether the financing bank and its branches have been, specified?
4. Whether the scheme has been approved by the competent authority?
5. Whether State Govt's clearance is obtained (for State Land Development Bank schemes)?
6. Background information and the status of the implementation of such schemes in the area, percentage failure of plants installed in the past in the area and the reasons for failure - Results of evaluation of the programme, if any, done
7. Criteria for selection of the beneficiaries and the size of plant with regard to adequacy of land for plant and connected pits, distance from the point of use of manure, nearness to cattle sheds, number of cattle owned, availability of water for use of plant, etc.
8. Designs, specifications and quality of material for the plant size
9. Designs, specifications and quality of material for the plant size
10. Financial returns: estimated benefits for the plant size, rates of returns and repay capacity
11. Lending terms: rate of interest, loan maturities (inclusive of grace period), down payment, nature of security, source and extent of subsidy, if any, available
12. Arranging for fabrication of gas holder, frame and burners, and quality control measures
13. Arrangements for procurement of the equipment and supply to the beneficiaries
14. Agency providing the technical support: Whether certified by the competent agency regarding the technical feasibility of the programme?
15. Availability of staff for technical appraisal of the proposal and furnishing technical certificate on completion and installation of the plant

16. Arrangements for supervision and monitoring of scheme implementation
17. Government support: (a) Technical guidance (b) Whether necessary budget provision has been made/proposed?

Appendix X: Do's and Dont's for floating-drum plant

Dos	Don'ts
1. Select the size of Biogas plant, depending on the number of animals you have.	1. Do not install a bigger size biogas plant if you don't have sufficient dung for it.
2. Install the plant as close to animal shed. Usually, the distance between the plant and the kitchen is kept at 9.1m (30ft) in the case of a 2m^3 (70ft^3) plant, 13.7m (45ft) for a 3m^3 (105ft^3) plant, 18.2m (60ft) for a 4m^3 (140ft^3) plant, 22.8m (75ft) for a 8m^3 (28.0ft^3) plant, and so on.	2. Do not install the plant at a distance more than that recommended. Otherwise you will have to provide additional pipeline which will add to your cost.
3. Make sure that the plant is installed in an open place, and gets plenty of sunlight during the day time round the year.	3. Do not install plant under a tree, inside the house or in any other shady place.
4. After construction, the digester should be cured for 10–12 days.	4. Do not use the digester immediately after construction without proper curing, otherwise cracks will develop.
5. The outer side of the digester wall must be compacted with soil.	5. The soil around the outer wall of digester should not be loose. Else, the digester may get damaged due to slurry pressure.
6. Feed the plant with cattle dung and water mixed in right proportion; add 4 parts of dung to 5 parts of water to make a homogeneous mixture.	6. Do not add more than required quantity of either dung or water - doing so might affect gas production efficiency.
7. While filling biogas plants with partition wall make sure that it is filled equally on both sides of the central partition wall, side by side up to the guide frame.	7. Do not fill the slurry unequally on either side – it may cause the central wall to collapse.
8. Make sure that the dung and water mixture is free from soil or sand.	8. Do not allow any soil or sand particle to enter into the digester – it may choke the inlet pipe at the bottom.
9. When digester is full with homogeneous dung-water mixture, place the gas holder on the central pipe of the guide frame. Keep the heavy duty gate valve closed.	9. Do not keep the gate valve loose. Else, gas will escape from the plant unutilised.
10. Rotate the gas holder once or twice every day in order to break the scum.	10. Do not allow scum to form in digester. Else, gas production will stop/reduce.
11. Release the first full drum of gas to the atmosphere to avoid any explosion. Biogas and oxygen mixture is explosive.	11. Don't use the initial gas produced. It may cause explosion due to the presence of O_2.
12. As soon as gas starts accumulating in gas holder, fit a pipeline from it to kitchen. Gas produced should be used regularly.	12. The gas should not be burnt directly from the gas outlet even for testing purpose.

Contd.

13. For efficient cooking, keep the approved burner preferably in the kitchen.	13. Do not light a burner in the open. Else, there can be considerable loss of heat.
14. Open the gas regulator cock at the time of its actual use.	14. Do not leave the gas regulator open when the burner is not in use.
15. Adjust flame by turning the air regulator till a blue flame; it gives maximum heat.	15. Do not use the gas if the flame is yellow. Adjust the flame till it is blue in colour.
16. Drain out any condensed water collected in the gas supply pipe every 4–5 days.	16. Do not allow water accumulation in gas pipe. Else, required gas pressure will not be maintained and the flame will sputter.
17. Clean gas holder from outside periodically with fresh water to avoid crust formation.	17. Do not keep the gas holder dirty.
18. Store the slurry in a proper pit so that the liquid content may not leak through.	18. Do not make slurry pit more than 0.9m (3ft) deep.
19. Check the outlet pipe periodically during the summer to avoid clogging.	19. Do not allow the slurry to dry or cake at the end of the outlet pipe.
20. Repaint the gas holder (if of MS sheet) from outside every year to prevent rusting. This will also prolong its life.	20. Do not allow gas holder (if of MS sheet) to rust; it develops holes resulting in gas leakage.

Appendix XI: Cost-benefit analysis of KVIC plant having an installed biogas generation capacity of 3 m^3/day (amount in Rs.)

a. Capital cost	
Gas holder and frame	4500
Piping and stove	1750
Civil engineering construction (tank, inlet and outlet, etc.)	10000
Total	16250
b. Annual expenditure	
The interest on investment @ 9% p.a.	1462.50
Depreciation on gas holder and frame @ 10% p.a.	450
Depreciation on piping and stove @ 5% p.a.	87.50
Depreciation on structure @ 5%	500
Cost of painting, yearly once	350
Total	2850
c. Annual income	
Gas 3m^3 daily @ Rs. 2.30/m^3 for 315 days (assuming initial 50 days HRT)	2173.50
Manure (6 tonnes compost) with refuse 15 tons @ Rs. 200/tonne	3000
Total	5173.50
d. Net annual income (b - c)	**2323.50**

Note: The net annual income of approximately Rs. 2300/- shows that the capital investment of Rs. 16250/- can be recouped in about seven years. There are also incidental advantages of improvement in hygienic, absence of smoke and soot in gas burning, convenience in burning, cooking material self-sufficiency and increased organic richness of manure for agriculture.

Annexure-II: Prominent Global Entities in Biogas R&D and Commercialisation

Disclaimer: Below is provided a representative list of some enterprises/ organisations enjoying global prominence that are actively engaged in biogas development as core focus, and a gist of their activities. For the ease of readers, they are arranged in alphabetical order of the primary country of activity. The list is neither exhaustive nor has any bearing what-so-ever in branding specific company/entity. Any such incidence interpreted would be a mere coincidence.

Australia

1. Paul Harris of the Agricultural Engineering at The University of Adelaide is a specialist expert in designing anaerobic digestion (both liquid and solid techniques), and integrated biosystems. In addition to lecturing, he involves heavily in disseminating biogas information in communities.
2. *Utilitas* provides design, drawing, construction, commissioning , operation and maintenance of biogas plants. Technologies available for solid waste and wastewater biogasification include CSTR, UASB, mCAL-U (modified covered anaerobic lagoon Up-flow). The company is designing a 665kWe biogas plant for Australia's second largest piggery; one 200kWe biogas plant for another piggery is completed. The plants will also produce a high quality biofertiliser.

Africa

3. *Avenam Links Intl Ltd* is a Renewable Energy company in Nigeria, incorporated to bring affordable biogas digester and generator technology to Nigeria. The company has designed BioDisc® treatment plants to provide an engineered package solution to meet a wide range of applications and maintain discharge qualities.

Austria

4. *Ennox* is an international manufacturer and developer of biogas components for wastewater treatment, sewage and landfill plants. Founded in 2011, the company supplies biogas technology and equipment in Europe and has a team of biogas professionals and technicians.
5. *Entec* focuses on the design of mid- and large-scale anaerobic digestion plants and offers design and supervision of construction, commissioning and operational monitoring of biogas plants.

6. *EnvironTec*, headquartered in Fussach at Lake Constance in West Austria, specialises in developing, producing and marketing gas technology components for a variety of fermentation gas applications in the bio, sewage and landfill gas areas. The company also undertakes the planning, conceptualisation and delivery of complete conduits and systems as well as their maintenance and inspection on behalf of the customer.

Canada

7. *Bio-en Power Inc.* designs, builds and operates anaerobic digesters, and focuses on developing anaerobic digestion facilities with environmentally friendly energy producing technology.
8. *Electrigaz* is a biogas engineering firm specialising exclusively in the designing, planning and realisation of biogas solutions for farms, agro-food industries and municipalities.
9. *PlanET* Biogas Solutions, incorporated in October, 2006, has designed and constructed twelve anaerobic digester facilities throughout Canada. PBS is an affiliate of PlanET Biogastechnik GmbH, Vreden, Germany.
10. *Greenlane* provides innovative biogas technology to biogas producers for environment friendly biomethane production. Greenlane Biogas claims to have developed biogas upgrading technique (from raw biogas to vehicle quality fuel), the biogas typically generating a mixture of 65% CH_4 and 35% CO_2.
11. *Yield Energy Inc.*, formed in 2007 with a focus towards converting urban waste streams into energy and value-added product, specialises in planning, designing, construction, commissioning and operation of renewable energy. It is active in biogas facilities in the North America based on the anaerobic digestion (AD) process of food waste and other organic materials.
12. *CH Four Biogas*, headquartered in Ottwa, Vancouver, specialises in the design, installation, optimization and maintenance of anaerobic digestion systems for the agricultural, industrial and municipal sectors, and offers complete system engineering and design, construction management and system optimisation and maintenance services.

China

13. *Fenghuo*, founded in 1989, manufactures a range of biogas systems from domestic units right through to large scale agricultural installations, a full range of biogas appliances. They also manufacture 700W to 140kW plus generator sets. The products are exported to over twenty countries and utilised by SNV Cambodia and United Nations in East Timor.

14. *Amoco*, founded in 1982 in Singapore, expanded the business into China in 1992 with the establishment of branches in Chengdu, Suzhou, Xuzhou, Kunming, and Guangzhou. The company designs, manufactures and sells their ISO-9000 accredited membrane products and engineering around the world.

Denmark

15. *ComBigaS* is located in Hemmet and collaborates with a number of suppliers and designs biogas systems for use in the Europe and Africa. They extend support from planning, installation, start-up to operations.

Germany

16. *Agraferm Technologies AG* based in Pfaffenhofen, Germany, designs and builds anaerobic digestion plants. It provides turnkey agricultural and industrial biogas plants in Europe, and operates internationally. Its services include project planning and construction as well as biological and technical services.

17. *Biogas-Ost* plan builds and maintains economical biogas plants. The company caters to customers from mainly agriculture, industry and the communities.

18. *Eisenmann*, headquartered in Germany, has offices/contacts in India, Brazil, China, France, Italy, Spain, Mexico, USA, Russia and England. Eisenmann biogas plants are double digesters. The primary digester is a plug flow digester with a continuous horizontal agitator shaft and the secondary digester is a stirred tank reactor.

19. *UTEC* is an independent engineering consultant group involved in developing and implementing biogas technology, since 1980. They are not bound to any manufacturing or technology company. Their extensive experience help avoid mistakes and unpleasant surprises.

20. *Nawaro* has offices in Leipzig, Krackow and Gustrow. Their bioenergy plants bring together the entire process of energy production based on biomass – from cultivation to the supply and processing of energy crops, from fermentation to biogas feed-in or the combustion of biogas to generate electricity and heat. On-site professional operator teams assure reliable operation round the clock using modern technology. The company assists the farmers, a biogas user, an investor, municipalities or a public authority.

21. *PlanET Biogastechnik GmbH,* Vreden, Germany.

Hong Kong

22. *Along Environ Tech Ltd* (AEL), founded in Hong Kong in 2006, has six years experience in the supply and installation of biogas plant and equipment in Norway, Ukraine, Kenya, South Africa, Thailand, Malaysia, Indonesia, Australia, and Brazil. Their 5500m^2 plant manufactures membrane products, biogas power generation units and purifying systems.

India

23. *Biotech*, with offices in Thiruvananthapuram, Kozhikkod and Ernakulam, provides all necessary technical advice to implement biowaste treatment programmes to generate bioenergy.
24. *Panse Consultants* specialises in biogas design and general environmental services, waste management services and project services.
25. *Synod Bioscience*, headquartered in Cochin, Kerala, develops waste management solutions for residential and commercial properties. It has designed, built and now operates power plants handling large quantities of waste, and generating other financial benefits. Product portfolio comprises domestic and commercial biogas plants, sewage treatment plants, and effluent treatment plants that have already been implemented in properties all over India.
26. *Urja Bio Systems*, established in 2006 and located in Pune, Maharashtra, has a team of qualified and trained engineers to design and install prefabricated biogas digesters, floating/fixed domes, biogas storage balloons, biogas scrubbing systems and biogas engines. They also offer biomass briquetting plants on turnkey basis.
27. The KISS installer.

Indonesia

28. *KIS Group*, with offices in Indonesia, India, Singapore, Malaysia, Brazil, Colombia and Africa, are arguably global leaders in biogas and biomass technologies. They have successfully installed and commissioned 330 projects in 17 countries. Their ZPHB™ (ZeroPond™, ZeroPollution™ and HigherBiogas™) technology is very popular and widely used across the globe.

Italy

29. *Biogas Engineering*, founded in 2005, specialises in the design and construction of plants for biogas production. It has installed 25 plants in the agricultural sector from 100kW through 1MW.

Malaysia

30. *Biodome*, a part of the Kirk Group, the United Kingdom, supports liquid and biogas storage across the globe. Their BIODOME® double membrane gas holders are used in wastewater, agricultural and municipal biogas systems throughout the world.

31. *Integrated Energy Industries Pte Ltd* has offices in Australia, Indonesia, Malyasia, Thailand and Singapore. It offers expertise in: Converting existing diesel engines to run on natural gas, biogas or bi-fuel; Generating and producing biogas from agricultural waste; Compressing and bottling biogas for use in vehicles and other prime movers; and Liquefying and storing biogas for use in vehicles, earth moving machinery and electrical generators.

Mexico

32. *Sistema Biobolsa* provides high quality biodigester system designed for small and medium farmers. Based in Condesa they also have a showroom and regional office in San Juan Tuxco, San Martin Texmelucan, Puebla.

Nepal

33. *Bageshwar Gobar Gas Company* (BGG), founded in 2006, employs 120 people and provides expertise in construction of small household biogas plants as well as larger (up to $50m^3$) units.

New Zealand

34. *Natural Systems Ltd*'s Ian Bywater has developed BioGenCool™, an on-farm energy system that transforms cow effluent into power that is used to heat and cool while providing greatly reduced electrical load. The system has been carefully researched and developed intended to how technology can deliver energy from renewable sources at low-cost and with benefits to the environment through a reduction in greenhouse gas emissions.

Netherlands

35. *Colsen* supplies several technologies in the field of biogas production, wastewater and digestate treatment. These have been successfully implemented in numerous projects for industrial and municipal customers.

North America

36. *American Biogas Council* is an industry association in the US that represents a full range of anaerobic digestion technologies and projects, including farm-based digesters, centralised facilities processing a variety

of municipal and industrial organic waste streams, and existing digesters at municipal wastewater treatment plants. They advocate for policy change in an effort to make industry investment more attractive.

37. *FirmGreen*, based in California, provides turnkey service to design, build, operate and maintain renewable energy (including biogas) projects. FirmGreen technology is used to process landfill gas into ultra-pure biomethane that can be compressed on-site for fuelling CNG vehicles.

38. *RCM Digesters*, headquartered in Oakland, CA, designs and builds anaerobic digester systems based on proprietary technology and expertise. The company provides relatively simple solutions for waste management and energy production for both family/farms and large regional waste centres. The privately held company has more than 100 RCM systems operating around the globe.

39. *Vahid Biogas*, headquartered in Oregon, USA, was established in 2008 by David Williams. The company consults worldwide and has undertaken projects in India, Singapore, Hawaii and Sri Lanka. Project development includes: yield calculation, design, sizing, feedstock, siting, economic benefit and financing options.

40. *Varec Biogas*, headquartered in California, began its business in the 1930s. It is a large supplier of biogas equipment and has many installations in municipal and industrial wastewater treatment plants, and sanitary landfills.

Singapore

41. *Biogas Helpline*, managed by Regen Energy, Singapore, is a professional body created by a team of biogas experts to 'promote and develop biogas'. This was set up as they found that, "a large part of potential beneficiaries are not aware of biogas". The Biogas Helpline was started to assist everyone in promoting and developing biogas, and help distantly placed people having difficulty with design, construction, operation or usage of biogas.

Spain

42. *Bentec Bioenergies* specialises in design, construction and delivery of 'turn-key' biogas plants for electricity production. Projects are developed in several phases including: first phase of analysis of economic viability, the intermediate phase of designing and planning, and the final phase of plant's construction.

Sweden

43. *AnoxKaldnes* has significant depth and expertise in biological processes and specialises in biopolymer production, biogas and biological wastewater treatment based on the MBBR™ technology. The company is a part of Veolia Water Solutions and Technologies, the technical subsidiary of Veolia Water, the water division of Veolia Environment, the largest environmental company in the world.

44. *Swedish Biogas International LLC* provides agricultural, municipal, and industrial anaerobic digestion technologies. The company specialises in integrating digestion technologies with existing WWTP facilities and provides consulting services for design and construction of digestion systems on order.

Thailand

45. *AsiaBiogas* is a large company which began in 1997 with a focus on providing energy solutions through a suite of technologies. Headquartered in Thailand, it has a regional presence with offices in Philippines and Indonesia. It currently operates in the niche small (ranging from 1–10MW) renewable energy power production systems. It has installed over 85 projects dealing with feedstock including palm oil effluent, animal manure and ethanol, over past 10 years.

46. *Renewable Cogen Asia* specialises in energy and environment, providing sustainable solutions in more than 30 countries. Its services cover Asia, Africa and other regions including customers from private sector, UN/international organisations, International funding agencies, and Government organisations. It expertises in biomass, biogas, MSW-based power/cogeneration plants, clean development mechanism, carbon foot printing.

47. *BioGasclean* specialises in biological desulfurisation of biogas and landfill gas. The company develops, manufactures and markets advanced fully automated bio-trickling filters for H_2S removal. BioGasclean has approx. 125 plants either in operation or under construction in 35 countries and supplies clean renewable fuel to more than 300MW gas engines.

Ukraine

48. *Zorg Biogas* has a full range of engineering services and has been constructing biogas plants since 2007. Zorg Biogas has designed, procured or constructed more than 55 biogas plants in 16 countries, catering from simple single-stage reactors for silage to complicated multi-stage reactor for pure chicken dung or distillery wastewater, wet or dry process.

United Kingdom

49. *BioGas Products Ltd* is a UK-based company specialising in process design and manufacture of anaerobic digesters and biogas utilisation equipment. They provide products and services to water utility, waste, agricultural and environmental sectors. They specialise in process design and construction of anaerobic digesters for UK farms, biogas cleaning, refining and storage, tank cover and lagoon liners, and onsite stainless steel fabrication.

50. *Bioplex* makes Portagester®, a new pre-composter and anaerobic fermenter to treat farm and horse stable manures; food and catering wastes; trade, garden and local authority wastes. The plant is relatively odourless, clean and compact, to an extent depending on the feedstock type being treated; sealed vessels prevent cross-contamination and disease risk from birds, rodents and insects.

51. *Kingdom Bioenergy Ltd*, established in September 2007, encourage the development of biomass energy, especially in the area of biogas technology. The company offers services in consulting, design, and research and development in the area of biogas technology as well in the broader area of biomass energy technologies.

52. *Vergas Ltd* specialises in design, manufacture and installation of flexible membrane biogas systems worldwide, with over 150 installations in the UK and elsewhere. The flexible membrane structures are typically for municipal treatment works, or solid waste processing facilities.

Zambia

53. *Southern BioPower* provides consultancy, construction, training services and feasibility studies for biogas-based waste management and energy (combined heat and power; CHP), including hybrid solutions combined with solar energy. The company constructs, supervises and provides skilled labour, quality control for biogas digesters, and turnkey project solutions in Zambia and other African countries.

Index

C

D

E

F

G

H

I

K

L

M

N

O

P

R

S

T